MICROBIOLOGY RESEARCH ADVANCES

ARCHAEA: STRUCTURE, HABITATS AND ECOLOGICAL SIGNIFICANCE

MICROBIOLOGY RESEARCH ADVANCES

Additional books in this series can be found on Nova's website under the Series tab.

Additional E-books in this series can be found on Nova's website under the E-books tab.

MICROBIOLOGY RESEARCH ADVANCES

ARCHAEA: STRUCTURE, HABITATS AND ECOLOGICAL SIGNIFICANCE

SAKURA Y. KATŌ
EDITOR

Nova Science Publishers, Inc.
New York

For permission to use material from this book please contact us:
Telephone 631-231-7269; Fax 631-231-8175
Web Site: http://www.novapublishers.com

Library of Congress Cataloging-in-Publication Data

Archaea : structure, habitats, and ecological significance / [edited by]
Sakura Y. Kato.
p. ; cm.
Includes bibliographical references and index.
ISBN 978-1-61761-932-8 (hardcover)
1. Archaebacteria. 2. Archaebacteria--Environmental aspects. I. Kato, Sakura Y.
[DNLM: 1. Archaea. 2. Biodegradation, Environmental. 3. Ecosystem. QW 50]
QR82.A69A738 2010
579.3'21--dc22
2010034002

Published by Nova Science Publishers, Inc. † New York

Contents

Preface

The Archaea are a group of single-celled microorganisms. They have no cell nucleus or any other organelles within their cells. In the past they were viewed as an unusual group of bacteria and named archaebacteria but since the Archaea have an independent evolutionary history and show many differences in their biochemistry from other forms of life, they are now classified as a separate domain in the three-domain system. Archaea include micro-organisms adapted to a variety of environmental conditions, including some of the most hostile locations on Earth. This book presents current research in the study of Archaea from across the globe, including the Archaeal flagellum; the Archaea ribonuclease P; mechanosensitive channels in Archaea; and the physiology, genomics and phylogeny of nanoarchaeota. (Imprint: Nova Biomedical)

Chapter 1-The archaeal flagellum is a unique prokaryotic motility apparatus responsible for swimming that is found in all of the major cultured subgroupings of Archaea. While they are rotating structures like bacterial flagella, archaeal flagella nevertheless differ from their bacterial namesake in several fundamental ways. Indeed, archaeal flagella have more similarities to bacterial type IV pili, organelles responsible for a type of surface motility called twitching. Structurally, archaeal flagella are thinner than bacterial flagella and lack an internal channel large enough to allow passage of subunits. The flagellins in archaea are made as preproteins with type IV pilin-like signal peptides which are removed by a dedicated preflagellin peptidase homologous to the bacterial type IV prepilin peptidase. Both enzymes belong to a novel family of aspartic acid proteases with conserved aspartic acid residues that are necessary for signal peptide cleavage. Bacterial flagellins, on the other hand, are never made with signal peptides. Additional homologues have been identified between the archaeal flagella system and the type IV pilus system, including a conserved ATPase and a membrane component. A further posttranslational modification commonly found on archaeal flagellins is attachment of an N-linked glycan. This is again never seen in bacterial flagellins which can be, in rare cases, glycoproteins but always with the glycan attached in an O-linkage. Much recent data has accumulated on the assembly, biosynthesis and attachment of the N-glycan to archaeal flagellins and the important role that the glycan plays in flagella assembly and function.

For *Methanococcus* species, a number of glycosyltransferases and the oligosaccharyltransferase needed to assemble the glycan and attach it to the flagellin target protein have been identified by deletion and complementation analysis coupled to mass spectrometry comparison data of the glycan in wild type and various mutant strains.

Additional genes encoding biosynthetic enzymes and sugar modification enzymes have also recently be identified. The current model of assembly of archaeal flagella is one where subunits, following processing by the preflagellin peptidase and glycan addition, are added at the base of the structure, as in pilus assembly, rather than at the distal tip, as is known for bacterial flagella.

Chapter 2-temperature underground oil reservoir indicates their key contribution to the material cycle and energy flow in the subsurface hot habitat. Methanogens dominated the anoxic ecosystem with amount of petroleumhydrocarbons by catalyzing the last step of carbon cycle by producing methane. The diversity of archaeal community was analyzedby 16S rRNA gene librariesconstructed from two continental- and offshore, high-temperature, long-term water-flooded petroleum reservoir in China. The continentallibrary includes237 clones clustered in 28 phylotypes. The dominant members were affiliated with the order Methanomicrobiales belonged to four orders: Methanobacteriales, Methanococcales, Methanomicrobiales and Methanosarcinales. Most of the clones clustered with sequences previously described for methanogens. Theoffshore archaeal 16S rRNA gene clone libraryincluded220 randomly selected clones clustered with 28 phylotypes dominated by methanogen-like rRNA genes (*Methanothermobacter*, *Methanobacter*, *Methanobrevibacter* and *Methanococcus*)and with a lower percentage of clones belonging to *Thermoprotei*. The FISH results indicated that the quantity of archaea was much more than bacteria in the samples and they had close spatial distribution *in situ* formed compact aggregates. Some thermophilic methanogens detected had been previously isolated from a number of high-temperature petroleum reservoirs worldwide, thus they might adapt to the environments and be the common habitants of geothermally heated subsurface environments. The existence, diversity, abundance and distribution of ammonia-oxidation archaea (AOA) in the high temperature oil reservoirs were investigated by using real-time PCR and phylogenetic analysis based on *amoA* genes. The result demonstrated the AOA-like phylotypes were mainly clustered within two major clades of archaeal *amoA* sequences knownfrom water columns, sediments, and soils: clusters A and B, and few clones were related to the new genera:*Candidatus* '*Nitrosocaldus yellowstonii*'. The abundance of AOA -like *amoA* gene ranged from2.92×10^3 to 9.21×10^4 per ml production water. Statistical analysisshowed that *amoA*gene fell into five groups and the distribution of *amoA* was of significant positive correlation with the environmental factors temperature and recovery process.The study provided the evidence for the distribution of putative prokaryotic*amoA* gene in various oil reservoirs, which improved ourunderstanding of the nitrogencycling in oilfield ecosystem.

Chapter 3-Recent major progress in sequencing technology has yielded more than 80 complete archaeal genome sequences and thousands of partial archaea-derived genomic and transcriptome sequences from various environmental samples. As a result of this rapid increase in the amount of archaeal sequence data, an unexpected variety of non-coding RNAs (ncRNAs) has been detected in Archaea, including fragments of transfer RNA (tRNA), a new class of small nucleolar RNA (snoRNA), a clustered, regularly interspaced short palindromic repeat (CRISPR) RNA used as an antiviral defense, and enzymatically cleaved introns found in many types of precursor tRNA, ribosomal RNA (rRNA) and messenger RNA (mRNA). Furthermore, several RNA elements such as riboswitches and SECIS elements have been predicted to exist in the 5' and 3' untranslated regions of archaeal mRNA. In this review, we discuss the distribution, function, and processing machinery of these archaeal ncRNA/RNA elements and the diversity of archaeal species. The authors have revised the phylogeny of the domain Archaea by reconstructing the most recent phylogenetic tree based on 8488 archaeal 16S rRNA sequences collected from all possible databases. The most recent findings on archeal ncRNA are explained along with the phylogeny of archaeal species, providing a deeper understanding of the functional and evolutionary roles of all types of archaeal RNA.

Chapter 4-Comparative sequence analysis of 16S RNA molecules carried out by Carl Woese in the 1970s revealed that life on Earth can be divided into three primary lineages (referred to as domains): *Eukarya*, *Bacteria* and *Archaea*. Presently, most known archaeal representatives are included into two kingdoms: *Crenarchaeota* and *Euryarchaeota*, with remarkable diversity of organisms. *Archaea* can be found in some of the most extreme environments on the planet, but are not restricted to them. Although the important role that *Archaea* play in biogeochemical cycles, such as carbon, sulfur and nitrogen cycles, physiology and ecological significance of archaeal populations are still poorly known. This lack of knowledge is, in large part, due to the limitations of cultivation procedures. In addition, information about the archaeal community in tropical ecosystems is still scarce. In this chapter the authors present a survey of the archaeal community present in aquatic ecosystems of Brazil: Madeira River and Tucuruí Power Plant Reservoir in Amazon; Santos and São Vicente Estuary and mangroves in the Southeast coast. The study of archaeal groups in freshwater and saline tropical aquatic ecosystems in Brazil revealed highly diverse communities and suggested the presence of new archaeal groups. In Amazonian environments, *Crenarchaeota* representatives such as Marine Group I, 1.1.band *Thermoplasmata* were frequent, suggesting a possible role of *Archaea* in nitrogen and carbon metabolism. Methanogenic groups were also detected in sediment and water where anaerobic conditions prevailed from both regions studied. In Santos and São Vicente Estuarine System, higher abundance of methanogenic *Archaea* was found in a highly impacted site when compared to a less contaminated area. In Mangrove sediments, the detection of MCG, MBG-B and MBG-D as an important component of the microbial community indicates the possible role of these groups in carbon cycle.This work emerges as a valuable contribution to the understanding of the biogeography and diversity of *Archaea* Domain in these rich ecosystems, besides providing hints as to help characterizing the role and dynamics of archaeal communities in major processes such methane production in tropical environments, subjected or not to the influence of human activities.

Chapter 5-Temperature is an essential and variable environmental factor, which impacts the phenotype, habitat distribution and evolution of organisms. The pervasive influence of temperature on organisms requieres an ensemble of compensatory adaptations that span all levels of biological organization, from behaviour to fine-scale molecular structure. The Archaea stand out as the only group of organisms that have species capable of growing at temperatures ranging from 0°C to 110°C. Thus, the recent availability of complete genome sequences for a large number of archaea species provides a powerful tool to analyze the mechanisms of thermal adaptation at the molecular level. In this study the authors have used comparative genomics to analyze thermal adaptation in the universally conserved ribonucleoprotein: Ribonuclease P (RNase P). We have analyzed compositional and structural features of both, the RNA and protein components of the archaeal RNase P, to identify molecular trends related to thermal adaptation. Our results show a robust correlation between the stems GC content of the RNase P RNA and the optimal growth temperature (OGT) of the archaea species. Moreover, the authors have detected significant differencesin the composition and distribution of several amino acids in the RNase P protein components of hyperthermophilic and mesophilic archaea. Thus, this study reveals that natural selection has adapted the RNase P holoenzyme to the OGT of the archaea species by modifying both, its RNA and proteins subunits.

Chapter 6-Archaea, the third domain of life, include micro-organisms adapted to a variety of environmental conditions, including some of the most hostile locations on Earth. Osmotic regulation is an important issue for a microbe, and mechanosensitive (MS) ion channels are important proteins capable to respond as transducers of mechanical stress by forming non-selective pores in order to deal with drastic osmotic changes in the milieu. As molecular sensors, MS channels respond to increases in membrane tension. When the membrane tension is small, these proteins are in the closed configuration; as the tension grows, the open state becomes ever more frequent, until it dominates at mild or high tensions. It has been proposed that these changes correspond to different channel conformations in a defined protein–lipid interface. This commentary discusses some of the most important aspects in order to understand the function of MS channels in archaebacteria. Special attention is focused on how archaebacterial lipids are attractive targets to study the function of these proteins.

Chapter 7-The Nanoarchaeota, proposed as the fourth sub-division of the Archaea in 2002, are known from a single isolate, "Nanoarchaeum equitans", which exists in a symbiotic association with the hyperthermophilic Crenarchaeote, Ignicoccus hospitalis. The sequencing of both genomes has provided clues on their unusual and very specific association. The discovery of nanoarchaeal sequences in diverse geographical hyperthermophilic locations, and most importantly the identification of nanoarchaeal phylotypes in mesophilic halophilic environments suggest the possibility of either free-living nanoarchaeotes or associations with novel hosts. The "true" phylogenetic position of the Nanoarchaea is still a matter of debate, having been described originally as a separate phylum and later as a member of the phylum Euryarchaeota. The present review will discuss the recent discoveries on this hyperthermophilic organism and its association to Ignicoccus and will analyse the different views on the phylogenetic position of this unique organism.

Chapter 8-Prefoldin and group II chaperonins are molecular chaperones found in Archaea and eukaryotic cytosol.

Chaperonins are large double ring complexes of ~1000 kDa enclosing a central cavity in which non-native proteins can refold into their native form. Group II chaperonins act independently of a cofactor that corresponds to GroES of the group I chaperonins.

Instead, the helical protrusion at the tip of the apical domain forms a built-in-lid for the central cavity.

Chaperonins use ATPase cycling to promote the conformational change necessary to open and close the lid. Prefoldin can capture nascent polypeptides or non-native proteins with its coiled coil motifs and, subsequently, transfers it to a group II chaperonin for correct folding. Prefoldin also cooperates with eukaryotic chaperonins in the folding of actin and tubulin.

Because of its biological importance, the manner in which prefoldin and group II chaperonin cooperate to fold a non-native protein has been extensively studied.

Here, the authors review a series of our studies that describe the structural and biological characteristics of archaeal prefoldin and chaperonin. This chapter also reviews recent reports on archaeal and eukaryotic prefoldin and chaperonin.

Detailed interaction studies between prefoldin and chaperonin allow us to discuss their possible cooperative mechanisms.

Chapter 9-The discovery of nonthermophilic archaea (crenarchaeota and euryarchaeota) in the open ocean and coastal water using culture-independent molecular technologies was reported in two seminal papers in the early 1990's (DeLong, 1992; Fuhrman et al., 1992), and this corrected our previous biased view that crenarcheota only existed in extreme environments such as hydrothermal vents or terrestrial thermal springs. Since then, research on nonthermophilic archaea has been intensely pursued intensely and it is now clear that these organisms occur widely in low temperature marine and terrestrial environments (see reviews by DeLong, 1998; Schleper et al., 2005). Overall, in marine water columns, nonthermophilic archaea are dominated by Group I crenarchaeota and Group II euryarchaeota. In marine sediments, not only Group I crenarchaeota and Group II euryarchaeota are present but also a variety of other archaeal species, including the Marine Benthic Group crenarchaeota, Miscellaneous Crenarchaeotal Group, and an unidentified euryarchaeota group (Vetriani et al., 1999; Inagaki et al., 2006; Biddle et al., 2006). In addition, anaerobic methane-oxidizing archaea (ANME groups) widely occur in association with gas hydrates, cold seeps, and organic rich thermal sediments (e.g., Boetius et al., 2000; Orphan et al., 2001; Mills et al., 2003; Knittel et al., 2005) as well as in anoxic water bodies like the Black Sea and Cariaco Basin (Madrid et al., 2001; Durisch-Kaiser et al., 2005; Schubert et al., 2006). This section reviews recent observations of the occurrence and diversity, abundance, metabolic pathways, and contributions to carbon and nitrogen cycles of nonthermophilic archaea in the deep oceans and marine sediments.

In: Archaea: Structure, Habitats and Ecological...
Editor: Sakura Y. Katō, pp. 1-37

ISBN 978-1-61761-932-8

Chapter 1

The Archaeal Flagellum: A Novel Prokaryotic Motility Structure and the Role of Posttranslational Modifications in Its Assembly and Function

Ken F. Jarrell*

Department of Microbiology and Immunology, Queen's University, Kingston, Ontario, Canada K7L 3N6

Abstract

The archaeal flagellum is a unique prokaryotic motility apparatus responsible for swimming that is found in all of the major cultured subgroupings of Archaea. While they are rotating structures like bacterial flagella, archaeal flagella nevertheless differ from their bacterial namesake in several fundamental ways. Indeed, archaeal flagella have more similarities to bacterial type IV pili, organelles responsible for a type of surface motility called twitching. Structurally, archaeal flagella are thinner than bacterial flagella and lack an internal channel large enough to allow passage of subunits. The flagellins in archaea are made as preproteins with type IV pilin-like signal peptides which are removed by a dedicated preflagellin peptidase homologous to the bacterial type IV prepilin peptidase. Both enzymes belong to a novel family of aspartic acid proteases with conserved aspartic acid residues that are necessary for signal peptide cleavage. Bacterial flagellins, on the other hand, are never made with signal peptides. Additional homologues have been identified between the archaeal flagella system and the type IV pilus system, including a conserved ATPase and a membrane component. A further posttranslational modification commonly found on archaeal flagellins is attachment of an N-linked glycan. This is again never seen in bacterial flagellins which can be, in rare cases, glycoproteins but always with the glycan attached in an O-linkage. Much recent data has accumulated on the assembly, biosynthesis and attachment of the N-glycan to archaeal flagellins and the important role that the glycan plays in flagella assembly and function.

* E-mail: jarrellk@queensu.ca , Tel: 613-533-2456, Fax: 613-533-6796

For *Methanococcus* species, a number of glycosyltransferases and the oligosaccharyltransferase needed to assemble the glycan and attach it to the flagellin target protein have been identified by deletion and complementation analysis coupled to mass spectrometry comparison data of the glycan in wild type and various mutant strains. Additional genes encoding biosynthetic enzymes and sugar modification enzymes have also recently be identified.The current model of assembly of archaeal flagella is one where subunits, following processing by the preflagellin peptidase and glycan addition, are added at the base of the structure, as in pilus assembly, rather than at the distal tip, as is known for bacterial flagella.

Introduction

When Carl Woese first proposed the Archaea as a third domain of life, it was a concept based initially on sequence of small ribososmal RNA where trees made from 16S/18S rRNA sequences clearly showed that Archaea formed a third grouping distinct from Bacteria and Eucarya [1-3].

However, even in those early days of archaeal research, it was recognized that, as a group, archaea possessed other traits that clearly differentiated them from the other domains. These initially recognized characteristics included unusual cell walls that lacked murein, ether-linked membrane lipids, a DNA-dependent RNA polymerase of atypical subunit composition and an unusual resistance pattern to most bacterial-effective antibiotics [4,5].

Later studies revealed clearly that the archaeal flagellum was also a domain specific trait, different fundamentally from the bacterial counterpart in structure, composition and likely mode of assembly [6-8].

Distribution of Flagella in the Domain Archaea

Flagellation is a very common trait among all the major cultivated subgroups of archaea, including methanogens, extreme halophiles, thermoacidophiles and a variety of thermophilic and hyperthermophilic organisms [7]. Even in species that lack a cell wall, e.g., *Thermoplasma volcanium* and *Thermoplasma acidophilum*, flagella have been isolated and *fla* gene clusters identified [9,10]. As in bacteria, the number and distribution of flagella on the cell surface can vary from species to species (Table 1). While some archaea possess only a single flagellum, others possess a few and still others can produce close to 100 flagella per cell. Correlations between flagellation/motility and *fla* gene loci are found in all cases but a few interesting exceptions. One involves various *Methanosarcina* species in which complete genome sequences revealed the unexpected presence of apparently complete *fla* operons although all of *M. mazei, M. acetivorans* and *M. barkeri* are described as nonmotile [10,11]. However, in the case of one species of *Methanosarcina*, *M. baltica*, at least some members are flagellated [12,13]. It may be that under certain growth conditions, yet to be identified, all of these *Methanosarcina* species could assemble flagella. A similar observation was recently reported for *Haloferax volcanii*, originally described as nonmotile [14], but now found under certain growth conditions to make flagella and swim [15,16]. The other exceptions are *Pyrobaculum aerophilum* and *Methanopyrus kandleri*, where flagellation and/or motility has been reported [17,18] but a *fla* operon has not been found in either sequenced genome. Clearly, in these two cases, there is a disconnect between the reported flagellation/motility and lack of *fla* genes. It is possible, but seemingly unlikely, that these cells make flagella from a set of genes so distantly related or totally unrelated to known *fla* genes that they have not been identified by searches. It may also be that only certain members of the species have flagella or that the flagella observed are, in fact, some other surface appendage resembling flagella but composed of novel proteins. Recently, a new archaeal appendage, the Iho670 fiber, was identified on the surface on *Ignicoccus hospitalis* [19]. These structures were originally thought to be flagella and had a diameter similar to flagella, but were, in fact, completely novel. Obviously, *P. aerophilum* and *M. kandleri* are organisms in which the issue of flagellation and motility need to be revisited.

Table 1. Flagella distribution on selected archaea from a variety of habitats

Organism	Habitat	Flagella distribution
Methanogens		
Methanocaldococcus vulcanius	Hydrothermal vent	Three tufts
Methanococcus maripaludis	Salt marsh	Large bundle
Methanospirillum hungatei	Peat bogs	Polar tufts
Methanococcoides burtonii	Polar regions	Monotrichous
Extreme halophiles		
Halobacterium salinarum	Salted fish and hides	Mono-or bi-polar bundles
Haloarcula quadrata	Brine pool	One or more flagella, at one or more sites
Natronomonas pharaonis	Soda lakes	Polär tuft
Naltrialba magadii		Polar tuft
Thermoacidophiles		
Thermoplasma volcanium	Solfataras, hot springs	Monotrichous polar
Sulfolobus solfataricus	Solfataras, hot springs	One or more flagella
Metallosphaera sedula	Thermal pond	One or more flagellum
Hyperthermophiles		
Pyrococcus furiosus	Hydrothermal vents	Large bundle
Thermococcus kodakaerensis	Hydrothermal vents	Polar tufts
Archaeoglobus fulgidus	Hydrothermal vents	Monopolar polytrichous
Desulfurococcus fermentens	Freshwater hot springs	Monotrichous

While not studied extensively outside of specific extreme halophiles, archaea do possess a taxis system that allows them to swim to better environments [10,20-24]. Chemotaxis and phototaxis as well as aerotaxis have all been reported [22,25-29]. The chemotaxis system in archaea is very bacterial-like, despite the dissimilarity in their respective flagella. Indeed, the interaction of the chemotaxis system with the flagella in archaea has only recently been examined [24] and is of great interest since the point of interaction of the two systems in bacteria involves a phosphorylated form of CheY interacting with the switch protein FliM [30] and archaea have no discernible FliM component [23,31].The flagella system, especially in *Methanococcus, Halobacterium* and *Sulfolobus* has been used as a source of reporter proteins to study a variety of basic biological events relating to protein export and posttranslational modifications, especially signal peptide removal and N-linked glycosylation of flagellins.

General Features of Archaeal Flagella and their Relationship to Bacterial Flagella and Type IV Pili

Archaeal flagella are superficially similar to bacterial flagella in appearance and both organelles do the same job. Both are rotating structures and the rotation can be switched from clockwise to counterclockwise [32-34]. They are the main organelle responsible for swimming in the bacterial domain and, thus far, the only organelle responsible for swimming in the domain Archaea. Electron microscopy analysis revealed that archaeal flagella are typically thinner in diameter compared to most bacterial flagella; archaeal flagella are usually 10-12 nm in diameter while the diameter of bacterial flagella usually ranges from 18-22 nm [35]. Both structures appear in EM examination to consist of a filament with a cell proximal hook region (Figure 1). The anchoring structure in the bacterial flagellum is a well characterized basal body with associated rod and rings while in archaea this is an ill-defined knob-like structure [36-39]. The length of the hook region, well defined in bacteria, appears much more variable in archaea [40,41].

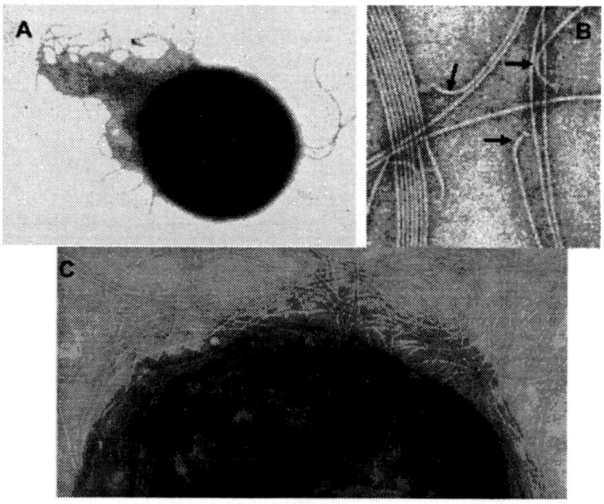

Figure 1. (A). Electron micrograph of a single M. voltae cell showing approximately 80 flagella. (B). Flagella isolation by detergent extraction showing filaments with hooks (arrows) ending in a featureless knob. (C). Close-up of the surface of a flagellated Methanococcus cell showing many flagella. Electron micrographs courtesy of Shin-Ichi Aizawa.

In both domains, the flagellum interacts with a very similar chemotaxis system that allows the cells to swim towards optimal growth environments. However, these superficial similarities belie the profound fundamental differences in the structures. With the advent of genome sequencing it became obvious that the two organelles did not share a genetic history. Although the genome sequences of many flagellated archaea have been completed, homologues of any bacterial genes that are involved in flagella structure or assembly are not found [23,31,42]. This includes genes for ring and rod components as well as mot proteins. Despite the fact that archaeal flagella have a hook and can switch rotation, there are no homologues of bacterial hook and switch protein genes. Other fundamental differences are found. Bacterial flagella rotation is well known to be driven by the protonmotive force or, in rarer cases, the sodium motive force [43,44].Recently, in archaea it was shown that rotation is dependent on ATP [45].

Finally, the structure of the archaeal flagellum [46], as well as the basic building blocks of the filament itself (the flagellins) are fundamentally different from that of flagella from bacteria.Interestingly, from a structural point of view the archaeal flagellum bears many similarities to a very different type of bacterial apparatus, the type IV pilus [8,47,48]. Type IV pili are involved, not in swimming motility, but in a type of surface motility caused twitching, which is driven by extension and retraction of the pilus by addition or removal of pilin subunits at the base of the structure [49,50].

Table 2. Comparisons of archaeal flagella, bacterial flagella and type IV pili

Trait	Archaeal flagella	Bacterial flagella	Type IV pili
Composition	Usually multiple flagellins some minor	Typically single flagellin sometimes multiple	One major pilin several minor
Signal peptides	Class 3 type on flagellins	None on flagellins	Class 3 type on pilins
Signal peptidase	FlaK/PibD	None	PilD
Glycosylatio	N-linked to flagellins	O-linked to flagellins, but rare	O-linked to pilins
Filament diameter	10-12nm	18-22 nm	6 nm
Filament structure	No central channel	Central channel	No central channel
Growth point	Presumed base	Distal tip	Base
Anchoring structure	Ill defined knob-like	Basal body, rod and rings	None observed
Hook region	Yes, variable length	Yes, defined length	No
Motility	Swimming	Swimming	Twitching (surface translocation)
Mechanism of movement	Rotation, switchable	Rotation, switchable	Extension/retraction
Force for movement	ATP	H^+/Na^+ gradient	ATP
Genes	Homologs to type IV pili, Not to bacterial flagella	No homologs to others	Homologs to archaeal flagella
Interaction with chemotaxis system	Yes	Yes	Sometimes

Critically, both type IV pili and archaeal flagella do not possess an internal channel large enough for subunits to pass through [46,51]. For bacterial flagella, this central channel which extends through the basal body, hook and filament is necessary for its mode of assembly where new subunits are sent through the hollow structure to assemble at the distal tip, i.e. at the point furthest from the cell surface [37]. This is accomplished by a specialized type III secretion system located at the base of the flagellum. The substrates for the type III secretion system are not made as preproteins with signal peptides. Archaeal flagellins were long ago recognized to be made with type IV pilin-like (or class III) signal peptides [52-54]. This signal peptide, like those of the type IV pilins, is cleaved by a specific peptidase, the prepilin peptidase for type IV pilins [55,56] and the prepilin peptidase homologue (preflagellin peptidase) FlaK in archaea [57-59]. Besides this common signal peptide, archaeal flagellins were noted to have significant amino acid sequence at the N-terminus to that of type IV pilins although, with rare exceptions, the archaeal flagellins lack the conserved +5 glutamic acid found in almost all type IV pilins [53].The similarities of archaeal flagellins to type IV pilins and the lack of a central channel in the archaeal flagella lead to the proposal that archaeal flagellins grow by addition of subunits to the base of the structure and not to the distal tip [6,60]. Comparisons of archaeal flagella with bacterial flagella and type IV pili are summarized in Table 2.

How Can One Differentiate among Flagellins, Pilins and Fibers?

It is becoming clear that Archaea as a group use type IV pilin-like proteins as structural units for a variety of surface appendages [61,62]. To date, this has been found first in archaeal flagella [8], later in type IV-like pili in *Methanococcus* [63,64] and *Sulfolobus* [65,66], and, most recently, in the Iho670 fibers of *Ignicoccus hospitalis* [19]. The Iho670 fibers were originally thought to be flagella but motility of these organisms was never observed and the main fiber protein has no identified homologues in the genetic databases and is certainly not a flagellin [19]. A short type IV pilus-like apparatus at the cell surface, termed the bindosome, may also be formed from sugar binding proteins in *Sulfolobus solfataricus* although EM evidence for this is lacking [62,67]. In all cases, the main structural proteins are made initially as preproteins with class III signal peptides that are removed by a homologue of the type IV prepilin peptidase in the bacterial type IV pilus system. These TFPP homologues identified to date have been designated FlaK (for flagellin processing in methanogens [57,58]), PibD (for processing of flagellins, pilins and sugar binding proteins in *S. solfataricus* and flagellins and pilin-like proteins in *H. volcanii* [15,59]) and EppA (for pilin processing in *M. maripaludis* [63]).

An important question is how one can differentiate among these different type IV pilin-like proteins. The class signal peptides of flagellins, pilins and fibers are often very similar, especially near the cleavage site, although the lengths can vary. However, this variation in signal peptide length is also observed solely within the flagellins themselves [60]. Archaeal flagellin proteins are typically 200-250 amino acids in length, although much longer examples can be found, for example, in *N. magadii* (395 amino acids [68]) and *T. kodakaraensis* (580 amino acids [69]). Pilins are smaller, typically in the 130-150 amino acid range in length. However, the major pilin of *M. maripaludis* (MMP1685) is only 72 amino acids in length, 62 after the signal peptide is removed. The Iho670 fiber protein is about 300 amino acids long. Flagellins can often be identified by several features. Usually, the flagellins are found in tandem at the beginning of operons that include some subset of the *fla* associated genes*flaC-J*. Sometimes the flagellins are located next to these *fla* associated genes but transcribed in the opposite direction, as in *Halobacterium* [70]. However, in isolated examples, there are additional flagellins located elsewhere on the chromosome separate from the flagellins located near the *fla* associated genes as in *Halobacterium* and *Haloarcula* [70-72]. One commonality found in all archaeal flagellins, but not pilins or the fiber protein, is a glycine residue located at position +3 relative to the cleavage site for FlaK (Figure 2) [60]. To date, this has been diagnostic for flagellins among the type IV pilin-like proteins. Among the type IV pilins, there is the presence of glutamic acid at position +5 in three of the four likely pilins of *M. maripaludis* (MMP0233, MMP0236, MMP1685), as found in bacterial type IV pilins, with the fourth putative pilin (MMP0237) having a aspartic acid at this position [63].

Surprisingly, both type IV pilins of *S. solfataricus* lack an acidic residue at the +5 position, instead having either leucine or phenylalanine [66]. Therefore, unlike type IV pilins of bacteria, which have an extremely conserved +5 glutamic acid, this is not a conserved feature of archaeal type IV pilins, even amongst the very few so far identified. Indeed, since several archaeal flagellins also have a +5 glutamic acid (e.g., in *Msp. hungatei, Mcc. burtonii, A. fulgidus* and *A. pernix* [60]), this position cannot be used to define a pilin vis-a-vis a flagellin. Furthermore, while a +1Q was shown to be important for pilin processing in *M. maripaludis* by EppA [63], outside of the methanococci where the flagellins do not have a +1 Q, several archaeal flagellins do possess this amino acid at the +1 position. Furthermore, this is not a conserved amino acid in the only other archaeal pilins studied, those of *S. solfataricus,* where the +1 position is held by leucine or isoleucine [66].

N-terminal alignment of type IV pilin-like proteins from selected archaea

		↓
Flagellins		
M. maripaludis FlaB1	MKITEFMKSKKG	ASGIGTLIVFIAMVLVAAVAA
H. salinarum FlaB1	MFEFITDEDERG	QVGIGTLIVFIAMVLVAAIAA
S. solfataricus Fla	MLKEYNKKVKRKG	LAGLDTAIILIAFIITASVLA
Pilins		
M. maripaludis MMP0233	MFKRFNRG	QISFEFSIIVLSILLISTITI
M. maripaludis MMP0236	MSKG	QVSVEFIVLFLALLVAVVVST
M. maripaludis MMP0237	MIKMLQLPFNKKG	QVSFDFIIAMLFLLLIFAFMG
M. maripaludis MMP1685	MKFLEKLTSKKG	QIAMELGILVMAAVAVAAIAA
S. solfataricus Sso0117	MMWLKA	ISSIFSTLIVVMITLSLIVPL
S. solfataricus Sso0118	MLQLMMKGGYKLKKRKG	LSSILGTVIVLAITLVLGGLL
Fibers		
I. hospitalis Iho670 fibers	MKIARKG	VSPVIATLLLILIAVAAAVLL

Figure 2. N-terminal alignment of type IV pilin-like proteins from selected archaea.

Archaeal Flagella Functions

In bacteria, the flagellum is not just a swimming apparatus. Bacterial flagella can be involved in swarming motility across surfaces, sense wetness and play a role in biofilm formation, for example [73-75]. Similarly, the flagella of archaea have been shown to perform multiple functions. While first and foremost the organelle used by archaea for swimming, recent studies have also indicated that in some species flagella can take on other important biological roles. In *Pyrococcus furiosus*, flagella have been shown to be involved in forming cable-like connections among cells and in adhesion to cell surfaces [76]. Interactions of *P. furiosus* with *Methanopyrus kandleri* cells can occur through flagella, resulting in a two component biofilm [77]. In *Sulfolobus*, flagella mediated swarming has been shown and in addition, flagella, along with pili, are both involved in surface adhesion [65]. In *Haloferax volcanii*, recently shown to be motile under certain growth conditions, the flagella were shown not to be involved significantly in surface adhesion [15]. It was suggested that perhaps extreme halophiles do not need flagella to adhere to surfaces because the high salt content of their environments would mask the repulsive forces between the charged surfaces. Instead, attachment here is mediated by other pilin-like proteins processed by a type IV prepilin peptidase-like enzyme.Swimming without flagella has been reported via different mechanism in bacteria [35]. No such nonflagella-driven swimming modes have yet to been reported in archaea.The possibility exists of twitching type movement across surfaces, as seen in bacteria mediated via extension and retraction of type IV pili, since several archaea have been reported to have type IV pili-like loci. However, it remains unclear whether these pili are retractile since, to date, only a single ATPase has been found associated with their formation [66]. In type IV pili systems where twitching is known to occur there are separate extension and retraction ATPases [78].

However, retraction has not been observed in all known type IV pili examples (e.g., those of *Vibrio cholerae* [79].

Structural Work on Archaeal Flagella

Although Gram positive basal body-like anchoring structures with two rings have been reported for *Msp. hungatei* and *M. thermolithotrophicus* flagella [41], other studies in *Methanococcus* and *Halobacterium* have found that flagella typically end in a rather featureless knob [39,80].

Since archaea have very different cell envelopes compared to bacteria [4,81], and, in particular, lack murein and almost universally an outer membrane, the rings found in bacterial flagella may not be typically associated with the archaeal flagellum anchoring structure. This may be particularly true of the many flagellated archaea that have a cell wall composed of only an S layer overlying the cytoplasmic membrane or, in the case of *Thermoplasma,* no wall component outside the membrane at all. Since, in bacteria, the stator component attaches the flagella to the murein layer that is lacking in archaea [82], how the archaeal flagella are attached to the cells is an intriguing but poorly studied area.

Electron microcopy evidence has been presented suggesting that a subcytoplasmic membrane polar-membrane like structure may aid in anchoring the flagella of at least some archaea [83,84]. This is best studied in extreme halophiles, but has also been observed in thin sections of methanococci [85]. In *H. salinarum*, thin sections revealed the presence of a plateau-like discoid lamellar structure that lies underneath the cytoplasmic membrane in the area of the cell where flagella are inserted. The structure can be up to 300 nm in diameter and is located 20 nm underneath the cytoplasmic membrane [83]. This structure has been suggested to be involved in additional anchoring of the flagella to the cell, a prediction made from early studies [86]. Whether this is the polar cap to which attached flagella were isolated by Kupper et al. is unclear [80].

Most archaea have multiple flagellins, the exceptions being members of the Sulfolobales which have only a single flagellin gene.In *Methanococcus,* there are 3-4 flagellins with two being the major structural proteins. Interestingly, one of the flagellins, FlaB3, in both *M. voltae* [40] and *M. maripaludis* appears to play the role of the hook, as deletion of *flaB3* in *M. maripaludis* results in flagella that lack the curved portion of the structure and complementation with a plasmid borne version of the gene restores the hook region [87].

A similar finding was reported for *H. salinarum* [88].Since there are genera of archaea that have 2 flagellins (e.g., *A. pernix*, *A. fulgidus, T. volcanium*), one flagellin may be the filament protein while the other may be the hook protein. What happens in *Sulfolobus* species that have only one flagellin gene is currently not known.

However, interestingly in *Sulfolobus* flagella, hooks have not been observed [89]. In fact, while archaeal flagellar hooks have clearly been observed in some archaeal flagella, they have not been reported in all species (e.g. *Haloarcula marismortui* [42] and in *S. solfataricus* [89]). Where observed, the hooks are unusual in being of great variations in length [40,41]. What features of the protein allow one of the flagellins to form a curved structure while the closely related other flagellins form a filament is not studied. Analysis of the three flagellins of *M. maripaludis* indicates that the two filament flagellins (FlaB1 and FlaB2) are more closely related to each other than to the hook flagellin (FlaB3). Nevertheless, FlaB3 is still over 40% identical and almost 70% similar to FlaB1 and FlaB2 as the amino acid level. Individual essential roles for all the flagellins of *M. maripaludis* are assumed since deletions of *flaB1* or *flaB2* lead to nonflagellated cells while, as mentioned above, deletion of *flaB3* leads to hookless filaments.

In most archaea, there are multiple flagellin genes and, where examined, all of the flagellins have been found in the isolated flagella. Thus, the five different flagellins of *H. salinarum* [90]and the three flagellins of *M. maripaludis* [91] and four flagellins of *M. voltae* [92] have been identified in purified flagella samples. Rarely, however, has the distribution of the flagellins within the filaments been addressed. FlaB3 of *M. maripaludis* has been identified by deletion analysis to be the hook [93] and in the case of *Natrialba magadii* flagella, two of four flagellins, FlaB4 and FlaB2, were located along the entire filament by immunoelectron microscopy [94].

The structural determinations of the flagella filaments from several archaea have been important in further showing the fundamental differences bewteen archaeal and bacterial flagella and have greatly supported the association of archaeal flagella and bacterial type IV pili that have previously been built on the common class III signal peptides and signal peptidases of archaeal flagellins and type IV pilins. The major work reported on filament structure has been done on two distantly related archaea, the halophilic euryarchaeote *H. salinarum* and the thermoacidophilic crenarchaeotes *Sulfololus shibatae* and *S. solfataricus* [46,47,95-97] The work on the structure of the archaeal flagella is of interest considering the extremes of pH, salinity and temperature that encompass archaeal habitats. Maintaining the flagella structure under these conditions may involve specific modifications to stabilize the component proteins, which may include glycosylation of the flagellin subunits [46,98]. Three dimensional reconstructions carried out on the flagella of *H. salinarum* and *S. shibatae* yielded similar structures suggesting there is a common archaeal flagellar architecture [46].

In archaea, a three start helical symmetry is observed, similar to pili and the reconstructions show that there is no central channel. A central channel is found in bacterial flagella and is necessary for assembly as the new subunits travel through this central cavity to assemble at the distal tip. In *H. salinarum*, a lower density central assembly domain is surrounded by three larger external domains. In *S. shibatae,* where the flagella diameters are larger than those of the halophiles, the different lies in the larger external domains, with the internal domains remaining conserved with the halophilic version. The internal domains are formed from the conserved hydrophobic N-termini of the flagellins, which are very similar in amino acid sequence to those of type IV pili, known to be involved in oligomerization [53,61].These amino acids form an alpha helix.Unpublished data from the Trachtenberg lab suggest that this conserved symmetry holds true for other archaeal flagella filaments as well [46].

Regulation

Relatively little is known about regulation of the *fla* genes. Most of what is known, comes from studies in methanogens. Mukhopadhyay et al. [99] showed that flagellation in *Methanocaldococcus jannaschii* was regulated by the partial pressure of H_2. Under conditions of H_2 excess such as high H_2 partial pressure and low cell density, there was little or no detectable levels of four of the *fla* operon proteins (FlaB2, FlaB3, FlaD and FlaE) and the cells were mostly nonflagellated. Flagella synthesis occurred when hydrogen became limiting. The molecular nature of this novel regulation of flagella remains unknown but a 10 nucleotide direct repeat was identified upstream of the first flagellin gene and it was suggested that this element may play a role in regulation of the operon [100].

In *M. maripaludis*, the response to various nutrient limitations was studied at the transcript level [101]. It was found that mRNA for *fla* genes responded to multiple factors, decreasing under leucine limitation but increasing when cells were hydrogen limited, as seen in *M. jannaschii*. Growth rate also affected mRNA for *fla* genes but in complex ways: rapid growth lead to increased mRNA if H_2 was limiting but to decreased mRNA if phosphate was limited. Subsequent quantitative proteomic analysis demonstrated decreased flagellins and chemotaxis proteins when cells were nitrogen starved but flagellins increased if cells were phosphate starved [102]. H_2 limitation did not affect flagellin levels, in contrast to that seen in the *M. jannaschii* proteomic study and the results of the *M. maripaludis* transcriptome analysis. No reports of the flagellation state of the cells under the various conditions were presented.

In addition, it has been observed, in many mutants of *M. maripaludis* carrying deletions in genes that affect flagella formation, that the entire *fla* operon appears not to be transcribed after serial passage in the lab setting [103]. This has been seen with deletions in, for example, *flaI* and *flaJ,* as well as genes involved in glycan formation. If the gene deletion results in a glycan of less than 2 sugars, the flagellins cannot be incorporated into flagella. These mutants apparently stop transcription of the operon and various proteins from the *fla* operon cannot be detected by western blots after serial passage of the strains.

The secondary mutations that occur during serial passage of the original, targeted deletions have not been identified but, in at least some cases, these have not occurred in the promoter region of the *fla* operon as the sequence of this region in wildtype cells and in mutants where the *fla* operon is not transcribed is identical (unpublished results).

In *Msp. hungatei*, it was shown that flagellation is dependent on growth temperature, with flagella made at 37°C but not at 25°C or 30°C [104]. In addition, with medium modifications that resulted in long chains of cells, the cells at the ends of the filament were nonflagellated.

Single cells and cells at the ends of short filaments were flagellated. Interestingly, in cells that were nonflagellated, similar amounts of flagellin protein could be detected in western blots to that found in flagellated cells suggesting that the regulation of flagellation here was posttranscriptional.

In a rare case where transcription of the *fla* genes was specifically followed under different growth conditions in a crenarchaeote, it was shown that the flagellin gene of *S. solfataricus* was strongly induced in stationary phase and under conditions of nutrition limitation [97].

In *H. salinarum*, a complex interaction of the *flgA* and *flgB* gene transcription was uncovered [105]. In wildtype cells, both transcripts for the *flaA* operon and the *flgB* operon were observed. In a *flgA* mutant, no *flgA* transcript was detected but the level of the *flgB* transcript decreased to background levels. In the *flgB* mutants, *flgB* transcript was not detected but the *flgA* mRNA level was unaffected. In a separate experiment, only the *flgA2* gene was deleted.

In this mutant, a shorter *flgA* transcript was detected, reflecting the fact that a portion of the *flgA* operon had been deleted. However, the mutant did not produce any detectable *flgB* transcript, again showing that transcription of the *flgB* operon was dependent on the *flgA* operon.

Genetics of Archaeal Flagellation

The first archaeal flagellin genes were cloned from *H. salinarum* by Gerl and Sumper [72]. They used a flagellin antibody probe to screen a *H. salinarum* genomic library constructed in an expression vector and were able to detect a flagellin gene fragment. Using this as a probe they were able to isolate a larger fragment encoding the multiple flagellin genes. A different approach was successfully employed in the first studies on methanogen flagellin genes. Here, the genes for flagellins were identified in *Methanococcusvoltae* from reverse genetics based on N-terminal sequence date from purified flagella [52]. These studies revealed a set of flagellin genes followed closely by a number of so-called *fla* genes which appeared likely to be all co-transcribed with the flagellins and thereby likely to be involved in flagellation. Later studies, in *Methanococcus* as well as other genera, did show that the flagellins are typically found at the beginning of an operon that almost always resulted in the co-transcription of these *fla* associated genes [69,100]. As genetic tools improved in various archaea, especially *Methanococcus*, these genes were systematically deleted or inactivated and their necessary involvement in flagella formation was proven [87,106,107]. In *Methanococcus,* the *fla* associated genes are *flaC-flaJ*. Examination of the sequenced genomes of most flagellated Euryarchaeota reveal the presence of a very similar *fla* operon with the same genes in the same order [10] (Figure 3).

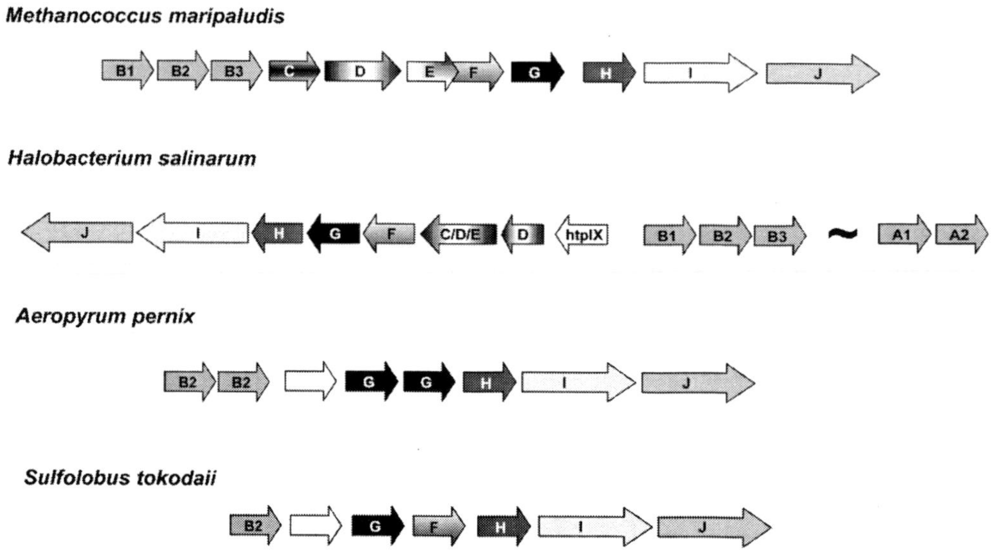

Figure 3. The fla gene region in selected archaea. Shown are a typical euryarchaeon, M. maripaludis; an atypical euryarchaeon, H. salinarum where flagellin genes are located next to, but transcribed in the opposite orientation from, the rest of the fla genes and a separate set of flagellin genes is located elsewhere on the chromosome. In addition, preceding the fla genes is a gene for a transducer protein (htpIX); a typical crenarchaeon A. pernix; an atypical crenarchaeon, S. tokodaii with a single flagellin gene. The white arrows in the crenarchaeote families do not have homologues in the databases.

Flagellated members of the Crenarchaeota however have a truncated version of the *fla* associated genes often missing the *flaCDE* genes [10]. In *M. maripaludis,* deletions in *flaD* and *flaE* could not be generated but deletions in *flaC* were made and such mutants were nonflagellated indicating an essential role for this protein in flagellation [87]. Obviously, in crenarchaeotes that do not have a *flaC* gene, the cells do not require this gene product for flagella formation.In *H. salinarum*, the *fla* associated genes are located next to flagellin genes but orientated in the opposite orientation so that co-transcription is not possible [70]. This organism is unusual as well in having multiple other flagellin genes that are located at a distance from the main *fla* locus [105]. Of the *fla* associated genes, *flaC, D, E, F, G* and *H* do not have strong similarities to other genes in the databases [60]. FlaH does share a RecA-like ATPase domain with distant homologues in both archaea and bacteria that are not involved in flagellation [108]. The Walker Box A for ATP binding but not the Walker Box B needed for ATP hydrolysis has been identified [100] but whether FlaH actually possesses ATPase activity has not been reported. Functions of FlaC, D, E, Fand G have not been reported although FlaCE and FlaD have recently been shown in *H. salinarum* to interact via other proteins with components of the Che system that may explain how the bacterial-like chemotaxis system of archaea interacts with the very different flagella found in archaea [24]. Unlike FlaC-H, FlaI and FlaJ have homologues in bacteria, specifically to proteins involved in type IV pilus biogenesis. FlaI is a readily identified homologue of ATPases required for type IV pilus extension and retraction, such as PilB/PilT [48,109].FlaJ is a homologue of the pilus membrane protein PilC/TadB [48]. Based on this, it has been suggested that FlaJ is the platform for assembly of archaeal flagella and that FlaI provides the energy for incorporation of new subunits [8,110]. Such a basic function in flagella assembly would explain while FlaI and FlaJ are found in genomes of flagellated archaea in the *fla* locus. Deletions of *flaI* and *flaJ* in different archaea, both euryarchaeota and crenarchaeota, result in nonflagellated cells [70,97,107]. Interestingly, homologues of *flaIJ* are found in *S. solfataricus* Ups pilus locus (*upsE* and *upsF*) and in the Bas locus believed to be involved in the bindosome structure assembly (*basE* and *basF*) [89].

Posttranslational Modifications of Flagellins

Archaeal flagellins are always synthesized initially as preproteins with class III signal peptides. This signal peptide must be removed in order for the flagellins to be incorporated into the filament [58]. For many archaeal flagellins, the apparent molecular weight estimated from SDS-PAGE migration does not closely correlate with the molecular weight deduced from the gene sequence. Often the flagellins migrate as though up to 10 kDa, or more, larger [7,42,68,111]. In several cases now, it has been shown that this is due to the attachment of glycans in N-linkage [91,92,112] and this posttranslational modification may be an almost universal feature of archaeal flagellins since analysis of the flagellins of most archaea reveals the presence of sequons that are used for N-linked glycan attachment [60,113]. While these can be found typically in the range of 2-6 sites per flagellin, in some cases there can be many more: for instance in *M. thermolithotrophicus* FlaB2 there are 16 such sites. The presence of the sequon does not necessarily mean glycan attachment occurs at each site but in *Methanococcus* most of the potential sites have been shown, in fact, to be occupied by glycan. The presence of at least some glycan attachment has been shown to be necessary for flagellin incorporation into filaments in *Methanococcus* [93,114]. These two important posttranslational events are discussed in detail below.

Signal Peptide Removal

One of the defining differences between archaeal flagella and bacterial flagella is that archaeal flagellins, which comprise the filament portion of the structure, are made as preproteins [52,53,59]. As first discovered over 20 years ago, the archaeal flagellins are made with the atypical, short signal peptides previously described for type IV pilins. These signal peptides typically are about 12 amino acids in length, although in some cases much longer or much smaller (as small as 4 amino acids) have been reported, mainly deduced from complete genome sequences [60]. Such signal peptides, termed class III signal peptides, typically end with a glycine at the -1 position and a basic amino acid, often lysine at position -2. These signal peptides are removed, not by the housekeeping signal peptidase I which removes signal peptides from most exported proteins, but by a type IV prepilin peptidase (TFPP) [57-59,115]. In archaea, site directed mutagenesis of both the TFPP and the flagellins have been done to help identify key residues for function.

Two different archaeal TFPPs and their substrates have been studied in some detail. These are FlaK of methanococci and PibD of *Sulfolobus*. FlaK appears to have a very limited substrate range and may in fact only process flagellins while the substrate range of PibD is much larger and includes flagellins, pilins and various sugar binding proteins which may all form a broadly similar pilus-like structure. In vitro assays, utilizing crude membrane fractions as a source of TFPP activity [116,117], were used to study the essential nature of several amino acids in the signal peptide and mature N-terminus of the flagellins of *M. voltae*. Site directed mutagenesis was used to replace various amino acids prior to their testing for cleavage by crude membrane fractions [118].

The highly conserved glycine found at position -1 relative to the cleavage site, could not be replaced by any of several amino acids tested, except for alanine where relatively poor processing was observed. In type IV pilins, it was also found that glycine was critical for proper processing of prepilins, with an alanine replacement leading to partial cleavage. Substitutions tested at the conserved -2 lysine position resulted in a complete loss of processing while replacing the -3 lysine resulted in partial processing. When the +3 glycine, conserved in every instance of an archaeal flagellin, was changed to valine, no processing was observed. More recently, it was discovered that *M. maripaludis* possessed a number of type IV pilin-like genes in addition to the pilin-like flagellin genes [63]. Three pilin-like genes were found in a cluster with another TFPP, termed EppA. Surprisingly, despite close similarities in their signal peptides, EppA was shown to process only pilins and FlaK only flagellins. For EppA cleavage, the presence of a +1 glutamine, found in all pilins but none of the flagellins, was important. Swapping the -2 to +2 amino acids of the pilins to flagellins resulted in flagellins that could then be now processed by EppA. However, swapping the -2 to +2 amino acids found in flagellins into pilins did not result in pilins that could be processed by FlaK. Thus, the amino acids located in the immediate vicinity of the cleavage site did not provide sufficient information for cleavage by FlaK, unlike the case with EppA.

However, the pilins do not have a glycine at position +3 and it may be that this is what determines the substrate specificity for FlaK since, in the case of *M. voltae* FlaB2, changing the +3 glycine to valine resulted in a lack of processing by FlaK [118].Recently, it was shown that FlaK also has a minimal signal peptide length needed for processing [119]. This was determined by generating a series of *M. voltae* preflagellins that have signal peptides ranging from the wildtype length of 12 amino acids to three amino acids while maintaining the wildtype positions at position -3 to -1. Preflagellins with signal peptides as short as 5 amino acids were processed but not ones with signal peptides shortened to 4 or 3. An exception was that a substrate containing a signal peptide of 6 was not processed. Similar studies have also been performed on substrates of PibD. A number of proposed substrates for PibD have very short signal peptides and it was unclear whether these would be processed by PibD.

However, even with a heterologous substrate (*Methanococcus* flagellins) it was shown that, unlike FlaK, PibD could process preflagellins with signal peptides as short as only 3 amino acids [119]. Site-directed mutagenesis studies of a PibD substrate, the sugar binding protein, GlcS, again showed the importance of a glycine at the -1 position but here the alanine substitution mutant protein was readily cleaved [59]. Various other substitutions at conserved positions in the substrate signal peptide revealed that PibD was much less stringent in its requirements for cleavage. This was in accordance was its extended range of substrates, compared to FlaK. PibD of *S. solfataricus,* however, may be exception in its substrate range given that even in other *Sulfolobus* species the genes for many of the PibD substrates are not present. While the activity of preflagellin signal peptide removal was readily followed in crude membrane fractions using an in vitro assay [116,117] based on one first developed for prepilin peptidase activity [120], it was ultimately determined that this activity resided in an enzyme designated FlaK [57,58]. The gene for FlaK is not usually found in the main cluster of *fla* genes that contains the flagellins, but is instead located at some distance form this locus. There are exceptions, as in *M. jannaschii,* where *flaK* is located immediately downstream of *flaJ* in the *fla* operon. Expression of *flaK* in *E. coli* and use of those membranes in the in vitro assay showed definitively that FlaK was the preflagellin peptidase.

Knockouts of *flaK* lead to nonflagellated cells confirming the essential role that preflagellin processing has in flagellin assembly into the filament. Deletion of the TFPP equivalent, *pibD,* in *Sulfolobus* has not been possible, likely reflecting its role in processes outside that of flagellation.

As with the substrates, site directed mutagenesis was used to determine critical amino acids necessary for catalysis of the archaeal TFPPs. Bacterial prepilin peptidases (i.e., PilD of *P.aeruginosa*) are members of COG 1989 and are a novel group of aspartic acid proteases [121]. Two aspartic acid residues have been shown to be critical to their catalytic activity. FlaK has several conserved aspartic acids and two of them align with the essential aspartic acids of the bacterial prepilin peptidase. As shown by site directed mutagenesis, these two aspartic acid residues, and not the other conserved ones, are necessary for FlaK activity [58]. This was subsequently shown to be the case for PibD as well [122], clearly indicating that the archaeal enzymes have the same mechanism of catalysis as the bacterial prepilin peptidases. A difference between the archaeal enzymes and the bacterial ones, however, is that the bacterial enzymes (like PilD) are bifunctional; not only do they remove the signal peptide but they subsequently methylate the resulting N-terminal amino acid, typically phenylalanine [55]. This is not the case with the archaeal TFPPs. In methanococci, there are both flagellins and pilins that are made with class III signal peptides. Separate signal peptidases cleave each; FlaK cleaves preflagellins and EppA cleaves the pilin-like preproteins [57,63]. In *Sulfolobus*, a single enzyme PibD cleaves a wide variety of substrates made with class III signal peptides [59]. Recently, the PibD equivalent in *Haloferax* was studied [15]. In this extreme halophile, only a single prepilin-like peptidase was found in the genome, although several different proteins with class III signal peptides were predicted. Comparisons of wildtype and PibD knockout strains showed that PibD is not only responsible for cleaving preflagellins but also several pilin-like proteins. Thus in *Haloferax*, as in *Sulfolobus*, a single enzyme is responsible for cleavage of multiple class III signal peptide-bearing proteins. Why *M. maripaludis* has two such enzymes with separate, designated substrates is currently unknown.

N-Linked Glycosylation of Flagellins

The second major posttranslational modification that flagellins undergo in archaea is glycosylation. Judging from the SDS-PAGE migration of archaeal flagellins relative to their mass deduced from gene sequences, it seems likely that most archaeal flagellins will ultimately be found to be glycoproteins. To date, only N-linked glycans have been reported on archaeal flagellins. In bacteria, flagellins may also be glycoproteins, although this seems to be a much rarer event than in archaea. However, in the case of all bacterial flagellin glycoproteins, there is a distinctive difference; the attached glycan is always O-linked to the protein [98,123-125]. Examination of archaeal flagellin amino acid sequences reveals that almost all have multiple N-linked sequons (N-X-S/T, where X is not proline). The flagellins of *Halobacterium salinarum* were the first to be shown to have attached glycan [112]. This glycan was originally reported as part of the glycan repertoire found on the S layer glycoprotein. The structure was revealed to be a sulfated oligosaccharide linked to asparagine via glucose, the first time such a linkage was reported in a glycoprotein. The glucose linking sugar is extended by 2-3 β-1,4 linked glucuronic acids, which are sometimes replaced with iduronic acid. Studies strongly suggested that the site of attachment of the glycan was located outside of the cell [111]. No genetic studies identifying genes involved in archaeal glycosylation were done, however, until almost 20 years later.

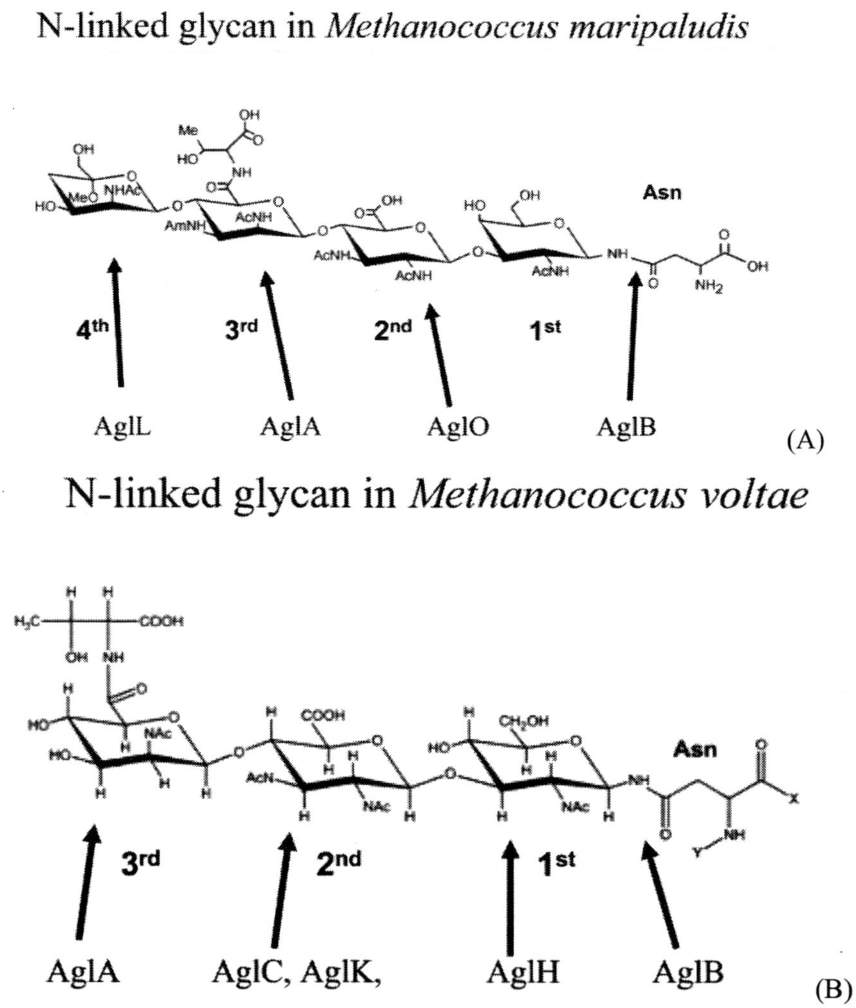

Figure 4. The N-linked glycans attached to the flagellins of M. maripaludis (A) and M. voltae (B). The step of involvement of glycosyltransferases, encoded by various agl genes, in transfer of the sugar units to assemble the glycan on a dolichol carrier, before transfer to the protein target by the oligosaccharyltransferase AglB, are indicated.

The first genes shown to be involved in archaeal N-linked glycosylation were reported almost simultaneously in *Methanococcus* [93] and *Haloferax* [126], but only in *Methanococcus* have the studies pertained to flagellins. Two *Methanococcus* species have been studied, initially *M. voltae* and subsequently, due to improved genetic techniques, the focus shifted to its relative *M. maripaludis*.The complete structures of the N-linked glycans of both *M. voltae* [92] and *M. maripaludis* [91] flagellins have been determined (Figure 4A, B). In the case of *M. voltae*, this is a trisaccharide (or in some versions of the strain, a tetrasaccharide [127]) that is also found on the S layer protein. The structure of the novel N-linked trisaccharide of mass 779 Da was β-Man*p*NAcA6Thr-(1–4)-β-Glc*p*NAc3NAcA-(1–3)-β-Glc*p*NAc with the linking sugar being N-acetylglucosamine as found in most eukaryotic glycoproteins. The second sugar is a di-acetylated glucuronic acid and the terminal sugar is a modified mannuronic acid with an attached threonine.

Genes involved in the assembly and transfer of the glycan have been identified. The oligosaccharyltransferase responsible for the transfer of the assembled glycan to the target protein was identified based on similarities to the Stt3 subunit that is the catalytic component of the oligosaccharyltransferase complex in eukaryotes [93]. This gene, designated *aglB*, encodes a protein that has the same catalytic motif (WWDYG) found in both Stt3 and in the bacterial version of the oligosaccharyltransferase, PglB. The archaeal enzyme is quite adept at transferring partial structures. In mutants that can only assemble a partial glycan, there is no evidence from western blot experiments that suggest any significant buildup of flagellins lacking the truncated glycans [114]. Disruption of *aglB* in *M. voltae* results in flagellins and S layer proteins that migrate as much smaller proteins that those in the wildtype cells, as expected if the *aglB* gene product was the oligosaccharyltransferase. Since the mutants were apparently healthy, the N-linked glycosylation system is not essential in *M. voltae*; nor is it essential in *H. volcanii* [126].

However, *aglB* mutant cells were nonflagellated, showing the importance of the glycan attachment for filament assembly.

The glycosyltransferases responsible for assembly of the trisaccharide were also identified. The first transferase which adds N-acetylglucosamine to a dolichol carrier could not be knocked out. However, it was shown that this gene (*aglH*) could complement a yeast conditional lethal mutant in *alg7*, which encodes N-acetylglucosamine-1-phosphate transferase [128]. Both eukaryotes and archaea use dolichol as a lipid carrier for assembly of the N-linked glycans, in contrast to bacteria which rely on undecaprenol. Other knockouts, coupled with mass spectrometry, revealed that AglA was responsible for adding the terminal sugar [93] while two proteins AglK and AglC were necessary for attachment for the second sugar [127]. Interestingly, the *aglA* knockout which resulted in a 2 sugar glycan was still flagellated while knockouts of *algK* or *aglC* which resulted in a single sugar glycan were nonflagellated. Thus, a minimum 2 sugar glycan was essential for incorporation of flagellins into the filament. In *M. maripaludis*, a tetrasaccharide was found attached to flagellins [91]. The structure was determined to be Sug-4-β-ManNAc3NAmA6Thr-4-β-GlcNAc3NAcA-3-β-GalNAc-Asn (where Sug is a novel monosaccharide unit (5*S*)-2-acetamido-2,4-dideoxy-5-*O*-methyl-α-l-*erythro*-hexos-5-ulo-1,5-pyranose). Significant differences to that of the *M. voltae* glycan were found. In *M. maripaludis*, the linking sugar to the protein is N-acetylgalactosamine, not N-acetylglucosamine. While the second sugar is the same, a diacetylated glucuronic acid, the third sugar is modified in *M. maripaludis*, in having an acetamidino group attached. The terminal sugar is a unique sugar which had previously been unknown in nature.Using newly developed genetic techniques, inframe deletions and complementation of targeted genes thought to be involved in the biosynthesis, assembly and attachment of the glycan were carried out. In this way, the oligosaccharyltransferase (AglB) as well as the glycosyltransferases involved in transfer of the last (AglL), the third (AglA) and the second sugar (AglO) were determined [114]. *M. maripaludis* contains many genes annotated as potential glycosyltransferases. Attempts to determine which one was responsible for the transfer of the first sugar to dolichol were unsuccessful, although some glycosyltransferase genes could not be successfully deleted. In addition, some genes involved in the biosynthesis of the sugars were also identified. MMP0350 is an acetyltransferase whose deletion leads to flagellins that migrate the same distance in western blots as knockouts in *aglO* (i.e. with a single sugar glycan) [103,114]. A simple explanation is that MMP0350 transfers one of the 2 acetyl groups to sugar 2 and that without this modification the second sugar is not added to the glycan. MMP0350 appears to transfer the acetyl group to the C-3 position as it has been shown in in vitro assays with purified overexpressed gene products that MMP1076 (GlcN-1P uridyltransferase/acetylase), acetylates the C-2 position of glucosamine-1-phosphate to generate *N*-acetylglucosamine. MMP1081 is responsible for adding the acetamidino group to sugar 3. Without this activity, the terminal sugar is not added. Finally, MMP1085 is a methyltransferase which adds a methyl group to the terminal sugar.

Mutants carrying deletions of *aglB,aglO* or MMP0350 are nonflagellated [103,114]. These mutants either lack a glycan altogether on their flagellins or the glycan is only a single sugar. Deletions of the other genes do not result in nonflagellated cells. However, the more truncated the attached glycan the less far these cells could migrate in semisolid medium. Thus, a minimum glycan size of 2 sugars is required for flagellin assembly, as seen also in *M. voltae*, but also a complete glycan is needed for maximum swimming capability [114]. A summary of the effects of specific *agl* gene deletions on glycan structure, flagellation and motility in *M. maripaludis* is presented in Table 3. Besides the genetic information available for the glycan components compiled by the Jarrell lab, in vitro assays have been done on a number of gene products expressed in *E. coli* that play a role in the biosynthesis of the acetamino sugars in methanococci by the Graham lab [129].

Several of the genes for these proteins are located very near to the genes for glycosyltransferases shown genetically to be involved in the glycan assembly. For example, MMP1077 converts glucosamine-6-phosphate to glucosamine-1-phosphate and MMP1076 acts in a bifunctional manner, both acetylating the 2-amino group as mentioned above and also facilitating the transfer to UTP to produce UDP-*N*-acetylglucosamine. MMP0705 (WecB) catalyses the isomerization of UDP-*N*-acetylglucosamine to UDP-N-acetylmannosamine while the neighboring gene MMP0706, encodes a UDP-ManNAc 6-dehydrogenase (WecC) which oxidizes UDP-N-acetylmannosamine to UDP-N-acetylmanno saminuronate, a direct precursor to the third sugar of the *M. maripaludis* glycan. Attempts to delete either MMP0705 or MMP0706 have so far been unsuccessful (unpublished data), indicating a possible essential role of these enzymes in the cell. However, since deletion of *aglB*, the oligosaccharyltransferase is possible, the essential role of MMP0705 and MMP0706 cannot be in the N-glycosylation system, but perhaps in component B biosynthesis. Component B is an essential cofactor for methanogenesis and mutations that prevent biosynthesis of its active form would be lethal. It has been reported that component B has an attached N-acetylmannosaminuronate β1-4 UDP-*N*-acetylglucosamine disaccharide [130] and disruption of the biosynthesis of its sugar moiety may lead to inactive component B and cell death. 2-acetamido sugars such as those discussed have also been found in *M. voltae* glycolipids [131] and these could prove essential for cell viability as well.

In *M. voltae,* all possible N-linked glycan sites in all four flagellins are occupied (one potential site in the minor flagellin FlaA was not covered in the peptide analysis) [92]. In *M. maripaludis*, the first potential N-linked glycan site in both FlaB1 and FlaB2 occurs at position N-26 of the mature protein (i.e. after signal peptide removal) and in both cases this site is unoccupied [91]. All of the other 9 sites found distributed in the three flagellins FlaB1, FlaB2 and FlaB3 were occupied. Strangely, some of the flagellins of *M. voltae* also have a potential N-linked site at position N-26 and, in this methanogen, that site is occupied by glycan.

Table 3. Effects of *agl* gene deletions on flagellin glycosylation and motility in *M. maripaludis*

Gene Deletion	Function	Effect on	Smaller	Flagella	Effect on motility
		glycan structure	MW flagellin		
aglL	4th GTase	Missing 4th sugar and Thr on 3rd sugar	Yes	Yes	Decreased cp to WT
aglA	3rd GTase	Missing 3rd and 4th sugars	Yes	Yes	Decreased cp to *aglL* mutant
aglO	2nd GTase	Missing 2nd, 3rd and 4th sugars	Yes	No	Nonmotile
aglB	OTase	No glycan	Yes	No	Nonmotile
MMP0350	Acetyl Transferase 2nd sugar	Missing 2nd, 3rd and4^{th} sugars	Yes	No	Nonmotile
MMP1081	Acetamidino transfer to 3rd sugar	Missing 4th sugar and acetamidinogroup of 3rd sugar	Yes	Yes	Decreased cp to WT
MMP1085	Methyl transferase	Missing methyl group of 4th sugar	Yes	Yes	ND
MMP1083	ND	ND	Yes	Yes	ND
MMP1094	ND	ND	Yes	Yes	ND

It has been assumed that a specific enzyme, a flippase, must be necessary for flipping the glycan assembled by glycosyltransferases on the cytoplasmic side of the cytoplasmic membrane to the external face of the cytoplasmic membrane where AglB transfers it to the target protein. An ATP-dependent flippase performs this function in bacteria [132] while the nature of the flippase in eukaryotes has recently become muddied, after doubts were expressed that the flippase here was Rft1, as originally proposed [133,134]. However, attempts to find this putative flippase in archaea have been unsuccessful.

Directed Studies on Archaeal Flagella

While much information has been gleaned from complete genome sequences concerning archaeal flagella and its widespread distribution and conservation throughout the domain, it has been directed studies on the flagella of several archaeal species that have defined the organelle as a unique motility apparatus. These studies have been done using biochemical, biophysical, electron microscopic and/or genetic approaches and all have contributed to our current understanding of this motility structure. These studies are reviewed below.

Extreme Halophiles

Halobacterium

The unusual flagella of archaea were first studied in any depth in *Halobacterium salinarum*. The filaments were shown to form a right handed helix, unlike the left handed helix typical of bacterial flagella [86].

Rotation of tethered *H. salinarum* cells demonstrated that the cells moved by rotation of flagella. *H. salinarum* cells swim forward by clockwise rotation of their flagella and backwards by counterclockwise rotation and the bundles do not fly apart when the rotation direction changes [86].

Later work showed that the flagella could undergo polymorphic transitions from normal to curly, straight and a ring by incubations at different temperatures or pHs [135].

The first archaeal flagellin genes were sequenced from *H. salinarum* [72] and they provided the first clues that archaeal flagella may be unique among prokaryotes as they bore no sequence similarity to bacterial flagellins. Two transcriptional units were identified [72,105], one encoding tandem genes *flgA1* and *flgA2* and a separate unit encoding three genes *flgB1*, *flgB2* and *flgB3*. The two gene groups are located at some distance on the chromosome from each other.

Nonetheless, all five gene products are found in purified flagella [90]. Three potential N-linked glycosylation sites were found in each flagellin. This was not unexpected as the flagellins were shown earlier to be glycoproteins and to carry an N-linked glycan that was also found on the S layer protein [112]. Mutational studies of the separate flagellin genes indicated specific roles for the various flagellins. Inactivation of both *flgA* genes resulted in short curved flagella, composed entirely of FlgB flagellins, that were mislocalized: instead of the normal polar location, these flagella were found at the sides and the poles of cells [105]. Inactivation of only *flgA2* lead to straight flagella located mainly at the poles of the cells. These flagella were composed entirely of FlgA1, as transcription of the *flgB* transcript was completely shut off. If the *flgB* flagellins were deleted, the cells made flagella composed of both FlgA flagellins but again the appearance was unusual, being spiral and located at the poles.

All flagellins appeared to be required for normal, functioning flagella. Based on this work, the authors proposed that the FlgA flagellins were incorporated first into the filaments followed by the cell proximal FlgB flagellins. Later evidence was presented suggesting that the hook region was composed of FlgB3 [88].

The *fla* associated genes are found on a separate transcript from both sets of flagellin genes [70]. This is due to their transcription being in the opposite orientation although the *fla* genes are located neighboring the *flgB* gene cluster. The *fla* associated genes *flaD-flaK* (*flaK* is not the preflagellin peptidase though), preceded by a halobacterial transducer gene, are all co-transcribed. Levels of mRNA for these genes were much lower than the mRNA from the flagellin structural genes. Within this cluster, *flaD, E, G, H, I* and *J* share high sequence homologies with *fla* associated genes from other Archaea. Inframe deletion of the ATPase, *flaI*, led to nonflagellated cells.

It was also in *H. salinarum* that the first connection of the archaeal specific flagella with the bacterial-like chemotaxis system was shown. In bacteria, this interaction is between CheY and the switch protein FliM where a phosphorylated CheY binds to FliM and results in a change in rotation of the flagella motor [30].

A homolog of *fliM* has not been found in archaea and how the chemotaxis system relays its message to archaeal flagella is a mystery.However, recently Oesterhelt's group identified three proteins in *H. salinarum* that interacted with the chemotaxis proteins CheY, CheD, and CheC2, as well as FlaCE and FlaD [24].

Strains deleted for these proteins failed to switch rotation of their flagella, and swam only clockwise and were unable to respond to different tactic signals. Another important finding was that flagella rotation was dependent on ATP, either directly or though an ATP-driven ion gradient, and not the proton motive force or sodium motive force used by bacterial flagella [45].

Haloferax

Haloferax volcanii had been considered a nonmotile organism [14] but analysis of the sequenced genome did reveal the presence of archaeal flagellin and *fla* associated genes [16]. Unlike the case in *Methanosarcina* where a similar situation is found, new results indicate that under certain growth conditions *H. volcanii* can swim in a flagella-dependent fashion, although critically no flagella were observed by electron microscopy [15]. Nonetheless, deletion of flagellin genes prevented swimming.

Deletion of *pibD* also lead to nonmotile cells, as expected, and the flagellins were unprocessed. This enzyme was also responsible for processing of type IV pilin-like proteins. The cells deleted for the flagellin genes were still able to adhere to surfaces by a nonflagellar type IV pilus-like structure.

Haloarcula

Like *Haloferax volcanii*, *Haloarcula marismortui* cells were considered nonmotile but recent studies have proven that conditions exist where these cells are flagellated and motile [42]. The genome sequence of *H. marismortui* revealed two typical flagellin genes (*flaB* and *flaA2*), only one of which (*flaB*) was associated with the *fla* operon [71]. The second was located on a different replicon.Isolated filaments were much thicker than typical archaeal flagella and, at 20 nm, were more like bacterial flagella in diameter [42]. SDS-PAGE revealed that the filaments were composed primarily of a single protein species of about

80kDa, much larger then either of the predicted flagellin proteins (approximately 45 kDa). Interestingly, none of the discrepancy between the SDS-PAGE determined mass and the gene deduced mass of the flagellins can be attributed to glycosylation (at least not to N-linked glycans) as the flagellin sequences did not contain any potential N-linked sites, a rarity for archaeal flagellins. Glycoprotein staining also was consistent with this, as the flagellins did not stain with Alcian blue or PAS. Two phenotypes of *H. marismortui* flagella were identified, one with filaments composed primarily of FlaB and the other with filaments composed primarily of FlaA2. The FlaA2 filaments were thicker (20-22 nm) than the filaments composed primarily of FlaB (16-18 nm). This observation is reminiscent of phase variation found in various Enterobacteria [136]. Interestingly, electron microscopic observations of isolated flagella or flagellated cells of *H. marismortui* and *H. salinarum* do not reveal the presence of a differentially staining hook region as seen in various methanogens [42].

Flagella and motility have also been studied in the square archaea, *Haloarcula quadrata* [137,138]. Like those of halobactera, the flagella of *H. quadrata* have right-handed helical form. The flagella were observed to rotate. These square archaea have either single filaments attached or giant flagella bundles [138]. Isolated flagella appeared to be composed of a single protein band of 73 kDa when examined by SDS-PAGE.

Natrialba

Four tandem flagellin genes were identified in *Natrialba magadii*, *flaB1*-*flaB4* encoding proteins with predicted molecular weights of 21.4, 26.5, 41.7 and 27.0 [68]. Transcriptional analysis revealed that all 4 flagellins and possible the first of the adjacent *fla* associated genes were co-transcribed. However, the flagellins identified by SDS-PAGE were much larger than the sizes predicted from the genes, at 105,000, 60,000, 59,000 and 45,000. *flaB1* corresponds to the 45,000 band; *flaB2* to 60,000, *flaB3* to 105,000 and *flaB4* to 59,000. These discrepancies between SDS-PAGE and deduced molecular weights are among the largest reported for archaeal flagellins which typically do run as larger than predicted due to glycosylation. In *N. magadii,* however, only the 60,000 and 45,000 flagellins are glycoproteins. *N. magadii* flagella dissociate into protofilaments when the salt levels fall below 8% [139]. The protofilaments are much thinner than the filaments themselves (3-5nm vs 10-12 nm). As the salt concentration is lowered further to 0, dissociation continues so that only protein aggregates are left. A rare attempt to monitor the distribution of the various flagellins along the flagellar filament was reported in *N. magadii* [94]. Antibodies that recognize only FlaB4 and FlaB2 uniformly bound along the length of filaments. However, other filaments, indistinguishable outwardly from labeled filaments went completely unlabelled, indicating the presence of at least two immunologically distinct protofilament types. Evidence was presented that parts of the *N. magadii* flagellin sequences may have been obtained through horizontal gene transfer from *E. coli* EPEC 1 group.

Sulfolobus

Flagella filaments were isolated from *Sulfolobus shibatae* and shown to have a slightly larger diameter, at 15 nm,than those of most typical archaea [9]. Analysis of the purified filaments by SDS-PAGE revealed two major flagellin bands at 31,000 and 33,000. Both bands stained positively as glycoproteins and both flagellins had identical amino acid sequences for at least the first 23 N-terminal positions. The genome sequence of *S. shibatae* is not available but since other *Sulfolobus* species possess only a single flagellin gene, this may be likely for *S. shibatae* too. If so, then the two flagellin bands observed by SDS-PAGE would likely represent the same flagellin protein glycosylated to different extents.

Flagellation has also been studied in *S. solfataricus*. Analysis of deletion mutants revealed that flagella, as well as pili, were necessary for attachment to various surfaces. In addition, a deletion mutant in *flaJ*, encoding the conserved membrane protein found in all *fla*operons, was nonflagellated [97]. *Sulfolobus* cells do not show reversed swimming or tumbling. Purified flagella had a diameter of 14.5 nm and formed a right handed helix, as also found in *Halobacterium* [86] and *Haloarcula* [42] A hook region has not been seen in *S. solfataricus* flagella [89]. The question of whether all archaea flagella contain a hook region or not is open to debate: given that the hook is composed of a "flagellin" in *Methanococcus*, this point is especially of interest in members of the Sulfolobales (*Sulfolobus* and *Metallosphaera*) where there is only one flagellin gene in the genome.

Thermoplasma

Flagellation in *Thermoplasma* species had been a somewhat controversial observation for years. This was due in part to *Thermoplasma* lacking a cell wall external to the membrane. Flagellation and swimming were observed in early studies of *Thermoplasma acidophilum* and then *T. volcanium* [140,141]. Later, flagella were isolated from *T. volcanium* and shown to consist primarily of a single 41 kDa band, which stained positively in PAS staining, indicative of glycoprotein [9]. N-terminal sequencing of this band revealed a very close sequence similarity to flagellins of *M. voltae*; 18 consecutive positions (amino acid positions 3-21) were identical in flagellins of the two organisms.Examination of the subsequent complete genome sequence revealed two flagellin genes of similar size (25-26 kDa) with very similar amino acid sequences at the N-termini. Whether the single 41 kDa band observed in SDS-PAGE analysis of purified filaments is a mix of the two flagellins is unknown. Only one flagellin is located next to the rest of the *fla* proteins while the second is located a distance away and not surrounded by any typical *fla* genes. Possibly, the one located within the *fla* operon is the major flagellin of the filament.

Methanococcus

Much of the work on the biochemistry, electron microscopy and genetics of flagellation in *Methanococcus* has been done on *M. voltae* and *M. maripaludis*. These studies have been detailed in the main body of the text above and are not repeated here. However, some additional work has been reported on other *Methanococcus* species.

Amplification of the flagellin genes from *M. vannielii* was done by inverse PCR, taking advantage of the highly conserved N-termini of tandem flagellin genes [142]. Three flagellin genes were obtained this way and FlaB1 and FlaB2 were found by N-terminal sequencing of the purified flagella to be the major structural proteins of 30.8 and 28.6 kDa. No detection of FlaB3 in the purified filaments was obtained. Examination of the subsequent complete genome sequence revealed only these three flagellin genes lying in tandem in a typical methanococcal *fla* operon. Since the flagella were obtained by shearing from the cells, if FlaB3 were the hook protein as in *M. maripaludis*, it would likely be present in only small amounts and could be missed.

Methanococcus deltae flagella were purified and shown to consist of two major flagellins of 34 and 32 kDa [143]. These flagellins stained positively as glycoproteins with PAS. Furthermore, the implied glycosylation was shown to be inhibited by the known glycosylation inhibitor, bacitracin [144]. Incubation with increasing concentrations of bacitracin lead to a progressive decrease in flagellin molecular weight detected in western blots, likely representing under or non glycosylated forms. Interestingly, these cells were nonflagellated, giving the first hint in methanogens that glycosylation was a necessary modification for flagellin assembly into filaments.

Flagellins of 32 and 27 kDa were found in purified flagella preparations of *M. jannaschii*, neither of which stained with PAS [143]. Subsequent genome sequencing revealed that this methanogen has a typical methanococcal *fla* operon with three flagellin genes located at the beginning of an operon including *flaC-J* but with the addition of *flaK* [145].

Methanospirillum

Methanospirillum cells have one of the most complex cell envelopes of all archaea [146-149]. Individual cells are surrounded by a cell wall consisting of an S layer and the individual cells are enclosed in a paracrystalline sheath [147]. The cells are separated from neighboring cells by complex spacer plugs and, at the ends of the sheaths, by end plugs. When grown as short filaments of a few cells or as single cells, the cells swam at speeds estimated at 8um/s. The flagella project through the end plugs and not the sheath [150]. Isolated flagella were sensitive to detergent, as is true of many archaeal filaments [151,152]. However, the filaments were stable to high temperatures and a broad pH range. In the case of *M. hungatei* flagella, two major flagellins of 24 and 25 kDa were found, both staining as glycoproteins [150]. Chemical deglycosylation lead to a reduced molecular weight supporting the glycoprotein staining result. Total amino acid compositions of the two flagellins revealed distinctive features of each, suggesting they were not differentially glycosylated forms of the same flagellin. N-terminal sequences of intact and cyanogen bromide cleaved flagellin fragments were used to piece together a long stretch of the N-terminus amino acid sequence of the flagellin [151]. Examination of the subsequent complete genome sequence revealed the presence of two tandem flagellin genes.

Methanosarcina

Like *H. volcanii* and *H. marismortui*, *Methanosarcina* is considered nonmotile but complete genome sequences of various species revealed an apparently full complement of flagellins and *fla* associated genes [11,153].Whereas the extreme halophiles have now been shown to be motile under certain growth conditions, such attempts in *Methanosarcina* remained ineffective. Whether these species can be motile given the right conditions remains unknown.

Methanoculleus and Methanothermus

Flagella preparations have been examined from other methanogens for molecular weight of the flagellins and their reaction in PAS staining. *Methanoculleus marisnigri* flagella were composed of two major flagellins, 31 and 25.5 kDa, which did not react with PAS [143].Those of *Methanothermus fervidus* were composed of a major smear centered on 34 kDa with minor components at 25 and 24 kDa. All of these stained positively with PAS [143].

Thermococcus/Pyrococcus

Transcriptional analysis of the five tandem flagellin genes of *Thermococcus kodakaraensis* was reported [69]. Multiple transcripts, all initiating upstream of *flaB1,* were found which would encode various combinations of all 5 flagellin genes. A transcript size of 5.4 kb was found that would extend from *flaB1* to *flaB5* but a much larger mRNA of 9.2 kb of strong intensity was also detected. Since *T. kodakaraensis* flagellins genes are followed by *flaC-J*, most, if not all, of the *fla* associated genes would be expected to be co-transcribed with the flagellin genes. From the subsequently published genome sequence, the beginning of *flaB1* to the end of *flaJ* can be estimated at approximately 12 kb.

Studies have recently begun of the flagella of *Pyrococcus furiosus*. *P. furiosus* is highly flagellated, with 50 or more flagella per cell. Purified filaments of *P. furiosus* were 9-10 nm in diameter, typical of most archaeal flagella and extremely stable to high temperature exposure [76].They were found to be composed primarily of a single glycoprotein of about 30 kDa. Besides a role in swimming, these flagella can associate to form cable-like structures that lead to cell to cell connections. In addition, the flagella can be used to adhere to a variety of solid surfaces as well as biotic surfaces [77].

Phylogenomics of the Archaeal Flagella System

Several labs have reported that searches of genetic databases have failed to detect any homologues of bacterial genes involved in flagella structure or assembly in archaeal genomes [23,31,42]. The phylogenetics of the archaeal flagella system have been investigated and homologues in type IV pilus and type II secretion systems have been identified. Indeed, after the initial finding of the sequence similarity of archaeal flagellins and type IV pilins at their

N-termini, the relationship between archaeal FlaI and type IV pilus system ATPases was the next evidence that suggested a close relationship between the two structures [109]. An in depth comparison of archaeal flagella, type II section and type IV pilus systems revealed the following homologues: a major and several minor prepilin-like proteins, a specific prepilin peptidase-like enzyme to cleave the unusual signal peptides from the prepilin-like proteins, an ATPase and an integral membrane protein with multiple transmembrane domains [48]. In the latter case, this is FlaJ of the archaeal flagella systems and type II secretion system protein F and type IV pilus protein C. Others have suggested that the sequence similarity is too weak for FlaJ to be considered a homologue of the others [108]. The archaeal flagella ATPase, FlaI, appears to function with the transmembrane protein FlaJ. Interestingly, however, the archaeal flagella ATPases are part of a large family that also includes archaeal ATPases that seemingly have nothing to do with archaeal flagella assembly and may, in fact, operate independently of a multispanning transmembrane protein homologue of FlaJ. The transmembrane protein probably arose from a gene duplication event in both bacteria and archaea [48]. The archaeal version, FlaJ, is about twice as big as the bacterial version suggesting that the bacterial version may function as an oligomer.

Phylogenomic studies have revealed that components of the archaeal flagella system have not been transferred frequently among different archaeal species and no clear evidence for any recent horizontal gene transfer of *fla* genes could be found [108]. However, it was suggested that an ancient horizontal gene transfer may have occurred involving an entire *fla* gene cluster from Crenarchaeota, such as members of the Sulfolobales, to a distantly related Methanomicrobia.

Assembly Model for Archaeal Flagellum

It was recognized early on in the study of archaeal flagella that perhaps the assembly of these structures may be different from those of bacterial flagella. Sumper's group [72] suggested a novel assembly model different from that of bacterial flagella may be required based on their observations that glycosylation of the flagellins occurred on the external face of the cytoplasmic membrane and so the flagellins would have to be translocated across the cytoplasmic membrane first [72]. It was Jarrell's group however that first recognized the similarity of archaeal flagellins and type IV pilins and suggested a model in which archaeal flagella grow by incorporation of subunits at the base of the structure rather than at the distal tip, i.e. a model like pili growth rather than like bacterial flagella growth [6]. This model was proposed based primarily on the discovery of the type IV pilin-like signal peptides on archaeal flagellins and the amino acid sequence similarity of archaeal flagellins and pilins at their mature N-termini [52,53]. The further discovery of genes with similarities to those in type IV pili systems, namely the homologous signal peptidases, ATPases and conserved membrane proteins all were consistent with this hypothesis [48,57,109]. Finally, the structural analysis of several archaeal flagella revealed again a similarity to type IV pili rather than bacterial flagella [46,47]. Critical among these discoveries was the fact that the archaeal flagella did not have a central channel with a large enough diameter to allow passage of new subunits, the key structural feature possessed by bacterial flagella that allows for the insertion of new subunits at the distal tip of the organelle. Without this ability to transport the newly synthesized subunits to the distal tip, the proposed mechanism of insertion of new subunits at the base seemed, by default, the only possibility.

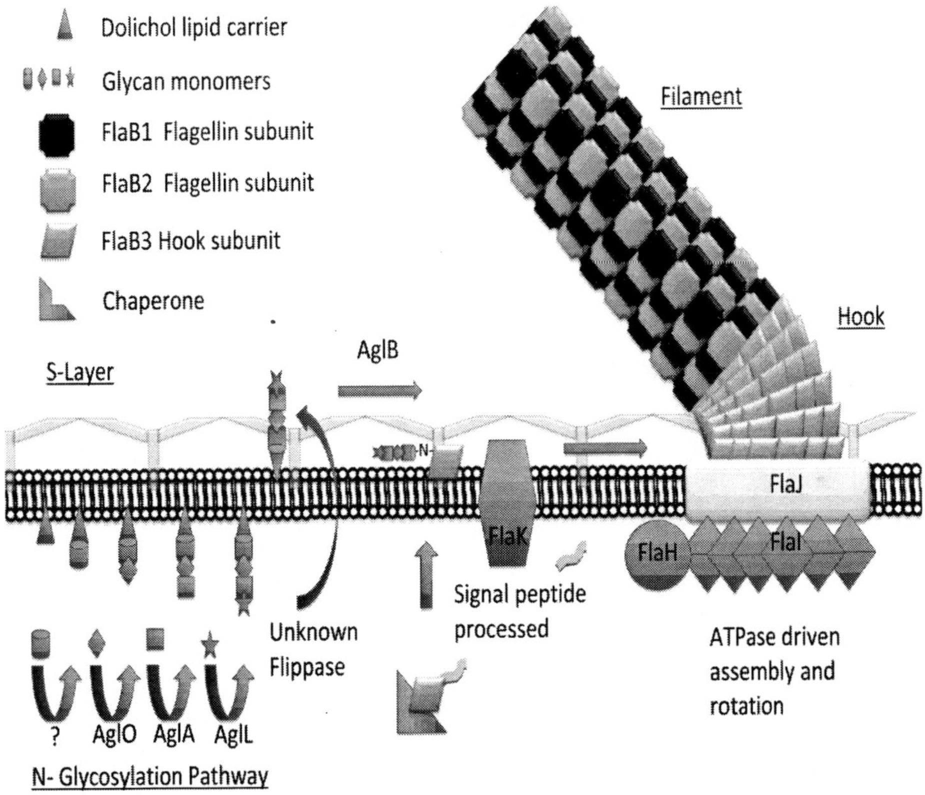

Figure 5. Model for assembly of flagella in M. maripaludis, with the N-linked glycosylation and signal peptide steps incorporated.

A working model of assembly of *M. maripaludis* flagella, incorporating signal peptide processing and N-linked glycosylation steps is presented in Figure 5. Flagellins are synthesized in the cytoplasm and transported to the cytoplasmic membrane, likely by chaperones, although early work on this important process has not been followed up [154].

Considering the importance of chaperones in assembly steps of bacterial flagella, including different chaperones for many different structural components, it would seem that further study of archaeal flagella chaperones is warranted [155-157].

Once at the membrane, the flagellins undergo two important posttranslational modifications necessary for the incorporation of the flagellins into the flagella filament. These are removal of the signal peptide by FlaK and transfer of the N-linked glycan by AglB to selected asparagines on the flagellins.These may be done virtually simultaneously in wildtype cells. Certainly, based on SDS-PAGE migration patterns in *flaK* mutants, flagellins that do not have their signal peptide removed can still be substrates for AglB and receive the N-linked glycan [58]. However, since *flaK* mutants are nonflagellated, removal of the signal peptide is an essential step for subsequent incorporation into the filament. Interestingly, the flagellins also need to be glycosylated to be incorporated into a stable flagella filament.

However, in mutants that render the glycan incomplete but with a minimum of the first two sugars intact, the flagellins are modified with the truncated glycan and incorporated. In mutants in which the glycan is composed of less than two complete sugars (or is completely nonglycosylated), the flagellins still receive the truncated glycan but these flagellins are not incorporated into filaments [93,114].

The significance of the flagellins needing at least some glycan structure attached is not understood. Perhaps, the assembly mechanism at the base of the flagella is recognizing the partial glycan. In *M. voltae*, the same glycan attached to flagellins is also found on the S layer protein and a significant decrease in the mobility of S layer is observed in stained whole cells lysates in mutants carrying mutations in *agl* genes involved in assembly or synthesis of the glycan [93]. This suggested that the glycan is assembled separately on its dolichol lipid carrier and then transferred by AglB to all external proteins that are to be N-linked glycoproteins. In *M. maripaludis*, the S layer protein from *agl* mutants does not run noticeably faster in SDS-PAGE compared to wildtype cells. However, the *M. maripaludis* S layer protein only contains a single potential N-linked site and it is currently unknown whether this is, in fact, occupied. Interestingly, a further extracytoplasmicly located protein, the pilin MP1685, is also an N-glycosylated protein. Its glycan however is slightly different in that it possesses a further hexose attached as a branch to N-acetylgalactosamine (the linking sugar in this archaeon).

It is assumed, based on homologies to type IV pilin proteins that the ATPase, FlaI, and the conserved membrane protein, FlaJ, may serve as the minimum platform for assembly [110], perhaps also including FlaH since *flaHIJ* always occur in tandem near the flagellin genes [8]. As such, FlaI and FlaJ may interact as their homologues do in pili systems [48,158], although no evidence for this has yet been presented. Since hookless mutants can be obtained in *M. maripaludis* by deleting *flaB3*, it was deduced that the filament is made first from FlaB1 and FlaB2 and then the hook is made from FlaB3 [93]. Hookless mutants would not be expected normally in bacterial flagella since the hook must be made first before the filament flagellins can be added.

Concluding Comments

Archaeal flagella have proven to be excellent model systems to study post translational modifications of protein in archaea. Use of this structure as a model system has been instrumental in identifying the preflagellin peptidase, still the best understood of the archaeal signal peptidases [115]. Since archaeal flagellins are only modified with N-linked glycans, rather than both N- and O-linked glycans as is the case in many archaeal S layers, they have proven readily amenable to genetic and biochemical studies of the N-linked glycans. It seems apparent that the archaea have a greater variety of N-linked glycan components, including novel sugars, compared to the bacterial systems and these may represent a large untapped resource for designer glycoproteins. Since archaea appear to assemble many different surface structures using the basic type IV pilin-like model, studies on archaeal flagella can contribute much to the study of these other appendages as well as to our understanding of the bacterial type IV pilus. Finally, it is clear that the archaea have found a novel mechanism for assembling a rotating swimming organelle that was not invented by bacteria. The novelty of this structure is apparent although much information on its components and assembly and regulation are yet to be revealed.

Acknowledgments

Research in my laboratory has been funded by the Natural Sciences and Engineering Research Council of Canada (NSERC). I wish to thank all of my former students and collaborators for the contributions they have made to our studies on flagellation in archaea.

References

[1] Woese CR, Fox GE. Phylogenetic structure of the prokaryotic domain: the primary kingdoms. *Proc.Natl.Acad.Sci.U.S.A.* 1977;74:5088-5090.

[2] Woese CR, Magrum LJ, Fox GE. Archaebacteria. *J.Mol.Evol.* 1978;11:245-251.

[3] Woese CR, Kandler O, Wheelis ML. Towards a natural system of organisms: proposal for the domains Archaea, Bacteria, and Eucarya. *Proc.Natl.Acad.Sci.U.S.A.* 1990;87:4576-4579.

[4] Kandler O, Konig H. Cell envelopes of archaebacteria. The bacteria. A treatise on structure and function. Orlando Florida: Academic Press Inc.; 1985. p. 413-457.

[5] Zillig W. Comparative biochemistry of Archaea and Bacteria. *Curr.Opin.Genet.Dev.* 1991;1:544-551.

[6] Jarrell KF, Bayley DP, Kostyukova AS. The archaeal flagellum: a unique motility structure. *J.Bacteriol.* 1996;178:5057-5064.

[7] Thomas NA, Bardy SL, Jarrell KF. The archaeal flagellum: a different kind of prokaryotic motility structure. *FEMS Microbiol.Rev.* 2001;25:147-174.

[8] Jarrell KF, VanDyke DJ, Wu J. Archaeal flagella and pili. In: Jarrell KF, editor. Pili and Flagella: Current Research and Future Trends. Norfolk, U.K.: Caister Academic Press; 2009. p. 215-234.

[9] Faguy DM, Bayley DP, Kostyukova AS, Thomas NA, Jarrell KF. Isolation and characterization of flagella and flagellin proteins from the thermoacidophilic archaea *Thermoplasma volcanium* and *Sulfolobus shibatae*. *J.Bacteriol.* 1996;178:902-905.

[10] Jarrell KF, Ng SY, Chaban B. Flagellation and chemotaxis. In: Cavicchioli R, editor. Archaea: molecular and cellular biology Washington , D.C.: ASM Press; 2007. p. 385-410.

[11] Galagan JE, Nusbaum C, Roy A, Endrizzi MG, Macdonald P, FitzHugh W, et al. The genome of *M. acetivorans* reveals extensive metabolic and physiological diversity. *Genome Res.* 2002;12:532-542.

[12] Singh N, Kendall MM, Liu Y, Boone DR. Isolation and characterization of methylotrophic methanogens from anoxic marine sediments in Skan Bay, Alaska: description of *Methanococcoides alaskense* sp. nov., and emended description of *Methanosarcina baltica*. *Int.J.Syst.Evol.Microbiol.* 2005;55:2531-2538.

[13] von Klein D, Arab H, Volker H, Thomm M. *Methanosarcina baltica*, sp. nov., a novel methanogen isolated from the Gotland Deep of the Baltic Sea. *Extremophiles* 2002;6:103-110.

[14] Mullakhanbhai MF, Larsen H. *Halobacterium volcanii* spec. nov., a Dead Sea halobacterium with a moderate salt requirement. *Arch.Microbiol.* 1975;104:207-214.

[15] Tripepi M, Imam S, Pohlschroder M. *Haloferax volcanii* flagella are required for motility but are not involved in PibD-dependent surface adhesion. *J.Bacteriol.* 2010 in press.

[16] Hartman AL, Norais C, Badger JH, Delmas S, Haldenby S, Madupu R, et al. The complete genome sequence of *Haloferax volcanii* DS2, a model archaeon. *PLoS One* 2010;5:e9605.

[17] Volkl P, Huber R, Drobner E, Rachel R, Burggraf S, Trincone A, et al. *Pyrobaculum aerophilum* sp. nov., a novel nitrate-reducing hyperthermophilic archaeum. *Appl.Environ.Microbiol.* 1993;59:2918-2926.

[18] Kurr M, Huber R, Konig H, Jannasch HW, Fricke H, Trincone A, et al. *Methanopyrus kandleri*, gen. and sp. nov. representsa novel group of hyperthermophilic methanogens, growing at 110°C. *Arch. Microbiol.* 1991;156:239-247.

[19] Muller DW, Meyer C, Gurster S, Kuper U, Huber H, Rachel R, et al. The Iho670 fibers of *Ignicoccus hospitalis*: a new type of archaeal cell surface appendage. *J.Bacteriol.* 2009;191:6465-6468.

[20] Szurmant H, Ordal GW. Diversity in chemotaxis mechanisms among the bacteria and archaea. *Microbiol.Mol.Biol.Rev.* 2004;68:301-319.

[21] Rudolph J, Nordmann B, Storch KF, Gruenberg H, Rodewald K, Oesterhelt D. A family of halobacterial transducer proteins. FEMS *Microbiol.Lett.* 1996;139:161-168.

[22] Rudolph J, Oesterhelt D. Deletion analysis of the *che* operon in the archaeon *Halobacterium salinarium*. *J.Mol.Biol.* 1996;258:548-554.

[23] Faguy DM, Jarrell KF. A twisted tale: the origin and evolution of motility and chemotaxis in prokaryotes. *Microbiology* 1999;145:279-281.

[24] Schlesner M, Miller A, Streif S, Staudinger WF, Muller J, Scheffer B, et al. Identification of Archaea-specific chemotaxis proteins which interact with the flagellar apparatus. *BMC Microbiol.* 2009;9:56.

[25] Nutsch T, Marwan W, Oesterhelt D, Gilles ED. Signal processing and flagellar motor switching during phototaxis of *Halobacterium salinarum*. *Genome Res.* 2003;13:2406-2412.

[26] Hou S, Larsen RW, Boudko D, Riley CW, Karatan E, Zimmer M, et al. Myoglobin-like aerotaxis transducers in Archaea and Bacteria. *Nature* 2000;403:540-544.

[27] Lindbeck JC, Goulbourne EA,Jr, Johnson MS, Taylor BL. Aerotaxis in *Halobacterium salinarium* is methylation-dependent. *Microbiology* 1995;141 (Pt 11):2945-2953.

[28] Storch KF, Rudolph J, Oesterhelt D. Car: a cytoplasmic sensor responsible for arginine chemotaxis in the archaeon *Halobacterium salinarum*. *EMBO J.* 1999;18:1146-1158.

[29] Sment KA, Konisky J. Chemotaxis in the archaebacterium *Methanococcus voltae*. *J.Bacteriol.* 1989;171:2870-2872.

[30] Bren A, Eisenbach M. The N terminus of the flagellar switch protein, FliM, is the binding domain for the chemotactic response regulator, CheY. *J.Mol.Biol.* 1998;278:507-514.

[31] Nutsch T, Oesterhelt D, Gilles ED, Marwan W. A quantitative model of the switch cycle of an archaeal flagellar motor and its sensory control. *Biophys.J.* 2005;89:2307-2323.

[32] Marwan W, Alam M, Oesterhelt D. Rotation and switching of the flagellar motor assembly in *Halobacterium halobium*. *J.Bacteriol.* 1991;173:1971-1977.

[33] Brown PN, Terrazas M, Paul K, Blair DF. Mutational analysis of the flagellar protein FliG: sites of interaction with FliM and implications for organization of the switch complex. *J.Bacteriol.* 2007;189:305-312.

[34] Berg HC, Anderson RA. Bacteria swim by rotating their flagellar filaments. *Nature* 1973;245:380-382.

[35] Jarrell KF, McBride MJ. The surprisingly diverse ways that prokaryotes move. *Nat.Rev.Microbiol.* 2008;6:466-476.

[36] Chevance FF, Hughes KT. Coordinating assembly of a bacterial macromolecular machine. *Nat.Rev.Microbiol.* 2008;6:455-465.

[37] Macnab RM. Type III flagellar protein export and flagellar assembly. Biochim.Biophys.*Acta* 2004;1694:207-217.

[38] Aizawa SI, Dean GE, Jones CJ, Macnab RM, Yamaguchi S. Purification and characterization of the flagellar hook-basal body complex of *Salmonella typhimurium*. *J.Bacteriol.* 1985;161:836-849.

[39] Kalmokoff ML, Jarrell KF, Koval SF. Isolation of flagella from the archaebacterium *Methanococcus voltae* by phase separation with Triton X-114. *J.Bacteriol.* 1988;170:1752-1758.

[40] Bardy SL, Mori T, Komoriya K, Aizawa S, Jarrell KF. Identification and localization of flagellins FlaA and FlaB3 within flagella of *Methanococcus voltae*. *J.Bacteriol.* 2002;184:5223-5233.

[41] Cruden D, Sparling R, Markovetz AJ. Isolation and ultrastructure of the flagella of *Methanococcus thermolithotrophicus* and *Methanospirillum hungatei*. *Appl.Environ.Microbiol.* 1989;55:1414-1419.

[42] Pyatibratov MG, Beznosov SN, Rachel R, Tiktopulo EI, Surin AK, Syutkin AS, et al. Alternative flagellar filament types in the haloarchaeon *Haloarcula marismortui*. *Can.J.Microbiol.* 2008;54:835-844.

[43] Yakushi T, Yang J, Fukuoka H, Homma M, Blair DF. Roles of charged residues of rotor and stator in flagellar rotation: comparative study using H^+-driven and Na^+-driven motors in *Escherichia coli*. *J.Bacteriol.* 2006;188:1466-1472.

[44] Blair DF. Flagellar movement driven by proton translocation. *FEBS Lett.* 2003;545:86-95.

[45] Streif S, Staudinger WF, Marwan W, Oesterhelt D. Flagellar rotation in the archaeon *Halobacterium salinarum* depends on ATP. *J.Mol.Biol.* 2008;384:1-8.

[46] Cohen-Krausz S, Trachtenberg S. The flagellar filament structure of the extreme acidothermophile *Sulfolobus shibatae* B12 suggests that archaeabacterial flagella have a unique and common symmetry and design. *J.Mol.Biol.* 2008;375:1113-1124.

[47] Trachtenberg S, Cohen-Krausz S. The archaeabacterial flagellar filament: a bacterial propeller with a pilus-like structure. *J.Mol.Microbiol.Biotechnol.* 2006;11:208-220.

[48] Peabody CR, Chung YJ, Yen MR, Vidal-Ingigliardi D, Pugsley AP, Saier MH,Jr. Type II protein secretion and its relationship to bacterial type IV pili and archaeal flagella. *Microbiology* 2003;149:3051-3072.

[49] Mattick JS. Type IV pili and twitching motility. *Annu.Rev.Microbiol.* 2002;56:289-314.
[50] Bradley DE. A function of *Pseudomonas aeruginosa* PAO polar pili: twitching motility. *Can.J.Microbiol.* 1980;26:146-154.
[51] Craig L, Volkmann N, Arvai AS, Pique ME, Yeager M, Egelman EH, et al. Type IV pilus structure by cryo-electron microscopy and crystallography: implications for pilus assembly and functions. *Mol.Cell* 2006;23:651-662.
[52] Kalmokoff ML, Jarrell KF. Cloning and sequencing of a multigene family encoding the flagellins of *Methanococcus voltae*. *J.Bacteriol.* 1991;173:7113-7125.
[53] Faguy DM, Jarrell KF, Kuzio J, Kalmokoff ML. Molecular analysis of archael flagellins: similarity to the type IV pilin-transport superfamily widespread in bacteria. *Can.J.Microbiol.* 1994;40:67-71.
[54] Bardy SL, Eichler J, Jarrell KF. Archaeal signal peptides--a comparative survey at the genome level. *Protein Sci.* 2003;12:1833-1843.
[55] Strom MS, Nunn DN, Lory S. A single bifunctional enzyme, PilD, catalyzes cleavage and N-methylation of proteins belonging to the type IV pilin family. *Proc.Natl.Acad.Sci.U.S.A.* 1993;90:2404-2408.
[56] Strom MS, Lory S. Kinetics and sequence specificity of processing of prepilin by PilD, the type IV leader peptidase of *Pseudomonas aeruginosa*. *J.Bacteriol.* 1992;174:7345-7351.
[57] Bardy SL, Jarrell KF. FlaK of the archaeon *Methanococcus maripaludis* possesses preflagellin peptidase activity. *FEMS Microbiol.Lett.* 2002;208:53-59.
[58] Bardy SL, Jarrell KF. Cleavage of preflagellins by an aspartic acid signal peptidase is essential for flagellation in the archaeon *Methanococcus voltae*. *Mol.Microbiol.* 2003;50:1339-1347.
[59] Albers SV, Szabo Z, Driessen AJM. Archaeal homolog of bacterial type IV prepilin signal peptidases with broad substrate specificity. *J.Bacteriol.* 2003;185:3918-3925.
[60] Ng SY, Chaban B, Jarrell KF. Archaeal flagella, bacterial flagella and type IV pili: a comparison of genes and posttranslational modifications. *J.Mol.Microbiol.Biotechnol.* 2006;11:167-191.
[61] Albers SV, Pohlschroder M. Diversity of archaeal type IV pilin-like structures. *Extremophiles* 2009;13:403-410.
[62] Ng SYM, Zolghadr B, Driessen AJM, Albers SV, Jarrell KF. Cell surface structures of Archaea. *J. Bacteriol.* 2008;190:6039-6047.
[63] Szabo Z, Stahl AO, Albers SV, Kissinger JC, Driessen AJM, Pohlschroder M. Identification of diverse archaeal proteins with class III signal peptides cleaved by distinct archaeal prepilin peptidases. *J.Bacteriol.* 2007;189:772-778.
[64] Wang YA, Yu X, Ng SYM, Jarrell KF, Egelman EH. The structure of an archaeal pilus. *J. Mol. Biol.* 2008;381:456-466.
[65] Zolghadr B, Klingl A, Koerdt A, Driessen AJ, Rachel R, Albers SV. Appendage-mediated surface adherence of *Sulfolobus solfataricus*. *J.Bacteriol.* 2010;192:104-110.
[66] Frols S, Ajon M, Wagner M, Teichmann D, Zolghadr B, Folea M, et al. UV-inducible cellular aggregation of the hyperthermophilic archaeon *Sulfolobus solfataricus* is mediated by pili formation. *Mol.Microbiol.* 2008;70:938-952.
[67] Zolghadr B, Weber S, Szabo Z, Driessen AJM, Albers SV. Identification of a system required for the functional surface localization of sugar binding proteins with class III signal peptides in *Sulfolobus solfataricus*. *Mol.Microbiol.* 2007;64:795-806.

[68] Serganova I, Ksenzenko V, Serganov A, Meshcheryakova I, Pyatibratov M, Vakhrusheva O, et al. Sequencing of flagellin genes from *Natrialba magadii* provides new insight into evolutionary aspects of archaeal flagellins. *J.Bacteriol.* 2002;184:318-322.

[69] Nagahisa K, Ezaki S, Fujiwara S, Imanaka T, Takagi M. Sequence and transcriptional studies of five clustered flagellin genes from hyperthermophilic archaeon *Pyrococcus kodakaraensis* KOD1. *FEMSMicrobiol.Lett.* 1999;178:183-190.

[70] Patenge N, Berendes A, Engelhardt H, Schuster SC, Oesterhelt D. The *fla* gene cluster is involved in the biogenesis of flagella in *Halobacterium salinarum. Mol.Microbiol.* 2001;41:653-663.

[71] Baliga NS, Bonneau R, Facciotti MT, Pan M, Glusman G, Deutsch EW, et al. Genome sequence of *Haloarcula marismortui*: a halophilic archaeon from the Dead Sea. *Genome Res*. 2004;14:2221-2234.

[72] Gerl L, Sumper M. Halobacterial flagellins are encoded by a multigene family. Characterization of five flagellin genes. *J.Biol.Chem.* 1988;263:13246-13251.

[73] Harshey RM. Bacterial motility on a surface: many ways to a common goal. *Annu.Rev.Microbiol.* 2003;57:249-273.

[74] Anderson JK, Smith TG, Hoover TR. Sense and sensibility: flagellum-mediated gene regulation. *Trends Microbiol.* 2010;18:30-37.

[75] Wang Q, Suzuki A, Mariconda S, Porwollik S, Harshey RM. Sensing wetness: a new role for the bacterial flagellum. *EMBO J.* 2005;24:2034-2042.

[76] Nather DJ, Rachel R, Wanner G, Wirth R. Flagella of *Pyrococcus furiosus*: multifunctional organelles, made for swimming, adhesion to various surfaces, and cell-cell contacts. *J.Bacteriol.* 2006;188:6915-6923.

[77] Schopf S, Wanner G, Rachel R, Wirth R. An archaeal bi-species biofilm formed by *Pyrococcus furiosus* and *Methanopyrus kandleri*. *Arch.Microbiol.* 2008.

[78] Burrows LL. Weapons of mass retraction. *Mol.Microbiol.* 2005;57:878-888.

[79] Li J, Lim MS, Li S, Brock M, Pique ME, Woods VL,Jr, et al. *Vibrio cholerae* toxin-coregulated pilus structure analyzed by hydrogen/deuterium exchange mass spectrometry. *Structure* 2008;16:137-148.

[80] Kupper J, Marwan W, Typke D, Grunberg H, Uwer U, Gluch M, et al. The flagellar bundle of *Halobacterium salinarium* is inserted into a distinct polar cap structure. *J.Bacteriol.* 1994;176:5184-5187.

[81] Kandler O, Konig H. Cell envelopes of archaea: Structure and chemistry. In: Kates M, Kusher DJ, Matheson AT, editors. The biochemistry of archaea. Amsterdam: *Elsevier*; 1993. p. 223-259.

[82] Kojima S, Imada K, Sakuma M, Sudo Y, Kojima C, Minamino T, et al. Stator assembly and activation mechanism of the flagellar motor by the periplasmic region of MotB. *Mol.Microbiol.* 2009;73:710-718.

[83] Speranskii VV, Metlina AL, Novikova TM, Bakeyeva LY. Disk-like lamellar structure as part of the archaeal flagellar apparatus. *Biophysics* 1996;41:167-173.

[84] Metlina AL. Bacterial and archaeal flagella as prokaryotic motility organelles. *Biochemistry* (Mosc) 2004;69:1203-1212.

[85] Koval SF, Jarrell KF. Ultrastructure and biochemistry of the cell wall of *Methanococcus voltae*. *J.Bacteriol.* 1987;169:1298-1306.

[86] Alam M, Oesterhelt D. Morphology, function and isolation of halobacterial flagella. *J.Mol.Biol.* 1984;176:459-475.

[87] Chaban B, Ng SY, Kanbe M, Saltzman I, Nimmo G, Aizawa SI, et al. Systematic deletion analyses of the *fla* genes in the flagella operon identify several genes essential for proper assembly and function of flagella in the archaeon, *Methanococcus maripaludis*. *Mol.Microbiol.* 2007;66:596-609.

[88] Beznosov SN, Pyatibratov MG, Fedorov OV. On the multicomponent nature of *Halobacterium salinarum* flagella. *Microbiology* (Russ.) 2007;76:435-441.

[89] Ellen AF, Zolghadr B, Driessen AJM, Albers SV. *Biogenesis of the archaeal cell surface.* Archaeain press.

[90] Gerl L, Deutzmann R, Sumper M. Halobacterial flagellins are encoded by a multigene family. Identification of all five gene products. *FEBS Lett.* 1989;244:137-140.

[91] Kelly J, Logan SM, Jarrell KF, Vandyke DJ, Vinogradov E. A novel N-linked flagellar glycan from *Methanococcus maripaludis*. *Carbohydr.Res.* 2009;344:648-653.

[92] Voisin S, Houliston RS, Kelly J, Brisson JR, Watson D, Bardy SL, et al. Identification and characterization of the unique N-linked glycan common to the flagellins and S-layer glycoprotein of *Methanococcus voltae*. *J.Biol.Chem.* 2005;280:16586-16593.

[93] Chaban B, Voisin S, Kelly J, Logan SM, Jarrell KF. Identification of genes involved in the biosynthesis and attachment of *Methanococcus voltae* N-linked glycans: insight into N-linked glycosylation pathways in Archaea. *Mol.Microbiol.* 2006;61:259-268.

[94] Pyatibratov MG, Leonard K, Tarasov VY, Fedorov OV. Two immunologically distinct types of protofilaments can be identified in *Natrialba magadii* flagella. *FEMS Microbiol.Lett.* 2002;212:23-27.

[95] Trachtenberg S, Galkin VE, Egelman EH. Refining the structure of the *Halobacterium salinarum* flagellar filament using the iterative helical real space reconstruction method: insights into polymorphism. *J.Mol.Biol.* 2005;346:665-676.

[96] Cohen-Krausz S, Trachtenberg S. The structure of the archeabacterial flagellar filament of the extreme halophile *Halobacterium salinarum* R1M1 and its relation to eubacterial flagellar filaments and type IV pili. *J.Mol.Biol.* 2002;321:383-395.

[97] Szabo Z, Sani M, Groeneveld M, Zolghadr B, Schelert J, Albers SV, et al. Flagellar motility and structure in the hyperthermoacidophilic archaeon *Sulfolobus solfataricus*. J.Bacteriol. 2007;189:4305-4309.

[98] Logan SM. Flagellar glycosylation - a new component of the motility repertoire? *Microbiology* 2006;152:1249-1262.

[99] Mukhopadhyay B, Johnson EF, Wolfe RS. A novel pH_2 control on the expression of flagella in the hyperthermophilic strictly hydrogenotrophic methanarchaeaon *Methanococcus jannaschii*. *Proc.Natl.Acad.Sci.U.S.A.* 2000;97:11522-11527.

[100] Thomas NA, Jarrell KF. Characterization of flagellum gene families of methanogenic archaea and localization of novel flagellum accessory proteins. *J.Bacteriol.* 2001;183:7154-7164.

[101] Hendrickson EL, Liu Y, Rosas-Sandoval G, Porat I, Soll D, Whitman WB, et al. Global responses of *Methanococcus maripaludis* to specific nutrient limitations and growth rate. *J.Bacteriol.* 2008;190:2198-2205.

[102] Xia Q, Wang T, Hendrickson EL, Lie TJ, Hackett M, Leigh JA. Quantitative proteomics of nutrient limitation in the hydrogenotrophic methanogen *Methanococcus maripaludis*. *BMC Microbiol.* 2009;9:149.

[103] VanDyke DJ, Wu J, Ng SY, Kanbe M, Chaban B, Aizawa SI, et al. Identification of putative acetyltransferase gene, MMP0350, which affects proper assembly of both flagella and pili in the archaeon *Methanococcus maripaludis*. *J.Bacteriol.* 2008;190:5300-5307.

[104] Faguy DM, Koval SF, Jarrell KF. Effect of changes in mineral composition and growth temperature on filament length and flagellation in the archaeon *Methanospirillum hungatei*. *Arch. Microbiol.* 1993;159:512-520.

[105] Tarasov VY, Pyatibratov MG, Tang SL, Dyall-Smith M, Fedorov OV. Role of flagellins from A and B loci in flagella formation of *Halobacterium salinarum*. *Mol.Microbiol.* 2000;35:69-78.

[106] Thomas NA, Pawson CT, Jarrell KF. Insertional inactivation of the *fla*H gene in the archaeon *Methanococcus voltae* results in non-flagellated cells. *Mol.Genet.Genomics* 2001;265:596-603.

[107] Thomas NA, Mueller S, Klein A, Jarrell KF. Mutants in *fla*I and *fla*J of the archaeon *Methanococcus voltae* are deficient in flagellum assembly. *Mol.Microbiol.* 2002;46:879-887.

[108] Desmond E, Brochier-Armanet C, Gribaldo S. Phylogenomics of the archaeal flagellum: rare horizontal gene transfer in a unique motility structure. *BMC Evol.Biol.* 2007;7:106.

[109] Bayley DP, Jarrell KF. Further evidence to suggest that archaeal flagella are related to bacterial type IV pili. *J.Mol.Evol.* 1998;46:370-373.

[110] Albers SV, Driessen AJM. Analysis of ATPases of putative secretion operons in the thermoacidophilic archaeon *Sulfolobus solfataricus*. *Microbiology* 2005;151:763-773.

[111] Sumper M. Halobacterial glycoprotein biosynthesis. *Biochim.Biophys.Acta* 1987;906:69-79.

[112] Wieland F, Paul G, Sumper M. Halobacterial flagellins are sulfated glycoproteins. *J.Biol.Chem.* 1985;260:15180-15185.

[113] Jarrell KF, Jones GM, Nair DB. Role of N-linked glycosylation in cell surface structures of Archaea with a focus on flagella and S layers. *Int. J.Microbiol.* 2010.

[114] Vandyke DJ, Wu J, Logan SM, Kelly JF, Mizuno S, Aizawa SI, et al. Identification of genes involved in the assembly and attachment of a novel flagellin N-linked tetrasaccharide important for motility in the archaeon *Methanococcus maripaludis*. *Mol.Microbiol.* 2009;72:633-644.

[115] Ng SY, Chaban B, VanDyke DJ, Jarrell KF. Archaeal signal peptidases. *Microbiology* 2007;153:305-314.

[116] Correia JD, Jarrell KF. Posttranslational processing of *Methanococcus voltae* preflagellin by preflagellin peptidases of *M. voltae* and other methanogens. *J.Bacteriol.* 2000;182:855-858.

[117] Bayley DP, Jarrell KF. Overexpression of *Methanococcus voltae* flagellin subunits in *Escherichia coli* and *Pseudomonas aeruginosa*: a source of archaeal preflagellin. *J.Bacteriol.* 1999;181:4146-4153.

[118] Thomas NA, Chao ED, Jarrell KF. Identification of amino acids in the leader peptide of *Methanococcus voltae* preflagellin that are important in posttranslational processing. *Arch.Microbiol.* 2001;175:263-269.

[119] Ng SY, VanDyke DJ, Chaban B, Wu J, Nosaka Y, Aizawa S, et al. Different minimal signal peptide lengths recognized by the archaeal prepilin-like peptidases FlaK and PibD. *J.Bacteriol.* 2009;191:6732-6740.

[120] Strom MS, Nunn DN, Lory S. Posttranslational processing of type IV prepilin and homologs by PilD of *Pseudomonas aeruginosa. MethodsEnzymol.* 1994;235:527-540.

[121] LaPointe CF, Taylor RK. The type 4 prepilin peptidases comprise a novel family of aspartic acid proteases. *J.Biol.Chem.* 2000;275:1502-1510.

[122] Szabo Z, Albers SV, Driessen AJM. Active-site residues in the type IV prepilin peptidase homologue PibD from the archaeon *Sulfolobus solfataricus. J.Bacteriol.* 2006;188:1437-1443.

[123] Schirm M, Kalmokoff M, Aubry A, Thibault P, Sandoz M, Logan SM. Flagellin from *Listeria monocytogenes* is glycosylated with beta-O-linked N-acetylglucosamine. *J.Bacteriol.* 2004;186:6721-6727.

[124] Szymanski CM, Logan SM, Linton D, Wren BW. *Campylobacter*--a tale of two protein glycosylation systems. *Trends Microbiol.* 2003;11:233-238.

[125] Twine SM, Paul CJ, Vinogradov E, McNally DJ, Brisson JR, Mullen JA, et al. Flagellar glycosylation in *Clostridium botulinum. FEBS J.* 2008;275:4428-4444.

[126] Abu-Qarn M, Eichler J. Protein N-glycosylation in Archaea: defining *Haloferax volcanii* genes involved in S-layer glycoprotein glycosylation. *Mol.Microbiol.* 2006;61:511-525.

[127] Chaban B, Logan SM, Kelly JF, Jarrell KF. AglC and AglK are involved in biosynthesis and attachment of diacetylated glucuronic acid to the N-glycan in *Methanococcus voltae. J.Bacteriol.* 2009;191:187-195.

[128] Shams-Eldin H, Chaban B, Niehus S, Schwarz RT, Jarrell KF. Identification of the archaeal *alg*7 gene homolog (encoding N-acetylglucosamine-1-phosphate transferase) of the N-linked glycosylation system by cross-domain complementation in *Saccharomyces cerevisiae. J.Bacteriol.* 2008;190:2217-2220.

[129] Namboori SC, Graham DE. Acetamido sugar biosynthesis in the Euryarchaea. *J.Bacteriol.* 2008;190:2987-2996.

[130] Sauer FD, Blackwell BA, Kramer JK, Marsden BJ. Structure of a novel cofactor containing N-(7-mercaptoheptanoyl)-O-3-phosphothreonine. *Biochemistry* 1990; 29:7593-7600.

[131] Ferrante G, Ekiel I, Sprott GD. Structural characterization of the lipids of *Methanococcus voltae*, including a novel N-acetylglucosamine 1-phosphate diether. *J.Biol.Chem.* 1986;261:17062-17066.

[132] Wacker M, Feldman MF, Callewaert N, Kowarik M, Clarke BR, Pohl NL, et al. Substrate specificity of bacterial oligosaccharyltransferase suggests a common transfer mechanism for the bacterial and eukaryotic systems. *Proc.Natl.Acad.Sci.U.S.A.* 2006;103:7088-7093.

[133] Frank CG, Sanyal S, Rush JS, Waechter CJ, Menon AK. Does Rft1 flip an N-glycan lipid precursor? *Nature* 2008;454:E3-4; discussion E4-5.

[134] Helenius J, Ng DT, Marolda CL, Walter P, Valvano MA, Aebi M. Translocation of lipid-linked oligosaccharides across the ER membrane requires Rft1 protein. *Nature* 2002;415:447-450.

[135] Alam M, Oesterhelt D. Purification, reconstitution and polymorphic transition of halobacterial flagella. *J.Mol.Biol.* 1987;194:495-499.

[136] Aldridge PD, Wu C, Gnerer J, Karlinsey JE, Hughes KT, Sachs MS. Regulatory protein that inhibits both synthesis and use of the target protein controls flagellar phase variation in *Salmonella enterica*. *Proc.Natl.Acad.Sci.U.S.A.* 2006;103:11340-11345.

[137] Oren A, Ventosa A, Gutierrez MC, Kamekura M. *Haloarcula quadrata* sp. nov., a square, motile archaeon isolated from a brine pool in Sinai (Egypt). *Int.J.Syst.Bacteriol.* 1999;49 Pt 3:1149-1155.

[138] Alam M, Claviez M, Oesterhelt D, Kessel M. Flagella and motility behaviour of square bacteria. *EMBO J.* 1984;3:2899-2903.

[139] Fedorov OV, Pyatibratov MG, Kostyukova AS, Osina NK, Tarasov VY. Protofilament as a structural element of flagella of haloalkalophilic archaebacteria. *Can.J.Microbiol.* 1994;40:45-53.

[140] Black FT, Freundt EA, Vinther O, Christiansen C. Flagellation and swimming motility of *Thermoplasma acidophilum*. *J.Bacteriol.* 1979;137:456-460.

[141] Segerer A, Langworthy TA, Stetter KO. *Thermoplasma acidophilum* and *Thermoplasma volcanium* sp. nov. from solfatara fields. *Syst. Appl. Microbiol.* 1988;10:161-171.

[142] Bayley DP, Florian V, Klein A, Jarrell KF. Flagellin genes of *Methanococcus vannielii*: amplification by the polymerase chain reaction, demonstration of signal peptides and identification of major components of the flagellar filament. *Mol.Gen.Genet.* 1998;258:639-645.

[143] Kalmokoff ML, Koval SF, Jarrell KF. Relatedness of the flagellins from methanogens. *Arch.Microbiol.* 1992;157:481-487.

[144] Bayley DP, Kalmokoff ML, Jarrell KF. Effect of bacitracin on flagellar assembly and presumed glycosylation of the flagellins of *Methanococcus deltae*. *Arch. Microbiol.* 1993;160:179-185.

[145] Bult CJ, White O, Olsen GJ, Zhou L, Fleischmann RD, Sutton GG, et al. Complete genome sequence of the methanogenic archaeon, *Methanococcus jannaschii*. *Science* 1996;273:1058-1073.

[146] Beveridge TJ, Stewart M, Doyle RJ, Sprott GD. Unusual stability of the *Methanospirillum hungatei* sheath. *J.Bacteriol.* 1985;162:728-737.

[147] Stewart M, Beveridge TJ, Sprott GD. Crystalline order to high resolution in the sheath of *Methanospirillum hungatei*: a cross-beta structure. *J.Mol.Biol.* 1985;183:509-515.

[148] Beveridge TJ, Southam G, Jericho MH, Blackford BL. High-resolution topography of the S-layer sheath of the archaebacterium *Methanospirillum hungatei* provided by scanning tunneling microscopy. *J.Bacteriol.* 1990;172:6589-6595.

[149] Beveridge TJ, Sprott GD, Whippey P. Ultrastructure, inferred porosity, and gram-staining character of *Methanospirillum hungatei* filament termini describe a unique cell permeability for this archaeobacterium. *J.Bacteriol.* 1991;173:130-140.

[150] Southam G, Kalmokoff ML, Jarrell KF, Koval SF, Beveridge TJ. Isolation, characterization, and cellular insertion of the flagella from two strains of the archaebacterium *Methanospirillum hungatei*. *J.Bacteriol.* 1990;172:3221-3228.

[151] Faguy DM, Koval SF, Jarrell KF. Physical characterization of the flagella and flagellins from *Methanospirillum hungatei*. *J.Bacteriol.* 1994;176:7491-7498.

[152] Faguy DM, Koval SF, Jarrell KF. Correlation between glycosylation of flagellin proteins and sensitivity of flagellar filaments to Triton X-100 in methanogens. *FEMS Microbiol.Lett.* 1992;69:129-134.

[153] Maeder DL, Anderson I, Brettin TS, Bruce DC, Gilna P, Han CS, et al. The *Methanosarcina barkeri* genome: comparative analysis with *Methanosarcina acetivorans* and *Methanosarcina mazei* reveals extensive rearrangement within methanosarcinal genomes. *J.Bacteriol.* 2006;188:7922-7931.

[154] Polosina YY, Jarrell KF, Fedorov OV, Kostyukova AS. Nucleoside diphosphate kinase from haloalkaliphilic archaeon *Natronobacterium magadii*: purification and characterization. *Extremophiles* 1998;2:333-338.

[155] Evans LD, Stafford GP, Ahmed S, Fraser GM, Hughes C. An escort mechanism for cycling of export chaperones during flagellum assembly. *Proc.Natl.Acad.Sci.U.S.A.* 2006;103:17474-17479.

[156] Aldridge P, Karlinsey J, Hughes KT. The type III secretion chaperone FlgN regulates flagellar assembly via a negative feedback loop containing its chaperone substrates FlgK and FlgL. *Mol.Microbiol.* 2003;49:1333-1345.

[157] Fraser GM, Bennett JC, Hughes C. Substrate-specific binding of hook-associated proteins by FlgN and FliT, putative chaperones for flagellum assembly. *Mol.Microbiol.* 1999;32:569-580.

[158] Craig L, Li J. Type IV pili: paradoxes in form and function. *Curr.Opin.Struct.Biol.* 2008;18:267-277.

In: Archaea: Structure, Habitats and Ecological... ISBN 978-1-61761-932-8
Editor: Sakura Y. Katō, pp. 39-68

Chapter 2

Methanogenes and Ammonia-Oxidation Archaeal Communities in High Temperature Oil Reservoirs

Hui Li[1] ***and Bo-zhong Mu***[2]
[1]School of Resources and Environmental Engineering, State Environmental Protection Key Laboratory of Environmental Risk Assessment and Control on Chemical Process, East China University of Science and Technology, Shanghai 200237, China
[2]Institute of Applied Chemistry, State Key Laboratory of Bioreactor Engineering and Institute of Applied Chemistry, East China University of Science and Technology, Shanghai 200237, PR China

Abstract

The widespread occurrence and diversity of archaea in high temperature underground oil reservoir indicates their key contribution to the material cycle and energy flow in the subsurface hot habitat. Methanogens dominated the anoxic ecosystem with amount of petroleum hydrocarbons by catalyzing the last step of carbon cycle by producing methane. The diversity of archaeal community was analyzedby 16S rRNA gene libraries constructed from two continental- and offshore, high-temperature, long-term water-flooded petroleum reservoir in China. The continentallibrary includes 237 clones clustered in 28 phylotypes. The dominant members were affiliated with the order Methanomicrobiales belonged to four orders: Methanobacteriales, Methanococcales, Methanomicrobiales and Methanosarcinales. Most of the clones clustered with sequences previously described for methanogens. The offshore archaeal 16S rRNA gene clone libraryincluded220 randomly selected clones clustered with 28 phylotypes dominated by methanogen-like rRNA genes (*Methanothermobacter*, *Methanobacter*, *Methanobrevibacter* and *Methanococcus*) and with a lower percentage of clones belonging to *Thermoprotei*. The FISH results indicated that the quantity of archaea was much more than bacteria in the samples and they had close spatial distribution *in situ* formed compact aggregates. Some thermophilic methanogens detected had been previously isolated from a number of high-temperature petroleum reservoirs worldwide, thus they might adapt to the environments and be the common habitants of geothermally heated subsurface environments. The existence, diversity, abundance and distribution of ammonia-oxidation archaea (AOA) in the high temperature oil reservoirs were investigated by using real-time PCR and phylogenetic analysis based on *amoA* genes. The result demonstrated the AOA-like phylotypes were mainly clustered within two major clades of archaeal *amoA* sequences known from water columns, sediments, and soils: clusters A and B, and few clones were related to the new genera: *Candidatus* '*Nitrosocaldus yellowstonii*'. The abundance of AOA -like *amoA* gene ranged from 2.92×10^3 to 9.21×10^4 per ml production water. Statistical analysis showed that *amoA* gene fell into five groups and the distribution of *amoA* was of significant positive correlation with the environmental factors temperature and recovery process. Our study provided the evidence for the distribution of putative prokaryotic *amoA* gene in various oil reservoirs, which improved our understanding of the nitrogen cycling in oilfield ecosystem.

Keywords: Methanogen; Ammonia-oxidation archaea (AOA); Ammonia-oxidation bacteria (AOB); Petroleum reservoir.

Introduction

Geothermally heated petroleum reservoirs are complex ecosystems comprising an as yet undetermined number of microorganisms and multi-phase fluids of oil, gas and water. Moreover, the deep subsurface petroleum reservoirs, as a rule, lack dissolved oxygen and are generally anoxic habitats, which are characterized by high temperature, high pressure and high salinity. Recently the microbial communities associated with subterranean petroleum-rich strata have attracted considerable attention from researchers (Bonch-Osmolovskaya EA et al., 2003; Grabowski et al., 2005; Ollivier et al., 2000). A great variety of anaerobic microorganisms have been recognized from a number of geographically distant oil reservoirs, including sulfate reducers, sulfidogens, methanogens, and manganese and iron reducers. Particularly, the methanogenic archaea are common in these unique ecological niches in which liquid petroleum hydrocarbons are the prevailing organic matters (Magot et al., 2005). The known thermophilic methanogens isolated from high-temperature oilfields were represented by *Methanococcus thermolithotrophicus*, *M. thermautotrophicus* and *M. thermoaggregans*. In oil-rich strata, hydrocarbons are degraded to methanogenic substrates such as acetate, CO_2 and H_2 by diverse bacteria and methanogenesis is a final degradation process of organic matter by methanogenic archaea in anoxic petroleum reservoirs environments, so it is important to understand the ecology of archaea in petroleum reservoir due to the scientific significance and the industrial application such as microbial enhanced oil recovery (Aitken et al., 2004; Head et al., 2003; Liu et al., 2005).

Nevertheless, the diversity and structure of archaeal communities of petroleum reservoirs were poorly understood and only a few pure cultures of methanogens from oil strata have been isolated. This might be attributed to that the selective enrichment cultivation had severe limitations when used as a single approach to thoroughly describe microbial diversity, and more than 90% of the microorganisms in nature including petroleum reservoirs were thought to have escaped traditional cultivation techniques. The application of molecular techniques to microbial ecology, in particular the use of the 16S rRNA as a molecular marker, could in part alleviate these limitations, revealing much of the microbial diversity that previously escaped identification (Kemp et al., 2004; Marchesi et al., 2001). In recent years, increasing application of culture-independent techniques allowed a more complete characterization of the microbial communities in subsurface petroleum reservoirs(Bonch-Osmolovskaya EA et al., 2003; Grabowski et al., 2005; Nazina et al., 2006). These studies had extended our view of archaeal communities in low-temperature, low-salinity and biodegraded petroleum reservoir of Western Canadian Sedimentary Basin and high-temperature, sulfur-rich oil-bearing formations in California and in Western Siberia. Analysis of the 16S rRNA genes showed that the archaeal assemblages inhabiting these anaerobic oil reservoirs contained: (1) mesophilic methanogens represented by *Methanosarcina*, *Methanohalophilus*, *Methanocalculus* and *Methanosaeta*; (2) thermophilic methanogens belonging to *Methanothermobacter*, *Methanococcus* and *Methanoculleus*; and (3) a number of previously unknown archaeal 16S rDNA sequences.

To detect the methanogen archaeal community in oil reservoir, we analyze the archaeal 16S rRNA gene diversity in two kinds of oil reservoirs located at Hebei province, China. One is the continental high-temperature, water-flooded petroleum reservoir in the J-12 Unit at Huabei Oilfield. The J-12 Unit was operated in primary production in March 1987. Water flooding of the reservoir began in May 1988, and from then on, it has been in continuous water flooding to the present time by use of recycled water from the production water from this reservoir. Another is an offshore high-temperature, water-flooded petroleum reservoir of the QH (Qinghuang) Unit. The QH Unit has been operated in primary production sinceOctober2001.Water flooding of the reservoir began in December 2003, and from that time on, it has been continuously flooded with recycled production water.

The exploration of petroleum reservoir is a topic of increasing importance in geomicrobiology because the global energy demand stimulates the research on carbon cycle related microbial processes in the petroleum reservoir, e.g., hydrocarbon biodegradation, microbial enhanced oil recovery (Head et al., 2003; McInerney et al., 2005; Youssef et al., 2009). Furthermore, the bioconversion of residual oil to methane may offer an innovative route to recover crude oil *in situ* through methanogenic conversion (Dolfing et al., 2008; Gieg et al., 2008; Jones et al., 2008; Gray et al., 2009).Recently, diverse microbes were observed in petroleum reservoirs worldwide by culture-independent methods(Bonch-Osmolovskaya et al., 2003; Nazina et al., 2004; Grabowski et al., 2005; Li et al., 2007; Nazina et al., 2007; Dahle et al., 2008; Agrawal and Lal 2009; Gittel et al., 2009; Kaster et al., 2009; Pham et al., 2009), including ammonia-producing and nitrate-reducing bacteria (NRB)*Sulfurospirillum* spp. (Eckford and Fedorak 2005; Hubert and Voordouw 2007), ammonia-oxidizing bacteria *Nitrosomonas* sp., *Nitrosopira* sp., *Clostridium* sp., *Denitrisovibrio* sp., and planctomycetes sp. (Orphan et al., 2000; Bonch-Osmolovskaya et al., 2003; Eckford and Fedorak 2005; Hubert and Voordouw 2007; Li et al., 2007; Pham et al., 2009). These microorganisms related to nitrogen cycling have significant implications in the petroleum industry, e.g., the utilizing of NRB to inhibit the sulfate reducing bacteria (SRB) to prevent the reservoirs souring by decreasing the production of hydrogen sulfite and using AOB to removal of ammonia in the production water. A large amounts of nitrate are routinely injected into the petroleum reservoirs in Shengli Oilfield, China to reduce the risk of pipeline corrosion and other negative effects including souring(Hubert et al., 2005). Several nitrogen cycle related microorganisms have been activated by the injected nitrate, including the nitrate-reducing interdependent with sulfite oxidation bacteria (NR-SOB) and heterotrophic NRB (hNRB) which compete the organic electron donors with SRB (Hubert and Voordouw 2007). Both of these NRB produce nitrite and ammonia during the nitrate-reducing process(Greene et al., 2003). Ammonia may stimulate the ammonia-oxidizing in the oil reservoirs by the thriving of AOB in production water (Orphan et al., 2000; Bonch-Osmolovskaya et al., 2003; Eckford and Fedorak 2005; Hubert and Voordouw 2007; Li et al., 2007; Pham et al., 2009). However, the presence or activity of AOA has not been detected in oilfield, although the AOA may likely to inhabit the geothermal oil reservoirs because of their natural ubiquity in the other high-temperature environments (Spear et al., 2007; Weidler et al., 2007; Perevalova et al., 2008; Reigstad et al., 2008; Weidler et al., 2008; Zhang et al., 2008; Pitcher et al., 2009; Wang et al., 2009). Up to now, ammonia-oxidation is related to two kinds of functional microorganisms AOA and AOB (Kowalchuk and Stephen 2001; Konneke et al., 2005). Although AOA was first discovered not long ago (Konneke et al., 2005),they have been found widely distributed among moderate temperature aquatic and terrestrial environments (Francis et al., 2005; Treusch et al., 2005; Hallam et al., 2006), and often amount number AOA in several terrestrial and marine systems (Leininger et al., 2006; Wuchter et al., 2006; Coolen et al., 2007; He et al., 2007; Lam et al., 2007; Adair and Schwartz 2008; Chen et al., 2008; Jiang et al., 2009). More recently, AOA was detected in the elevated temperature environments (Nunoura et al., 2005; Ehrhardt et al., 2007; Spear et al., 2007; Weidler et al., 2007; Perevalova et al., 2008; Reigstad et al., 2008; Weidler et al., 2008; Zhang et al., 2008; Pitcher et al., 2009; Wang et al., 2009), in particularly a moderately thermophilic and a thermophilic enrichment provide the direct evidence for AOA thriving in the hot habitats, which greatly extend the upper temperature limit of nitrification and document that the capacity for ammonia-oxidation is broadly distributed among *Crenarchaeota*. Furthermore, AOA had also been detected in the water columns of Eastern Tropical North Pacific, one of the largest pelagic oxygen minimum zones in the ocean (Francis et al., 2005) as well as in suboxic water columns of the Black Sea (Coolen et al., 2007), and Treusch et al. had suggested that AOA might be capable of ammonia-oxidation under anaerobic conditions (Treusch et al., 2005). Despite recent advances in our understanding of nitrogen cycling activities in hot habitats, our knowledge on high-temperature oil reservoir ecosystems has been very limited. So we initiated the study to detect AOA community in Shengli Oilfield in China and analyze effect of oil drilling operations on their distribution and abundance. Our aim is to unveil more of the microbial diversity, which in turn will improve our understanding of the complex community that inhabits the subterranean petroleum-rich system.

Materials and Methods

Sample Collection

The samples of production water were collected into sterile steel screw-cap bottles directly from a sampling valve at the pipe line of the well head. The bottles completely filled with the production water (oil/water mixture) were tightly sealed from contamination, and immediately transported to the laboratory and stored at 4°C before treatment. The in-situ temperature and pressure of the onshore oil reservoir are about 75°C and 18 MPa, respectively. The salinity of groundwater of the oil-bearing formation with depth of 1,500-

1,700 m in the reservoir is 16622 mg L^{-1}. The sandy oil-bearing horizon in the offshore oil reservoir is at about 1100-1300 m, and the temperature is about 65 °C *in situ*. The salinity of the formation water is about 4226 mg L^{-1}, and its pH is 7.1-7.6.

Nucleic Acid Extraction

The oil in water samples was removed by heating the samples to 70°C for 15 min, followed by phase separation in sterile separatory funnels. The microbial biomass was collected by centrifugation (15000*g*, at 4°C for 10 min). Total community DNA was extracted from the cell pellets using a lysozyme/proteinase K/sodium dodecyl sulfate (SDS) treatment, followed by standard phenol/chloroform extractions (Murray et al., 1998). A final purification was carried out using Wizard DNA Clean Up kit (Promega, USA).

Construction of 16S rRNA Gene Library

The DNA from the production water was used to construct the archaeal library. Archaeal small-subunit rRNA genes were amplified by PCR using a set of primers (109f, 5′-ACK GCT CAG TAA CAC GT-3′; 1041r, 5′-GGC CAT GCA CCW CCT CTC-3′) (Grosskopf et al., 1998; Kolganova et al., 2002). The PCR reaction mixture contained 1x Taq 2000 enzyme and reagents(Takara, Japan), 0.1 μM of each primer and 200 μM of each dNTP. The genomic DNA from the water sample was used as template for amplification with final reaction concentrations of 1.5-3 ngμL^{-1}. Amplification program was consisted of a3 min denaturing step at 95°C followed by 30 cycles of 30sec at 95°C, 30sec at 50°C and 90 sec at 72°C. To recondition the PCR product, the amplified reaction was diluted10-fold into fresh reaction mixture of the same composition cycled three times using the parameters specified above (Thompson et al., 2000). The PCR products from the same sample were pooled to minimize PCR bias. The pooled PCR products were purified by the PCR production purification kit (V-gene, China). The purified amplicons were cloned with pMD19-T vector kit (Takara, Japan) according to the manufacturer's instructions.

Generation of Prokaryotic *amoA* Clone Library

Total genomic DNA was extracted from centrifuged materials using the lysozyme/proteinase K/sodium dodecyl sulfate (SDS) treatment followed by standard phenol/chloroform extractions. Nucleic acids were purified by the DNA purification kit (Promaga, USA). Samples DNA were extracted several times, pooled and concentrated. PCR amplification of prokaryotic *amoA* gene was performed by two protocols. The crenarchaeal *amoA* fragments were amplified according to the previous study with specific primer set, Arch-amoAF (5'-STAATGGTCTGGCTTAGACG-3') and Arch-amoAR (5'-GCGGCCATCCATCTGTATGT-3') (Francis et al., 2005). The bacterial *amoA* fragments were amplified according to the previous study with specific primer set, amoA1F (5'-GGGGTTTCTACTGGTGGT-3') and amoA2R (5'-CCCCTCKGSAAAGCCTTCTTC-

3')(Rotthauwe et al., 1997). PCR products were purified by gel cut procedure, and then used to construct the clone libraries with pMD-18 T-vector (Takara, Japan).

Sequencing and Phylogenetic Analysis

The plasmid DNA was isolated from selected clones by AxyPrep-96 Plasmid kit (Axygen, USA). The rDNA inserts were sequenced on automated ABI 377 sequencer (Dye-Terminator Cycle Sequencing Ready Reaction FS Kit; PE Applied Biosystems) using M13 universal sequencing primers. The resulting sequences (approximately 450 bp) were determined for orientation and then preliminarily compared with those available in GenBank database of NCBI (http://www.ncbi.nlm.nih.gov) using the BLAST network service (Altschul et al., 1997) to determine their approximate phylogenetic affiliations. Partial sequences that differed by less than 3% were considered to the same phylotype (Stackebrandt et al., 1994). Finally, one to three sequences per phylotype were sequenced using M13 universal sequencing primers. The sequences (about 950 bp of archaeal rDNAs) retrieved were checked for possible chimeric artifacts using the CHIMERA-CHECK program (Cole et al., 2003) in the Ribosomal Database Project II (http://rdp.cme.msu.edu/) and Bellerophon (Huber et al., 2004).

The sequences with no chimeras were initially submitted to FASTA program (version 3) (Pearson et al., 1988) to determine their closest phylogenetic relatives. Sequences were aligned to their nearest neighbor using the CLUSTAL_X software(Thompson et al., 1997).

Phylogenetic trees were constructed based on the Kimura two-parameter model (Kimura et al., 1980) and neighbor-joining algorithm(Saitou et al., 1987) using the PHYLIP package (Felsenstein et al., 2005). Bootstrap analysis with 1000 replicates was applied to assign confidence levels to the nodes in the trees. The CONSENSE program was used to generate consense tree and the TREEVIEW program was used to obtain a phylogenetic tree (Page et al., 1996).

In Situ Hybridization

Microbial cells collection were washed in PBS buffer, fixed for 3 h at 4 °C in freshly prepared 4% paraformaldehyde-PBS (1 ml), washed with PBS and stored in PBS/ethanol (1:1, v/v). The probes of 16S rRNA-targeted oligonucleotide were used in this study: Hyr (5'-AGTTAGCCGGTGCTTCCT-3') for bacteria (Loy et al., 2002) and Met (5'-GGCCACGGTGCGACCGTTGTCG-3') for Methanogenics archaea (Kotelnikova et al., 2000).

More detailed information about the probes can be found at Probe-Base (http://www.microbial-ecology.net/probeBase) (Loy et al., 2003). The oligonucleotide probes were labeled with either fluorescein (Hyr) or Cy3 (Met). Fixed cells were spotted on microscopic slides, air-dried and subsequently dehydrated with increasing concentrations of ethanol (50%, 70%, 96%, 5 min each).

A volume of 10 μl of hybridization solution (20% deionized formamide, 0.9 M NaCl, 20 mM Tris–HCl (pH 7.2), 0.1% SDS, 5 ng of each fluorescent labeled probe) was added to each well, and the slides were incubated at 46 °C for 3 h. Slides were rinsed with washing buffer (0.18 M NaCl, 20 mM Tris–HCl (pH 7.2), 0.1%SDS) and incubated in washing buffer at

48 °C for 20 min. Slides were then rinsed with sterile water, air-dried, and covered with the antifading agent (Invitrogen, US) in 50% glycerol then examined by a Olympus model BX51T-32P01-FLB3TCS NT fluorescence microscope.

Statistical Analysis

Operational taxonomic units (OTUs) for prokaryotic *amoA* gene diversity analysis were defined by 3% differences in nucleotide sequences, as determined by using the nearest-neighbor algorithm in DOTUR program (Schloss and Handelsman 2005). DOTUR was also used to generate Chao, Shannon and Simpson index for each clone library. The coverage of clone libraries was calculated from the equation $C = \left[1 - \left(n_1/N\right)\right] \times 100$ (Good 1953). Community classification of the AOA and AOB assemblages was determined with principal coordinates analyses (PCA) (Lozupone et al., 2006). Correlations between the AOA and AOB communities and environmental factors were determined by the canonical correspondence analysis (CCA) using the software CANOCO (ver. 4.5, Microcomputer Power, Ithaca, NY, USA) (ter Braak and Šmilauer 2002).

Real-Time PCR Assay

Copy numbers of crenarchaeal *amoA* gene in each sample were quantified with qPCR analysis. Real-time PCR was performed by using a FastStart Universal SYBR Green Master PCR mix (Roche Applied Science, Mannheim, Germany) in 96-well optical plates placed in an ABI 7000 Sequence detection system (Applied Biosystems, Foster City, CA). Q-PCR assay was performed with the primer sets described above. Three replicates were analyzed for each sample. Amplifications were performed in a 25 μl volume containing 1μl of DNA template, 2.5 μl BSA (0.1%), 0.5 μl of each primer (20μM) and 12.5μl of the ready reaction mix. The PCR cycle was started with 2 min at 50°C and 3 min at 95 °C, followed by 45 cycles of 95°C for 30 sec, 30 s at 55 °C for AOB or 53 °C for AOA, 1min at 72°Cand plate read at 83 °C. Cycle thresholds were determined by comparison with standard curves constructed using several concentrations of positive clone.

The plasmid DNA concentration was determined on a Biophotometer (Eppendorf, Germany) and the copy number of target gene was calculated directly from the concentration of the extracted plasmid DNA. Ten-fold serial dilutions of a known copy number of the plasmid DNA were subjected to real-time PCR in triplicate to generate an external standard curve.

The PCR efficiency and correlation coefficients for standard curves were higher than 95% and R^2 values were greater than 0.99 for the curves. The specificity of the PCR amplification was determined by the melting curve and by gel electrophoresis.

Results

Methanogens Community in Onshore Oil Reservoir

A total of 320 white colonies were selected and PCR analysis of these clones using universal plasmid primers revealed 237 positive clones containing inserts of 16S rDNA fragments of the corresponding size. The clone library obtained was designated HBO-A (Huabei Oilfield Archaea), and the 237 clones were partially sequenced (around 450 bp) and then categorized on the basis of their sequence similarity (97% identity). Finally, twenty-five phylotypes were identified and one to three clones per phylotype was sequenced again (around 950 bp).

The decrease in the rate of phylotype detection shown on the curves indicated that the major part of the diversity in the libraries had been detected. This conclusion was further supported by calculating the coverage of the library, which was 94.4%.

Comparative analyses of the retrieved sequences showed that all the archaea were members of the phylum Euryarchaeota. Seventeen sequence types had high levels of similarity with their closest counterparts in the public databases (Table 1, Figure 1), whereas 8 phylotypes (32%) showed less than 96% sequence similarity to their nearest database entries and may belong to hitherto unknown phylotypes (Stackebrandt et al., 1994).

All the phylotypes were represented by sequences related to the hydrogenotrophic and acetoclastic methanogens of four orders: Methanobacteriales, Methanococcales, Methanomicrobiales and Methanosarcinales. Methanomicrobiales-related archaeal clones comprised the largest portion of the HBO-A library, which included 112 clones and represented 47% of the total library.

The most abundant sequence type (81clones) in this group was closely related to *Methanoplanus petrolearius* SEBR 4847 (>98%), a methanogenic strain isolated from an African offshore oil field. Two clones were closely related to *Methanocorpusculum parvum* IF9601b (AY057068), a strain isolated from waste anaerobic sludge digestor in Taiwan. Methanococcales is another group that was retrieved abundantly in the library, and 45 clones (19% of the library) clustered within this group (Table 1).

The majority of the sequences in this group (29 clones) was closely related to *Methanococcus infernus* (>98%), a hyperthermophilic lithotrophic methanogen isolated from a deep-sea hydrothermal vent (Jeanthon et al., 1998).

Methanobacteriales-related clones were the most frequent phylotype in production well enrichments(Orphan et al., 2000). In this study, forty-three clones (18% of the total clone population) were placed into Methanobacteriales, and most of these phylotypes were >97% identical to previously determined rDNA sequences (Table 1).

Within this group, 22 clones were identified as *Methanothermobacter thermautotrophicus* GC-1 (>97%) and 14 clones were closest to *Methanobacterium subterraneum* A8p DSM 11074 (>98%), an alkaliphilic, eurythermic and halotolerant methanogen isolated from deep granitic groundwater [21].

Table 1. Distribution of sequence types from the onshore oilfield archaeal 16S rRNA gene library

Order	Type sequence	Closest cultivated species (accession numbers)	Similarity (%)[a]	No. of clones (%)[b]
Methanobacteriales	HBO-A9	*Methanothermobacter wolfeii* DSM 2970(AB104858)	98	1(<1)
	HBO-A6	*Methanobacterium subterraneum* A8p DSM 11074(X99044)	98	14(5.9)
	HBO-A24	*Methanobacterium subterraneum* C2BIS DSM 11075(X99045)	97	5(2.1)
	HBO-A4	*Methanothermobacter thermautotrophicus* Delta H(AE000666)	94	1(<1)
	HBO-A10	*Methanothermobacter thermautotrophicus* GC-1 (AY196661)	97	22(9.2)
Methanococcales	HBO-A12	*Methanococcus infernos*(AF025822)	98	29(12.2)
	HBO-A23	*Methanococcus maripaludis* S2(BX957219)	98	2(<1)
	HBO-A17	*Methanococcus thermolithotrophicus* DSM 2095(M59128)	99	1(<1)
	HBO-A25	*Methanocaldococcus* sp. Mc-1-85(AB095160)	95	11(4.6)
	HBO-A13	*Methanococcus jannasch0ii* DSM 2661(M59126)	97	1(<1)
	HBO-A3	*Methanococcus voltae* PS(U38461)	94	1(<1)
Methanomicrobiales	HBO-A8	*Methanocorpusculum parvum* IF9601b(AY057068)	98	2(<1)

Table 1 (Continued)

	HBO-A15	*Methanoculleus thermophilus* DSM 3915(M59129)	99	1(<1)
	HBO-A19	*Methanoplanus petrolearius* SEBR 4847(U76631)	98	81(34.2)
	HBO-A2	*Methanomicrobium mobile* DSM 1539(AY196679)	95	1(<1)
	HBO-A22	*Methanocalculus halotolerans* P2F9705(AF411470)	87	11(4.6)
	HBO-A26	*Methanoculleus bourgensis* CB1(AB065298)	95	1(<1)
	HBO-A27	Archaeon HR3812-Enrichment-018(DQ058818)	76	15(6.3)
Methanosarcinales	HBO-A1	*Methanosarcina lacustris* MM(AY260431)	99	1(<1)
	HBO-A7	*Methanosarcina thermophila* DSM 1825(M59140)	95	1(<1)
	HBO-A28	Deep-sea hydrothermal vent clone pISA16(AB019758)	98	29(12.2)
	HBO-A11	*Methanosaeta concilii*(X16932)	97	1(<1)
	HBO-A16	*Methanosarcina barkeri* CM1(AY196682)	93	1(<1)
	HBO-A20	*Methanosarcina siciliae* T4/M(U20153)	97	1(<1)
	HBO-A5	*Methanosarcina siciliae* C2J (U89773)	98	3(1)

[a] Percentage of 16S rDNA similarity to its closest relative.
[b] Proportion of clones in the library.

Five phylotypes were placed into the genus *Methanothermobacter* which were demonstrated to have the ability to consume hydrogen and carbon dioxide as basic substrates. In addition, thirty-seven clones (15.6% of the total library) were placed into Methanosarcinales.

Twenty-nine clones in this group were closely related to archaea clone pISA16 (>98%) which was detected in deep-sea hydrothermal vent environment (Takai and Horikoshi, 1999). And three clones were closely related to *Methanosarcina siciliae* C2J, an aceticlastic strain of *Methanosarcina siciliae* isolated from marine canyon sediments(Elberson et al., 1997).

On the other hand, there were eight phylotypes distantly related to known cultured relatives in the databases. Phylotype HBO-A25 representing 11 clones was related (<95%) to *Methanocaldococcus* sp. Mc-1-85 (AB095160), a strain detected in a hydrogen-based, subsurface lithoautotrophic microbial ecosystem beneath an active hydrothermal field. Phylotype HBO-A22 representing 11 clones was distantly related to *Methanocalculus halotolerans* P2F9705 (87.4%), a strain isolated from an estuarine environment (Lai et al., 2002).

Phylotype HBO-A27 representing 15 clones was remotely related (76.7%) to archaeal clone HR3812-Enrichment-018 (DQ058818), a cultivation of methanogens from shallow marine sediments at Hydrate Ridge.

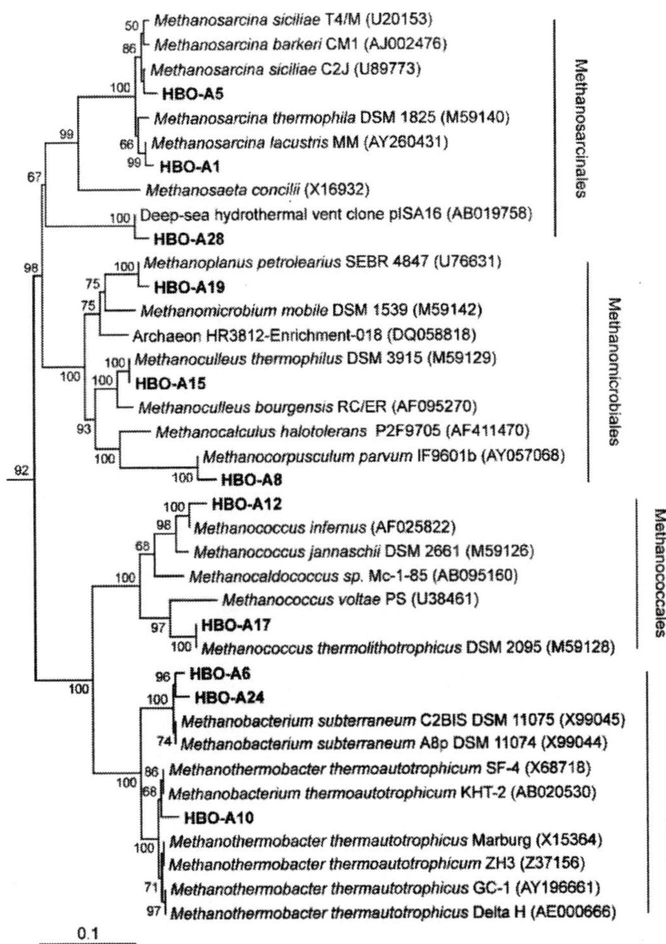

Figure1. Phylogenetic tree of methanogenic phylotypes of HBO-A library (showed in bold) and their closely related sequences from EMBL database. Putative divisions are listed to the right. The topology showed was that obtained with the Neighbor-joining method. Bootstrap values (n=1000 replicates) of 50% are reported. The scale bar represents the number of changes per nucleotide position.

In addition, diverse archaeal lineages, each represented by a few clones related to archaeal members, were found in this petroleum formation (Table 1). They were all less than three clones, thus not major components of the community.

Methanogens Community in Offshore Oil Reservoir

To determine the archaeal diversity retrieved from the offshore petroleum reservoir, an archaeal 16S rRNA gene library was constructed with the community DNA extracted from the production water. A total of 220 clones were partially sequenced (around 450 bp) and categorized on the basis of their sequence similarity (97% identity).

The clone libraries obtained were designated QHO-A (QH Oilfield Archaea). In total, 28 phylotypes were identified\ from the archaeal library. Using the formula of Good, the coverage of archaeal libraries were 92.3%, which indicated that the major part of the diversity in the libraries was detected.

At least one clone per phylotype was sequenced (around 950 bp), and the corresponding type sequences were used to construct phylogenetic trees.

Of the 28 phylotypes in total, half-sequence types had relatively high levels of similarity with their closest counterparts in public databases(Table 2, Figure 2), whereas 14 phylotypes (50%) showed less than 96% sequence similarity to their nearest database entries and may belong to hitherto unknown phylotypes (Stackebrandt and Goebel, 1994).

Table 2.Distribution of dominant sequence types from the offshore oilfield archaeal 16S rRNA gene library

Domain, Phylum	Type Sequence[a]	Closest cultivated species	Similarity (%)[b]	No. of clones(%)[c]
Euryarchaeota	QHO-A15	*Methanobacterium formicicum* Fcam	99	36(9.3)
	QHO-A7	*Methanothermobacter thermautotrophicus* GC-1	99	58(15)
	QHO-A11	*Methanobacterium subterrane um*C2BIS	99	23(5.9)
	QHO-A25	*Methanococcus maripaludis* S2	93	8(2.1)
	QHO-A2	*Methanobrevibacter* sp. Mc30	94	1(<1)

[a] Only partial16S rRNA gene phylotypes were listed.
[b]Percentage of 16S rRNA gene similarity to its closed relative.
[c] Proportion of clones in the library.

Comparative analysis of the retrieved sequences showed that sequences were representatives of *Methanobacteria*. Methanogens were phylogenetically related to the representativesof the genus *Methanothermobacter*, *Methanobrevibacter*, *Methanococcus* and *Methanothermobacter*. The phylotype QHO-A15, representing 36 clones, was closely related

(>99%) to *Methanobacterium formicicum* Fcam, a culturable methanogen isolated from ricefield soils(Joulian *et al.*, 1998).

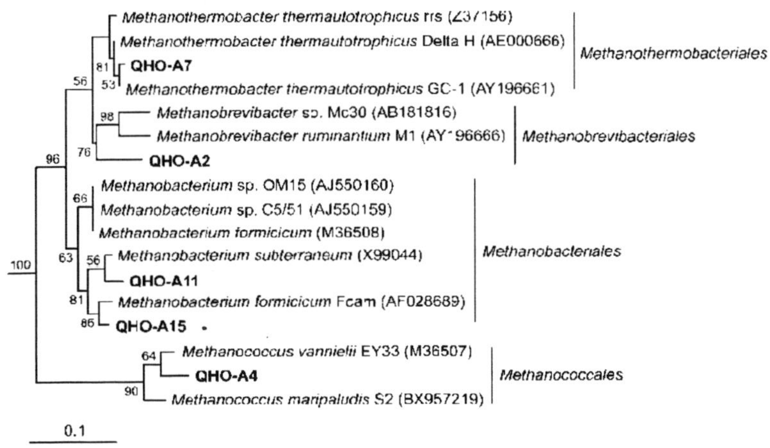

Figure2. Phylogenetic tree of methanogens 16S rRNA gene phylotypes of QHO-A sequence types (shown in bold) and closely related sequences from EMBL database. Putative divisions are listed to the right. The topology shown was obtained with the Neighbor-joining method. Bootstrap values (n = 1000 replicates) of 50% are reported. The scale bar represents the number of changes per nucleotide position. *Thermoproteus tenax* (M35966) and *Desulfurococcus mobilis* (X06188) were used as outgroups.

The most abundant sequencetype in this group, QHO-A7, displaying 58 clones, was closely related (>99%) to *Methanothermobacter thermautotrophicus* GC-1, which was similar to the phylotype found in Dagang oilfield (Nazina *et al.*, 2006).

The phylotype QHO-A11, including 23 clones, was closely related (>99%) to *Methanobacterium subterraneum*C2BIS, an alkaliphilic, eurythermic and halotolerant methanogen isolated from deep granitic groundwater(Kotelnikova *et al.*, 1998). Within *Methanococcus* most sequences were distantly related (<93%) to *Methanococcus maripaludis* S2.

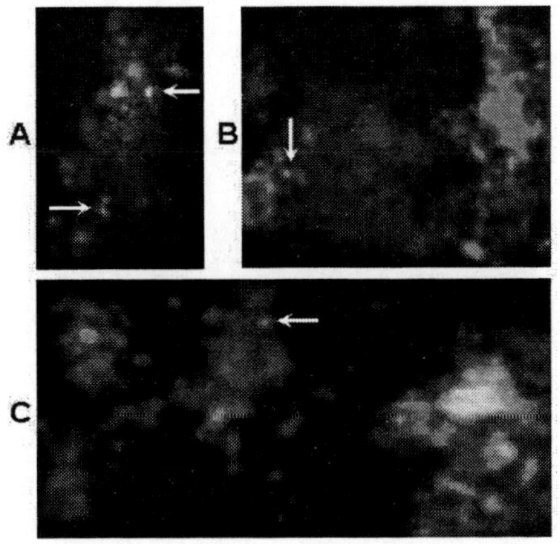

Figure3.In situ hybridization of bacteria and methanogenic archara in the water sample Hua from oil reservoir. They were simultaneously hybridized with the fluorescein-labeled probe Hyr and Cy3-labeled probe Met. Arrows indicate methanogenic-like cells (A, B and C) and hydrocarbon-degrading-like cells (A). Photographs are overlays of several microscopic images.

Methanogens Distribution in Situ

The relative abundance identification as well as spatial organization of methanogens archaeal populations in the syntrophic consortia was elucidated by fluorescence in situ hybridization. Simultaneous hybridizations with the fluorescein-labeled probe for bacteria (Hyr) and the Cy3-labeled probe for the methanogenics archaea (Met) demonstrated that these organisms formed dense aggregates (Figure 3). For the spacial distribution, no layered structure was observed and methanogens were randomly distributed around the syntrophic bacteria, in a close association. Although the amount of the different partners was difficult to quantify in all samples, we observed that methanogens seemed to dominate (Figure 3).

Presence and Diversity of Archaeal amoA Gene in Oilfield

The formations of the Shengli Oilfield lie on average 1200-2000 m below the ground surface with *insitu* temperature between 55 and 75 °C. The sampled wells at this oilfield covered three kinds of petroleum recovery processes: primary oil production at wells S6 and S7 and the formation water mainly form groundwater in the strata; water flooding for 2-10 years at wells S1, S2, S3, S4, S5, S8 and S9 to maintain the strata pressure and allow the further production of residual oil; and polyacrylamide (PAM) solution injection for two years at well S5. Overall, the production water was carbonaceous type with slightly alkaline pH values between 7.3 to 8.1 under ambient conditions (Table 3). The presence of *amoA* gene was confirmed in 9 out of the 17 water samples from Shengli Oilfield. The crenarchaeal *amoA* gene libraries were then constructed and a total of 800 clones were screened by PCR and positive clones ranging from 32 to 50 per well sample were sequenced (Table 4).

Table 3. Environmental characteristics of the production water samples from Shengli Oilfield

No.	Well no.	Depth, m	Temp[a], °C	pH	WF[b]	CF[b]	Content, $mg\ l^{-1}$				
							NO_3^-	NH_4^+	NO_2^-	S_2^-	Fe^{2+}
S1	YXD-9X5	2000	75	7.7	5	0	7.18	37.3	0.027	0.07	0.05
S2	YXD-10X5	1900	75	7.8	10	0	9.67	33.2	0.030	0.02	0.01
S3	C7-J9	1300	55	7.6	10	0	8.23	15.1	0.010	0.04	0.02
S4	C6-15	1200	55	7.5	5	0	5.16	11.7	0.012	0.06	0.07
S5	C26-G10	1300	55	8.1	10	2	10.2	39.2	b.d.l.	0.04	0.03
S6	C91-13	1200	55	7.3	0	0	2.09	7.15	0.092	0.72	0.11
S7	CN13-13	1500	65	7.4	0	0	1.13	5.31	0.087	0.35	0.10
S8	Luo801-10	1600	65	7.6	5	0	6.37	21.7	0.008	0.06	0.06
S9	Luo801-19	1600	65	7.7	10	0	9.12	27.6	b.d.l.	0.04	0.02

[a,]The average temperature of the oil bearing horizons of the unites.
[b,]WF, waterflooding; CF, chemical flooding.
[b.]d.l, below detection limit.

Table 4. Diversity and predicted richness of crenarchaeal and bacterial *amoA* gene retrieved in the production water samples from Shengli Oilfield

No.	Well no.	No. of sequences		No. of OTUs [a]		Coverage [b]		Shannon		Simpson		Chao1	
		AOA	AOB	AOA	AOB	AOA	AOB	AOA	AOB	AOA	AOB	AOA	AOB
S1	YXD-9X5	41	32	8	4	1.00	1.00	1.89	1.21	0.16	0.32	8.00	4.00
S2	YXD-10X5	37	32	7	5	1.00	1.00	1.81	1.51	0.16	0.22	7.00	5.00
S3	C7-J9	46	39	10	7	0.98	0.97	2.14	1.76	0.12	0.17	10.0	7.00
S4	C6-15	45	40	11	7	0.98	0.98	2.32	1.75	0.09	0.18	12.0	7.00
S5	CN13-13	37	32	7	5	1.00	1.00	1.74	1.40	0.19	0.28	7.00	5.00
S6	C91-13	35	32	7	5	1.00	1.00	1.78	1.52	0.17	0.22	7.00	5.00
S7	CN13-13	37	36	7	5	1.00	1.00	1.76	1.49	0.18	0.23	7.00	5.00
S8	Luo801-10	46	39	9	7	0.98	0.97	2.15	1.75	0.11	0.18	10.0	7.00
S9	Luo801-19	42	38	8	7	0.98	0.97	1.88	1.77	0.16	0.17	8.00	7.00

[a]·Unique OTUs were determined using the DOTUR program based on the 3% sequences difference.

[b]·The coverage, Shannon-Weiner, Simpson and Chao1richness estimators were calculated using the OTUs data.

The operational taxonomic units (OTUs) were calculated for all clones in each well sample using the nearest-neighbor algorithm in DOTUR program (Schloss and Handelsman 2005)based on 3%differences of nucleotide acid (Table 4).The coverage value of all libraries ranged from 0.97 to 1.00, indicating that the majority the phylotypes in the samples was detected (Table 4).The Chao1 diversity estimator was from7 to 12of crenarchaeal *amoA* OTUs and the Shannon index ranged from 1.74 to 2.32of crenarchaeal *amoA* OTUs(Table 4).. The archaeal *amoA* gene diversity of samples S3, S4 and S8was the highest while samples S1,S2, S6 and S7 the lowest.

Phylogenetic Affiliation of Crenarchaeal amoA Gene in Oilfield

Ammonia-oxidation crenarchaeal diversity in Shengli Oilfield was detected by*amoA* gene clone libraries constructed with the community DNAretrieved from the production water samples from the underground oil strata. Screening of randomly selected 360 sequences without chimera against the FASTA database indicated that the clones matched sequences related to crenarchaeal genera:*Candidatus* Nitrosopumilus, *Candidatus* Nitrososphaera and *Candidatus* Nitrosocaldus and the environmental phylotypes. Comparative analysis of *amoA*-like sequences retrieved in this study showed the presence of 24 OTUs categorized on the basis of sequence similarity (97% identity) and most of the sequence types had relatively high levels of similarity with their closest counterparts in the public databases. And the phylogenetic trees of *amoA*-like sequences were constructed to compare oilfield crenarchaeal *amoA* sequences to each other and tothose recovered from environment (Figure 4).The crenarchaeal *amoA* genes from oilfields were branched within the two known major clusters of archaeal *amoA* sequences: cluster A,cluster B (Francis et al., 2005) and cluster C including the *amoA* gene sequence of *Candidatus*'Nitrosocaldus yellowstonii' isolated in Yellow Stone National Park hot springs recently (de la Torre et al., 2008). Clone libraries from the different geographic locations and recovery processes of the oilfield contained very different numbers of the major subclusters (Figure 4).

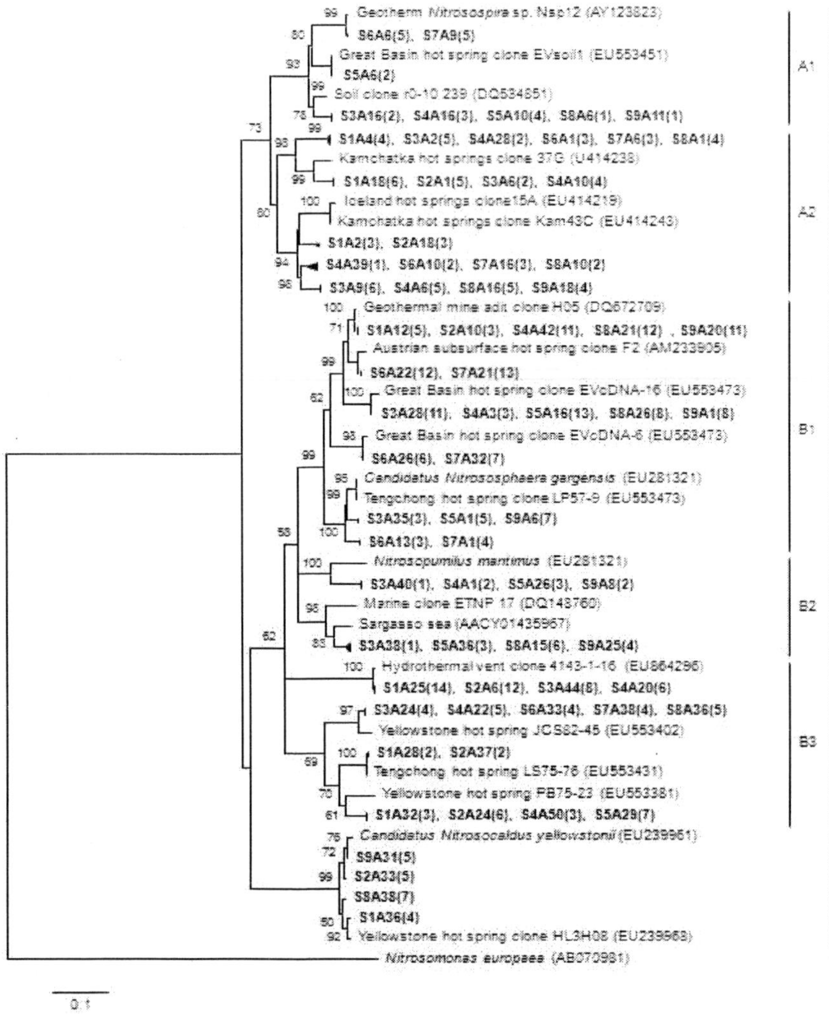

Figure 4. Phylogenetic tree of the AOA-like *amoA* sequences from Shengli Oilfield shown in boldface. and the referred sequences in the EMBL database with the putative divisions listed to the right. The topology shown was calculated with the Neighbor-joining method. Bootstrap values n = 1000 replicates of ≥ 50% were reported near the corresponding nodes. The scale bar represented 0.10 nuclear acid substitutions per nucleotide position. The *amoA* sequences of *Nitrosomonas europaea* AB070981 was used as outgroup.

Cluster A contained the phylotypes closely related to *amoA* sequences typically found in soil and terrestrial hot springs, representing one-fifth of the OTUs fell into two major subgroups A1 andA2 (Figure 4).

Groups A1 formed a clade containing sequences from moderate thermal springs and soil habitats. This group was comprised mainly of *amoA* sequences from S3, S4 and S5 oilfields with a temperature of 55 °C (Figure 4), making up 7% of all oilfield phylotypes.

The sequences were 97% to 99% identicalto *amoA* sequences recovered from geothermal mine adit (Spear et al., 2007), Eagleville hot spring (Zhanget al., 2008) and topsoil (Leininger et al., 2006). Group A2 sequences were from wells S1, S2, S8 and S9 with temperature between 65 °C and 75 °C and 10% sequences of the oilfield clones. This group was a large monophyletic clade with 97% to 99% identities to *amoA* sequences recovered from 84 to

85°C terrestrial hot springs at Hveragerdi in southwest Iceland in the Kamchatka Peninsula of Far East Russia(Reigstad et al., 2008).

Cluster B contained the phylotypes closely related to *amoA* sequences typically found in marine environment, hot springs, *Nitrosopumilus maritimus* and *Candidatus*'Nitrososphaera gargensis' and more than half of the oilfield OTUs fell into three major groups, namely B1 to B3 (Figure 4) with a range of temperature gradient and different recovery processes. Group B1 sequences were retrieved almost exclusively from lower temperature water samples from wells S3, S4 and S7 with a temperature of 55 °C (Figure 4). This group had a large number of sequences recovered from hot springs, geothermal mine adit and *Candidatus*'Nitrososphaera gargensis'(Hatzenpichler et al., 2008) and contained 20% phylotypes of the oilfield clones. These sequences were 97% to 99% identical to *amoA* sequences recovered from 45.6°C thermal mineral spring from deep rock fissures in the Alps near Salzburg, Austria (Weidler et al., 2007), a mat-like biofabrics in two kinds of speleothem from a 49.6 °C subsurface geothermal mine adit near Glenwood Springs (Spear et al., 2007), 41°CEagleville hot spring in Great Basin spring and 57 °C Little Pond hot spring in Tengchong area of China(Zhang et al., 2008). Group B2 also comprised sequences from lower temperature oilfield (Figure 4) with 10% oilfield phylotypes.

The sequences were 97% to 99% identical to *amoA* sequences of *Nitrosopumilus maritimus* (Konneke et al., 2005) and recovered from marine sample from the northwest Sargasso Sea, at the Bermuda Atlantic and the oxygen minimum zones of the Eastern Tropical North Pacific(Francis et al., 2005).Interestingly, group B3, composed of two sequences retrieved from the 75 °C wells S1 and S2, were closely related to the *amoA* sequence detected in the out layers of an active deep-sea chimney with 316°C black smoker vent in the Mothra field(Wang et al., 2009).

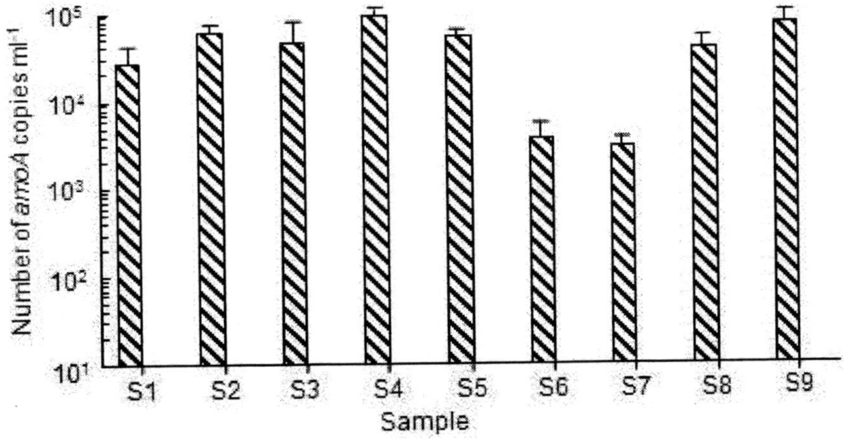

Figure 5. Abundance of crenarchaeal and bacterial amoA gene copy numbers in production water from Shengli Oilfield. Mean values and standard deviations were calculated from triplicate assay within a single qPCR setup.The number on the top of the columns represented the ratios of AOA:AOB.

Group B3 also contained sequences from 65°C oilfields and included 11% of the phylotypes closely related (98% identities) to the *amoA* sequence recovered in terrestrial hot springs of Yellowstone National Park and Tengchong in China(Zhang et al., 2008). Cluster C contained the sequences retrieved from above 65°C wells S1, S2, S8 and S9, accounting for one-sixth of oilfield phylotypes and closely related (98% identities)to the *amoA* gene sequence of *Candidatus* 'Nitrosocaldus yellowstonii' which was the highest temperature autotrophic ammonia-oxidationcrenarchaeote growing upto 74°C(de la Torre et al., 2008).

This group appeared to be exclusive to the geothermal habitats with 97% to 99% identities to the *amoA* gene sequence from 78.9 °Csediment of Heart lake and microbial mat of 95 °C Obsidian Pool Fumarole from Yellowstone National Park (de la Torre et al., 2008), which was obviously separated from the clusters A and B (Figure 4).

Quantification of AOA in Oil Reservoirs

The abundance of AOA in oilfield production water was estimated by using real-time PCR assay targeting *amoA* gene. Result showed that the copy numbers of AOA *amoA* gene ranged from2.92 $\times 10^3$ to 9.21 $\times 10^4$ per ml production water (Figure 5). The highest AOA population size was detected in the wells S4 and S9, followed by the wells S2 and S5, and then by the S1 and S8 (Figure 5). No significant trend in AOA population was delineated along the different samples. The lowest abundance of AOA was detected in the primary production oilfields S6 and S7 with 3.71 $\times 10^3$ and 2.92 $\times 10^2$ copies per ml formation water (Figure 5).

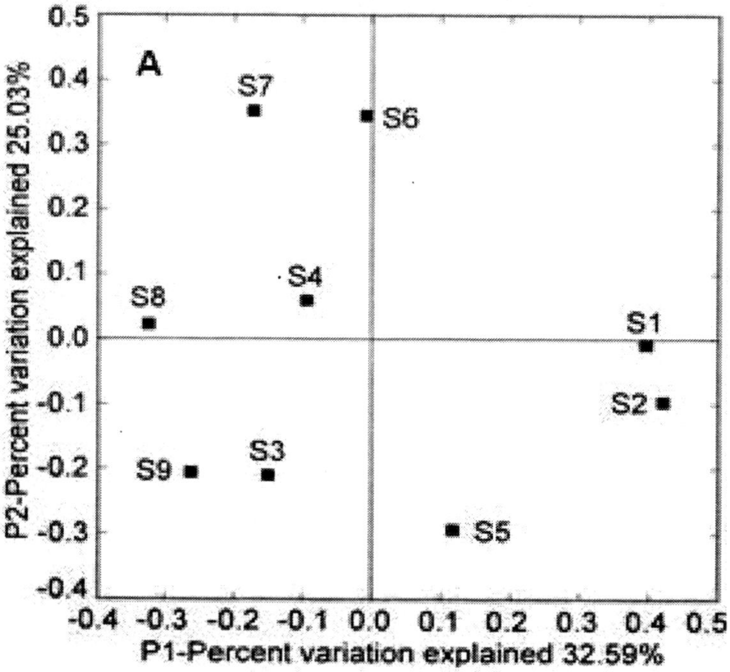

Figure 6. PCA ordination diagrams of ammonia-oxidizing prokaryotic AOA assemblages calculated with *amoA* gene sequences from production water of Shengli Oilfield. Shown are the plots of the first two principal coordinate axes P1 and P2. for PCA and the classifications of the AOA assemblages designated with the sampling wells. in response to these axes.

Ammonia-Oxidizing Archaeal Communities Classification

Significant difference of the ammonia-oxidizing archaeal (AOA) assemblages in the oilfields were detected via the PCA analyses ($p < 0.005$) (Lozupone et al., 2007) basing on the phylogenetic tree of crenarchaeal *amoA* sequences. Crenarchaeal assemblage fell into five groups, S6 and S7 for no waterflooding oilfields, S4 and S8 for lower temperature and moderate waterflooding oilfields, S3 and S9 for moderate temperature but longest water flooding oilfields, S1 and S2 for highest temperature oilfield and S5 for the chemical flooding oilfield (Figure 6). Wells S1 and S2 could be separated from the rest along the first principle coordinate (P1) which explained 32.6% of the total AOA and 55.2% of the total AOB community variability among all the sampled oilfields.

Among the 7 samples with temperature 55-65 °C, S5 representing chemical flooding oilfield showed a different classification scheme from other samples along the second principle coordinate (P2) which explained 25.0% of the total AOA community variability (Figure 6).

The other six samples with moderate temperature were grouped into three classifications along the P2 axe (Figure 6), consistent with the water flooding. S6 and S7 representing the no water flooding oilfields with the prokaryotic phylotypes from underground geothermal water system were grouped together. S4 and S8 with moderate water flooding while S3 and S9 with longest water flooding were separated into two groups.

Geographic Distribution of Ammonia-Oxidizing Archaeal Communities

To detect the correlations between the distribution of AOA community and the environmental variables of the oilfields, the canonical correspondence analysis (CCA) were conducted based on the crenarchaeal *amoA* sequences and the physiochemical parameters of the sampling wells Shengli Oilfield.

Both the temperature and recovery process had the most significant impact on the composition and structure of the AOA assemblages in the oilfield water samples (Figure 7).

In total, the CCA axe in crenarchaeal *amoA* gene deduced schema explained 52.6% (Figure 7) of the cumulative variations of the correlation between the environmental factors and the AOA community distribution (Figure 7).The distribution of S1 and S2 with the highest temperature was positive correlated with temperature (Figure 7).

In addition, the recovery process contributed significant effects on the distribution of the AOA community. S3 and S9 oilfields with the longest water flooding showed the highest positive correlation with the WF factor (Figure 7) and the S4, S8 oilfields also positively related to the water flooding factor, but the S6 and S7 oilfields, without water flooding process, showed the a negative relationship with the water flooding factor (Figure 7). The AOA community in S5 oilfield showed a strong association with the chemical flooding. None of the other environmental factors contributed significantly ($p > 0.500$) to the AOA community, while the combination of these environmental factors provided an additional 35.1% of the total CCA explanatory power.

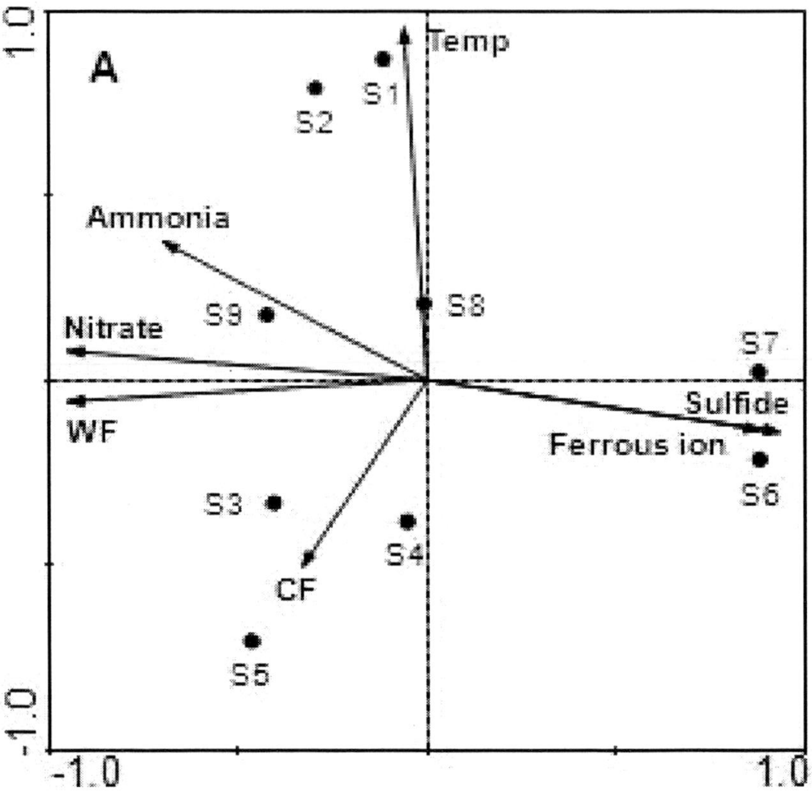

Figure 7. CCA ordination plots for the first two dimensions to show the relationship between the AOA communities designated with the sampling wells with environmental parameters analyzed using AOA amoA gene sequence from the production water of Shengli Oilfield. Correlations between environmental variables and CCA axes are represented by the length and angle of arrows environmental factor vectors in bold. Abbreviation: Temp, temperature; WF, waterflooding; CF, chemical flooding.

Discussion

Methanogens Community in Onshore Oil Reservoir

In the present study, molecular analysis was used to characterize the diversity of archaeal community in the production water from the high-temperature, water-flooded petroleum reservoir at Huabei oilfield. Although the level of effort expended in this study did not exhaustively sampled archaeal diversity, we can conclude that it is sufficient to capture a large fraction of archaea in the petroleum reservoir. Interestingly, the phylotypes retrieved all belong to methanogenic archaea, which indicates that the archaeal diversity is low in the high-temperature petroleum reservoir. This apparently low diversity might attribute to the extreme conditions of the subsurface petroleum reservoir environment. In fact, it is widely recognized that the microbial diversity is often low in the hyperthermal environments (Reysenbach et al., 2000). Although the PCR and DNA extraction used in investigating the microbial ecosystems would introduce biases (Suzuki et al., 1996; Von Wintzingerode et al., 1997), they are the primary sources of information available to assess the phylogenetic richness and complexity of microbial communities. And compared to other molecular techniques, rRNA approach has a great advantage that the generation of sequence data can be used to design group-specific probes and primers for further studies such as microarray and real-time PCR which have not been used to characterize the microbial diversity in the petroleum reservoir.

Geothermally heated oil reservoirs, in which liquid hydrocarbons are the prevailing organic matter, are unique ecological niches for thermophilic methanogens. In this study the Methanogen-related clones displayed a large amount of 16S rRNA gene sequence variability. Considering the fact that these phylotypes were detected from multiple oil reservoir environments worldwide (Bonch-Osmolovskaya et al., 2003; Li et al., 2006), they might be the indigenous microorganisms of petroleum reservoirs. However, many phylotypes related to mesophilic methanogen such as *Methanoplanus petrolearius* were clearly detected. The reasons mainly are that the oil reservoir is an open system with continuously flooded operation. Together with the tons of unsterilized water recycled from injection wells to production wells, numbers of microorganisms would be introduced into the petroleum reservoir. Some of them may reside in cooler portions of the reservoir or along walls or openings of production-well tubing, which may result in detection of mesophilic microorganisms in production water.

Many Methanomicrobiales-related clones can utilize formate and alcohols alternatively to H_2+CO_2, whereas Methanosarcinales-related ones grow generally by dismutating methyl compounds or with acetate as sole energy source. Here, we have shown that they are present in the production water and could participate in biogeochemical cycles in oil reservoir. Obviously, these phylotypes are consistent with the physicochemical conditions of oil bed, which indicates that they may be the indigenous microorganisms in petroleum reservoir systems and likely to have contributed to methane production in most petroleum reservoirs. The prevalence of the methanogens phylotypes in the archaeal library led us to conclude that thermophilic methanogens may dominate over the subsurface oil reservoir ecosystem.

Our study revealed the phylogenetic diversity of archaea in previously poorly characterized subsurface environment. We believe additional insights into the underlying microbiology in petroleum reservoir ecosystems are important for stimulating them at optimal efficiencies. Indeed, our data suggest that petroleum reservoir systems are far more complex than previously estimated and remain largely unexplored. The apparent abundance and diversity of these hydrogenotrophic and acetoclastic methanogens sequence types implied that their corresponding organisms may play significant but unknown roles in the anaerobic processes in the petroleum reservoir.

In anoxic environment of oil reservoir, because of the thermodynamic considerations, mineralization of alkanes was performed by the cooperation of at least two different metabolic groups including hydrocarbons-degrading bacteria and methanogenics achaea.

These two groups usually formed closely syntrophic associations *in situ*, so it was quite difficult to isolate the pure cultures from environmental samples. Here, we used FISH method to characterize the bacteria and methanogenics archaea in the syntrophic associations from oil reservoir sample, which overcame the culture limitation and showed the spacial distribution of them objectively. The results showed the two groups in close consortia indeed, same as the obvious study (Grabowski et al., 2005; Sekiguchi et al., 2000). This was perhaps due to the close contact between the different partners could favor optimal metabolite transfer in the syntrophic associations.

Normally, bacteria adhering in the insoluble alkanes substrates metabolized the hydrocarbon to formate, acetate and so on. These small molecular matters were the carbon and energy source of the methanogens, so it was reasonable in our observation that the methanogens archaea were randomly distributed around the bacteria, which contributed to methanogens both absorbed the substrates and remove the end products limitations for bacteria.

Bacteria and methanogenics archaea formed the compacted consortia *in situ*. However, the multi-discipline researches will have to be preformed to systematically state the ecological distribution and role of the functional populations in the underground oil reservoirs. We believe that the result provides insight into the diversity of microbial communities in petroleum reservoirs, which in turn will improve our understanding of efficient use in, e.g. the microbial enhanced oil recovery.

Methanogens Community in Offshore Oil Reservoir

Molecular analysis was used to characterize the microbial diversity in production water from a high-temperature, water-flooded petroleum reservoir at the QH offshore oilfield. Although many efforts have been made to define the microbial diversity of petroleum reservoirs in recent years (Bonch-Osmolovskaya *et al.*, 2003; Grabowski *et al.*, 2005; Li *et al.,* 2006; Orphan *et al.*, 2000), the present study still revealed many 16S rRNA gene sequences that have not been previously reported from such extreme environments. We conclude that the level of effort expended in this study has not been sufficient to exhaustively sample microbial diversity, but that it is successful in characterizing a large fraction of the oilfield system.

The phylotypes belong mainly to methanogenicarchaea, which indicates that the phylotype richness is low in the high-temperature petroleum reservoir, as shown in our previous work about the archaeal community int an onshore oil reservior (Li *et al.*, 2006). We note that the apparently low diversity might be attributed to the extreme conditions of the subsurface petroleum reservoir environment. In fact, it is widely recognized that microbial diversity is often low in hyperthermal environments, as emphasized by studies in a particular hot spring (Reysenbach *et al.*, 2000).On the other hand, geothermally heated oil reservoirs, in which liquid hydrocarbonsare the prevailing organic matter, are unique ecological niches for thermophilic anaerobic hydrocarbondegraders and methanogens. We have studied the bacterial community of a high-temperature petroleum reservoir and have shown that some clones have a number of ecologically beneficial featuresadapting them to the habitat (Li *et al.*, 2006). Within the archaeal library, 16S rRNA gene types belonged to the genera *Methanothermobacter*, *Methanobacter*, *Methanobrevibacter* and *Methanococcus*, which is consistent with the results of studies of methanogens from high-temperature oilfields obtained by other authors (Mueller and Nielsen, 1996; Nilsen and Torsvik, 1996). The prevalence of the methanogen phylotypes in the association studied led us to conclude that thermophilic methanogens may dominate the subsurface ecosystem in these oil reservoirs.

Although this petroleum reservoir system is characterized by high temperature (65 °C), many phylotypes related to mesophilic microorganisms were clearly detected. In fact, the oil is an open system, since it is continuously flooded with water. Together with the tons of unsterilized water recycled from injection wells to production wells, numbers of archaea would be introduced into the petroleum reservoir. Some of them may reside in cooler portions of the reservoir or along walls or openings of production-well tubing, which may result in detection of mesophilic archaea in production water. In addition, intensive use of molecular techniques in investigating microbial ecosystems would introduce biases that may increase or decrease the frequency of the particular phylotype found in the16S rRNA gene library, which may not necessarily represent the true proportion of corresponding species in the original environment (Suzuki *et al.*, 1998). Basically, the PCR and DNA extraction are two acknowledged sources of a methodological bias (Suzuki and Giovannoni, 1996; von Wintzingerode *et al.*, 1997), but they are primary sources of information we have available to assess the phylogenetic richness and complexity of microbial communities. Analysis of this information provides insight into the likely complexity of archaeal communities in petroleum reservoirs and inspiration for future studies.

Ammonia-Oxidization Archaeal Community in Oil Reservoirs

The observation of crenarchaeal *amoA* gene described here confirmed and substantiated the occurrence of AOA in the geothermal oilfield environment. Our result showed that the condition of limited DO and high temperature potentially associated with geothermal oilfield should not deter AOA from proliferating in these habitats. Statistical analyses indicated that majority of the crenarchaeal *amoA* gene diversity in these oilfields was revealed. Most of the phylotypes discovered in oilfields were related to the cultured thermophilic AOA or the *amoA* ecotypes found in the thermophilic habitats, e.g. geothermal mine adit, Great Basin hot spring, Kamchatka hot spring, Iceland hot spring, Austrian subsurface hot spring, Tengchong hot spring, hydrothermal vent and Yellowstone hot spring(Spear et al., 2007; Weidler et al., 2007; Perevalova et al., 2008; Reigstad et al., 2008; Weidler et al., 2008; Zhang et al., 2008; Pitcheret al., 2009; Wang et al., 2009). In particular, two cultured thermophilic AOA strains *Nitrososphaera gargensis* and *Candidatus* 'Nitrosocaldus yellowstonii'(de la Torre et al., 2008; Hatzenpichler et al., 2008) related phylotypes were detected in oilfields, indicating that these thermophilic AOA might be ubiquitous in the hot habitats all over the world. However, not all *amoA* gene alleles from oilfield were related to the high temperature *amoA* phylotypes, while these sequences included few clones suggesting they were only the minority of the AOA population in the oilfield water samples. Interestingly, very little clones were related to *Nitrosopumilus maritimus* which was reported to be inhibited by organic substrates even at very low concentrations (Konneke et al., 2005), indicating that these clones were not active in oil with plenty of hydrocarbons and small molecular organic acid. Although the amount of carbonate in the formation water of oil reservoir could supply the adequate carbon source for their growth, the sequences related to *Candidatus* 'Nitrosocaldus yellowstonii' were minority in the clone, indicating that the acetate or other small organic acids might inhibit their growth.

The distribution of AOA was not random with respect to the physicochemical conditions of oil reservoir. Compared to the conventional AOA habitats(Avrahami et al., 2003; Park and Noguera 2004; Erguder et al., 2009), the salinity, pH and ammonia concentration in oilfield were essentially constant in Shengli Oilfield while the temperature and DO of oil reservoir were especially high and limited to the normal ammonia-oxidation. The recovery process was another important factor affecting the AOA community in different oil field, which could explain the discrepancy of AOA distribution in different oilfields by the influence of injected water and chemical(Li et al., 2007; Dahle et al., 2008).

Originally, the subsurface oil reservoir was exclusively reducing environment. At the secondary oil recovery process, oxygen and chemical were injected with the water flooding to dramatically change the indigenous chemical conditions of the oil strata *in situ*. In addition, the volume of injection water was so large that many kinds of microorganisms together with nutrients were introduced into the oil reservoirs(Korenblum et al., 2010), which subsequently changed the microbial community distribution. Interestingly, the ammonia and nitrate concentration showed no distinct correlation with the AOA assemblage distribution, which might be due to the ammonia and nitrate concentration were in excess to the AOA assemblage under the water flooding condition. More importantly, our results suggest that AOA might have broader ecophysiological adaptation potential than previously estimate.

Our study demonstrated the occurrence of AOA-like *amoA* in the production water, suggesting the existence of putative AOA in the oil reservoir. However the knowledge of the activity of AOA-like strain was very limited under the oil reservoir condition. Our future effort will focus on the enrichment of the AOA-like members in a laboratory scale bioreactor to reveal whether these organisms are indeed capable of catalyzing ammonia-oxidation process in the oil field ecosystem.

Our study allows a better comprehension of the nitrogen cycle in the oil reservoir and will provide the basis for further exploration AOA-like organisms in the subterranean environment.

Acknowledgments

This study was financially supported by the National Natural Science Foundation of China (40871223, 41003031, 41073055), the National Environmental Protection Public Welfare Science and Technology Research Program of China (200909089, 2010467032, 201109013), the Specialized Research Fund for the Doctoral Program of Higher Education (20110074130002) the Fundamental Research Funds for the Central Universities (WB1014004), the Major Science and Technology Program for Water Pollution Control and Treatment (2009ZX07101-009-03).

References

Adair KL and Schwartz E (2008) Evidence that ammonia-oxidizing archaea are more abundant than ammonia-oxidizing bacteria in semiarid soils of northern Arizona, USA. *Microb. Ecol.* 56:420-426.

Agrawal A and Lal B (2009) Rapid detection and quantification of bisulfite reductase genes in oil field samples using real-time PCR. *Fems.Microbiol. Ecol.* 69:301-312.

Aitken CM, Jones DM, Larter SR (2004) Anaerobic hydrocarbon biodegradation in deep subsurface oil reservoirs. *Nature* 431:291-294

Altschul SF, Thomas LM, Alejandro AS, Zhang JH, Zhang Zh, Webb M, David JL (1997) Gapped BLAST and PSI-BLAST: a new generation of protein database search programs. *Nucleic Acids Res.*25:3389-3402.

Avrahami S, Liesack W, Conrad R (2003) Effects of temperature and fertilizer on activity and community structure of soil ammonia oxidizers. *Environ. Microbiol.* 5:691-705.

Bonch-Osmolovskaya EA, Miroshnichenko ML, Lebedinsky AV, Chernyh NA, Nazina TN, Ivoilov VS, Belyaev SS, Boulygina ES, Lysov YP, Perov AN, Mirzabekov AD, Hippe H, Stackebrandt E, L'Haridon S, Jeanthon C (2003) Radioisotopic, culture-based, and oligonucleotide microchip analyses of thermophilic microbial communities in a continental high-temperature petroleum reservoir. *Appl. Environ. Microbiol.*69:6143-6151.

Chen XP, Zhu YG, Xia Y, Shen JP, He JZ. 2008. Ammonia-oxidizing archaea: important players in paddy rhizosphere soil? *Environ. Microbiol.* 10:1978-1987.

Cole JR, Chai B, Marsh TL, Farris RJ, Wang Q, Kulam SA, Chandra S, McGarrell DM, Schmidt TM, Garrity GM, Tiedje JM (2003) The Ribosomal Database Project (RDP-II):previewing a new autoaligner that allows regular updates and the new prokaryotic taxonomy. *Nucleic Acids Res.* 31:442-443.

Coolen MJL, Abbas B, van Bleijswijk J, Hopmans EC, Kuypers MMM, Wakeham SG, Damste JSS. 2007. Putative ammonia-oxidizing Crenarchaeota in suboxic waters of the Black Sea: a basin-wide ecological study using 16S ribosomal and functional genes and membrane lipids. *Environ. Microbiol.* 9:1001-1016.

Dahle H, Garshol F, Madsen M, Birkeland NK (2008) Microbial community structure analysis of produced water from a high-temperature North Sea oil-field. Antonie Van Leeuwenhoek *Int. J. Gen. Mol. Microbiol.* 93:37-49.

de la Torre JR, Walker CB, Ingalls AE, Konneke M, Stahl DA (2008) Cultivation of a thermophilic ammonia oxidizing archaeon synthesizing crenarchaeol. *Environ. Microbiol.* 10:810-818.

Dolfing J, Larter SR, Head IM. 2008. Thermodynamic constraints on methanogenic crude oil biodegradation. *Isme J.* 2:442-452.

Eckford RE, Fedorak PM (2005) Applying a most probable number method for enumerating planktonic, dissimilatory, ammonium-producing, nitrate-reducing bacteria in oil field waters. *Can. J. Microbiol.* 51:725-729.

Elberson MA, Sowers KR (1997) Isolation of an aceticlastic strain of Methanosarcina siciliae from marine canyon sediments and emendation of the species description for Methanosarcina siciliae. *Int. J. Syst. Bacteriol. 47*:1258-1261.

Ehrhardt CJ, Haymon RM, Lamontagne MG, Holden PA (2007) Evidence for hydrothermal Archaea within the basaltic flanks of the East Pacific Rise. *Environ. Microbiol.* 9:900-912.

Erguder TH, Boon N, Wittebolle L, Marzorati M, Verstraete W (2009) Environmental factors shaping the ecological niches of ammonia-oxidizing archaea. *Fems Microbiol. Rev.* 33:855-869.

Felsenstein J (2005) PHYLIP: Phylogeny inference package, Version 3.65. University of Washington, Seattle.

Francis CA, Roberts KJ, Beman JM, Santoro AE, Oakley BB (2005) Ubiquity and diversity of ammonia-oxidizing archaea in water columns and sediments of the ocean. *PNAS* 102:14683-14688.

Gieg LM, Duncan KE, Suflita JM (2008) Bioenergy production via microbial conversion of residual oil to natural gas. *Appl. Environ. Microbiol.* 74:3022-3029.

Gittel A, Sorensen KB, Skovhus TL, Ingvorsen K, Schramm A (2009) Prokaryotic Community Structure and Sulfate Reducer Activity in Water from High-Temperature Oil Reservoirs with and without Nitrate Treatment. *Appl. Environ. Microbiol.* 75:7086-7096.

Good IJ (1953) The population frequencies of species and the estimation of population parameters. *Biometrika*40:237-264.

Grabowski A, Blanchet D and Jeanthon C (2005) Characterization of long-chain fatty-acid-degrading syntrophic associations from a biodegraded oil reservoir. *Res. Microbiol.* 156: 814-821

Grabowski A, Nercessian O, Fayolle F, Blanchet D, Jeanthon C (2005) Microbial diversity in production waters of a low-temperature biodegraded oil reservoir. *FEMS Microbiol. Ecol.* 54:424-433

Gray ND, Sherry A, Larter SR, Erdmann M, Leyris J, Liengen T, Beeder J, Head IM (2009) Biogenic methane production in formation waters from a large gas field in the North Sea. *Extremophiles* 13:511-519.

Greene EA, Hubert C, Nemati M, Jenneman GE, Voordouw G (2003) Nitrite reductase activity of sulphate-reducing bacteria prevents their inhibition by nitrate-reducing, sulphide-oxidizing bacteria. *Environ. Microbiol.* 5:607-617.

Grosskopf R, Janssen PH, Liesack W (1998) Diversity and structure of the methanogenic community in anoxic rice paddy soil microcosms as examined by cultivation and direct 16S rRNA gene sequence retrieval. *Appl. Environ.Microbiol.* 64:960-969.

Hallam SJ, Konstantinidis KT, Putnam N, Schleper C, Watanabe Y, Sugahara J, Preston C, de la Torre J, Richardson PM, DeLong EF (2006) Genomic analysis of the uncultivated marine crenarchaeote Cenarchaeum symbiosum. *PNAS* 103:18296-18301.

Hatzenpichler R, Lebedeva EV, Spieck E, Stoecker K, Richter A, Daims H, Wagner M (2008) A moderately thermophilic ammonia-oxidizing crenarchaeote from a hot spring. *PNAS* 105:2134-2139.

Head IM, Jones DM, Larter SR (2003) Biological activity in the deep subsurface and the origin of heavy oil. Nature 426:344-352.Hubert C, Nemati M, Jenneman G, Voordouw G (2005) Corrosion risk associated with microbial souring control using nitrate or nitrite. *Appl. Microbiol. Biotech.* 68:272-282.

Hubert C, Voordouw G (2007) Oil field souring control by nitrate-reducing Sulfurospirillum spp. that outcompete sulfate-reducing bacteria for organic electron donors. *Appl. Environ. Microbiol.* 73:2644-2652.

Huber T, Faulkner G., Hugenholtz P (2004) Bellerophon: a program to detect chimeric sequence in multiple sequence alignments. *Bioinformatics* 20:2317-2319.

Jeanthon C, L'Haridon S, Reysenbach AL, Vernet M, Messner P, Sleytr UB, Prieur D (1998) Methanococcus infernus sp. nov., a novel hyperthermophilic lithotrophic methanogen isolated from a deep-sea hydrothermal vent. *Int. J. Syst. Bacteriol.* 48:913-919.

Jiang HC, Dong HL, Yu BS, Lv G, Deng SC, Berzins N, Dai MH (2009) Diversity and Abundance of Ammonia-Oxidizing Archaea and Bacteria in Qinghai Lake, Northwestern China. *Geomicrobiol.J.* 26:199-211.

Jones DM, Head IM, Gray ND, Adams JJ, Rowan AK, Aitken CM, Bennett B, Huang H, Brown A, Bowler BFJ, Oldenburg T, Erdmann M, Larter SR (2008) Crude-oil biodegradation via methanogenesis in subsurface petroleum reservoirs. Nature 451:176-181.

Joulian C, Ollivier B, Patel BKC and Roger PA (1998) Phenotypic and phylogenetic characterization of dominant culturable methanogens isolated from ricefield soils. *FEMS Microbiol. Ecol.* 25: 135-145.

Kaster KM, Bonaunet K, Berland H, Kjeilen-Eilertsen G, Brakstad OG (2009) Characterisation of culture-independent and -dependent microbial communities in a high-temperature offshore chalk petroleum reservoir. Antonie Van Leeuwenhoek *Int. J. Gen. Mol. Microbiol.* 96:423-439.

Kemp PF and Aller JY (2004) Bacterial diversity in aquatic and other environments: what 16S rDNA libraries can tell us. *FEMS Microbiol. Ecol.* 47: 161-177.

Kimura M (1980) A simple method for estimating evolutionary rates of base substitutions through comparative studies of nucleotide sequences. *J. Mol. Evol.*16:111-120

Kolganova TV, Kuznetsov, BB, Turova TP (2002) Selection and testing of oligonucleotide primers for amplification sequencing of archaeal 16S rRNA genes. *Mikrobiologiya* 71:283-285

Konneke M, Bernhard AE, de la Torre JR, Walker CB, Waterbury JB, Stahl DA (2005) Isolation of an autotrophic ammonia-oxidizing marine archaeon. *Nature* 437:543-546.

Korenblum E, Valoni E, Penna M, Seldin L (2010) Bacterial diversity in water injection systems of Brazilian offshore oil platforms. *Appl. Microbiol.Biotech.* 85:791-800.

Kotelnikova S, Macario AJ and Pedersen K (1998) Methanobacterium subterraneum sp. nov., a new alkaliphilic, eurythermic and halotolerant methanogen isolated from deep granitic groundwater. *Int. J. Syst. Bacteriol.* 48: 357-367.

Kowalchuk GA, Stephen JR (2001) Ammonia-oxidizing bacteria: A model for molecular microbial ecology. *Annual Rev. Microbiol.* 55:485-529.

Lam P, Jensen MM, Lavik G, McGinnis DF, Muller B, Schubert ÇJ, Amann R, Thamdrup B, Kuypers MMM (2007) Linking crenarchaeal and bacterial nitrification to anammox in the Black Sea. *PNAS* 104:7104-7109.

Lai MC, Chen SC, Shu CM, Chiou MS, Wang CC, Chuang MJ, Hong TY, Liu CC, Lai LJ, Hua JJ (2002) Methanocalculus taiwanensis sp. nov., isolated from an estuarine environment. *Int. J. Syst. Evol. Microbiol.* 52:1799-1806.

Leininger S, Urich T, Schloter M, Schwark L, Qi J, Nicol GW, Prosser JI, Schuster SC, Schleper C (2006) Archaea predominate among ammonia-oxidizing prokaryotes in soils. *Nature* 442:806-809.

Li H, Yang SZ, Mu BZ, Rong ZF, Zhang J (2006) Molecular analysis of the bacterial community in a continental high-temperature and water-flooded petroleum reservoir. *FEMS Microbiol. Lett.* 257:92-98.

Liu JF, Ma LJ, Mu BZ, Liu RL, Ni FT, Zhou JX (2005) The field pilot of microbial enhanced oil recovery in a high temperature petroleum reservoir. *J. Petrol. Sci. Eng.* 48:265-271.

Loy A, Lehner A, Lee N, Adamczyk J, Meier H, Ernst J, Schleifer KH and Wagner M (2002) Oligonucleotide microarray for 16S rRNA gene-based detection of all recognized lineages of sulfate-reducing prokaryotes in the environment. *Appl. Environ. Microbiol.* 68:5064-5081.

Loy A, Horn M and Wagner M (2003) ProbeBase-an online resource for rRNA-targeted oligonucleotide probes. *Nucleic Acids Res.* 31: 514-516.

Lozupone C, Hamady M, Knight R (2006) UniFrac-an onlinetool for comparing microbial community diversity in a phylogenetic context. *BMCBioinform* 7:371-384.

Lozupone CA, Hamady M, Kelley ST, Knight R (2007) Quantitative and qualitative beta diversity measures lead to different insights into factors that structure microbial communities. *Appl. Environ. Microbiol.* 73:1576-1585.

Magot M (2005) Indigenous microbial communities in oil fields. In: Ollivier B, Magot M eds. Petroleum Microbiol. ASM Press, Washington, pp 21–34.

Marchesi JR, Weightman AJ, Cragg BA, Parkes RJ, Fry JC (2001) Methanogen and bacterial diversity and distribution in deep gas hydrate sediments from the Cascadia Margin as revealed by 16S rRNA molecular analysis. *FEMS Microbiol. Ecol.* 34:221-228.

McInerney MJ, Nagle DP, Knapp RM (2005) Microbially enhanced oil recovery: past, present, and future. In Petroleum Microbiol. Ollivier B, and Magot M. eds.. Washington, DC, USA: American Society for Microbiol Press, pp 215–237.

Murray AE, Preston CM, Massana R, Taylor LT, Blakis A, Wu K, DeLong EF (1998) Seasonal and spatial variability of bacterial and archaeal assemblages in the coastal waters near Anvers Island, Antarctica. *Appl.Environ. Microbiol.*64:2585-2595.

Mueller RF, Nielsen PH (1996) Characterization of thermophilic consortia from two souring oil reservoirs. *Appl. Environ. Microbiol.* 62:3083-3087.

Nazina TN, Shestakova NM, Grigoriyan AA, Mikhailova EM, Tourova TP, Poltaraus AB, Feng CX, Ni FT and Belyaev SS (2006) Phylogenetic Diversity and activity of anaerobic microorganisms of high-temperature horizons of the Dagang oil field (P. R. China).*Microbiology* 75: 55-65.

Nilsen RK, Torsvik T (1996) Methanococcus thermolithotrophicus isolated from North Sea oil field reservoir water. *Appl. Environ. Microbiol.*62: 728-731.

Nunoura T, Hirayama H, Takami H, Oida H, Nishi S, Shimamura S, Suzuki Y, Inagaki F, Takai K, Nealson KH, Horikoshi K (2005) Genetic and functional properties of uncultivated thermophilic crenarchaeotes from a subsurface gold mine as revealed by analysis of genome fragments. *Environ.Microbiol.* 7:1967-1984.

Orphan VJ, Taylor LT, Hafenbradl D, Delong EF (2000) Culture-dependent and culture-independent characterization of microbial assemblages associated with high-temperature petroleum reservoirs. *Appl. Environ.Microbiol.* 66:700-711.

Park HD, Noguera DR. 2004. Evaluating the effect of dissolved oxygen on ammonia-oxidizing bacterial communities in activated sludge. *Wat. Res.* 38:3275-3286.

Page RDM (1996) TREEVIEW: an application to display phylogenetic trees on personal computers. *Comput. Appl. Biosci.* 12:357-358.

Perevalova AA, Kolganova TV, Birkeland NK, Schleper C, Bonch-Osmolovskaya EA, Lebedinsky AV (2008) Distribution of Crenarchaeota representatives in terrestrial hot springs of Russia and Iceland. *Appl. Environ. Microbiol.* 74:7620-7628.

Pearson WR, Lipman DJ (1988) Improved tools for biological sequence comparison. *Proc. Natl. Acad. Sci.* 85:2444-2448.

Pham VD, Hnatow LL, Zhang SP, Fallon RD, Jackson SC, Tomb JF, DeLong EF, Keeler SJ (2009) Characterizing microbial diversity in production water from an Alaskan mesothermic petroleum reservoir with two independent molecular methods. *Environ. Microbiol.* 11:176-187.

Pitcher A, Schouten S, Damste JSS (2009) In Situ Production of Crenarchaeol in Two California Hot Springs. *Appl. Environ. Microbiol.* 75:4443-4451.

Polz MF, Cavanaugh CM (1998) Bias in template to product ratios in multitemplate PCR. *Appl. Environ. Microbiol.* 64:3724-3730.

Reysenbach AL, Ehringer H, Hershberger K (2000) Microbial diversity at 83°C in Calcite Springs, Yellowstone National Park: another environment where the Aquificales and "Korarchaeota" coexist. *Extremophiles* 4:61-67.

Saitou N, Nei M (1987) The neighbor-joining method: a method for constructing phylogenetic trees. Mol Biol Evol4:406-425. Rotthauwe JH, Witzel KP, Liesack W (1997) The ammonia monooxygenase structural gene amoA as a functional marker: Molecular fine-scale analysis of natural ammonia-oxidizing populations. *Appl. Environ. Microbiol.* 63:4704-4712.

Schloss PD, Handelsman J (2005) Introducing DOTUR, a computer program for defining operational taxonomic units and estimating species richness. *Appl. Environ. Microbiol.* 71:1501-1506.

Sekiguchi Y, Kamagata Y, Nakamura K, Ohashi A and Harada H (2000) Syntrophothermus lipocalidus, gen. nov., sp. nov., a novel thermophilic, syntrophic, fatty-acid-oxidizing anaerobe which utilizes isobutyrate. *Int. J. Syst.Bacteriol.* 50:771-779.

Spear JR, Barton HA, Robertson CE, Francis CA, Pace NR (2007) Microbial community biofabrics in a geothermal mine adit. *Appl. Environ.Microbiol.* 73:6172-6180.

Stackebrandt E, Goebel BM (1994) Taxonomic note: a place for DNA-DNA reassociation and 16S rRNA sequence analysis in the present species definition in bacteriology. *Int. J. Syst. Bacteriol.*44:846-849.

Suzuki MT, Giovannoni SJ (1996) Bias caused by template annealing in the amplification of mixtures of 16S rRNA genes by PCR. *Appl. Environ.Microbiol.* 62:625-630.

Suzuki MT, Rappe MS and Giovannoni SJ (1998) Kinetic bias in estimates of coastal picoplankton community structure obtained by measurements of small-subunit rRNA gene PCR amplicon length heterogeneity. *Appl. Environ.Microbiol.*64: 4522-4529.

Takai K, Horikoshi K (1999) Genetic diversity of archaea in deep-sea hydrothermal vent environments. *Genetics* 152:1285-1297.

ter Braak CJF and Šmilauer P (2002) CANOCO Reference Manual and CanoDraw for Windows User's Guide: Software for Canonical Community Ordination version 4.5. *Microcomputer Power Ithaca NY*, USA.

Thompson JD, Gibson TJ, Plewniak F, Higgins DG (1997) The CLUSTAL_X windows interface: Flexible strategies for multiple sequence alignment aided by quality analysis tools. *Nucleic Acids Res.*25:4876-4882.

Thompson JR, Marcelino LA, Polz MF (2000) Heteroduplexes in mixed-template amplifications: formation, consequence and elimination by "reconditioning PCR". *Nucleic Acids Res.*30:2083-2088.

Treusch AH, Leininger S, Kletzin A, Schuster SC, Klenk HP, Schleper C (2005) Novel genes for nitrite reductase and Amo-related proteins indicate a role of uncultivated mesophilic crenarchaeota in nitrogen cycling. *Environ.Microbiol.* 7:1985-1995.

Van Hamme JD, Singh A, Ward OP (2003) Recent advances in petroleum microbiology. *Microbiol. Mol. Biol. Rev.*67:503-49.

Von Wintzingerode F, Gobel UB, Stackebrandt E (1997) Determination of microbial diversity in environmental samples: pitfalls of PCR-based rRNA analysis. *FEMS Microbiol. Rev.*21:213-229.

Wang SF, Xiao X, Jiang LJ, Peng XT, Zhou HY, Meng J, Wang FP (2009) Diversity and abundance of ammonia-oxidizing archaea in hydrothermal vent chimneys of the Juan de Fuca Ridge. *Appl. Environ.Microbiol.* 75:4216-4220.

Weidler GW, Gerbl FW, Stan-Lotter H (2008) Crenarchaeota and their role in the nitrogen cycle in a subsurface radioactive thermal spring in the Austrian central Alps. *Appl. Environ. Microbiol.* 74:5934-5942.

Weidler GW, Dornmayr-Pfaffenhuemer M, Gerbl FW, Heinen W, Stan-Lotter H (2007) Communities of archaea and bacteria in a subsurface radioactive thermal spring in the Austrian Central Alps, and evidence of ammonia-oxidizing Crenarchaeota. *Appl. Environ. Microbiol.* 73:259-270.

Wuchter C, Abbas B, Coolen MJ, Herfort L, van Bleijswijk J, Timmers P, Strous M, Teira E, Herndl GJ, Middelburg JJ, Schouten S, Sinninghe Damste JS (2006) Archaeal nitrification in the ocean. *PNAS* 103:12317-12322.

Youssef N, Elshahed MS, McInerney MJ (2009) Microbial processes in oil fields: culprits, problems, and opportunities. *Advan. Appl. Microbiol.* 66:141-251.

Zhang CL, Ye Q, Huang ZY, Li WJ, Chen JQ, Song ZQ, Zhao WD, Bagwell C, Inskeep WP, Ross C, Gao L, Wiegel J, Romanek CS, Shock EL, Hedlund BP (2008) Global occurrence of archaeal amoA genes in terrestrial hot springs. *Appl. Environ. Microbiol.* 74:6417-6426.

In: Archaea: Structure, Habitats and Ecological...
Editor: Sakura Y. Katō, pp. 69-94

ISBN 978-1-61761-932-8

Chapter 3

Diversity, Function, and Processing of Archaeal Non-Coding RNAs

Kosuke Fujishima and Akio Kanai*
Institute for Advanced Biosciences, Keio University, Tsuruoka, Yamagata, Japan 9970017

Abstract

Recent major progress in sequencing technology has yielded more than 80 complete archaeal genome sequences and thousands of partial archaea-derived genomic and transcriptome sequences from various environmental samples. As a result of this rapid increase in the amount of archaeal sequence data, an unexpected variety of non-coding RNAs (ncRNAs) has been detected in Archaea, including fragments of transfer RNA (tRNA), a new class of small nucleolar RNA (snoRNA), a clustered, regularly interspaced short palindromic repeat (CRISPR) RNA used as an antiviral defense, and enzymatically cleaved introns found in many types of precursor tRNA, ribosomal RNA (rRNA) and messenger RNA (mRNA). Furthermore, several RNA elements such as riboswitches and SECIS elements have been predicted to exist in the 5' and 3' untranslated regions of archaeal mRNA. In this review, we discuss the distribution, function, and processing machinery of these archaeal ncRNA/RNA elements and the diversity of archaeal species. We have revised the phylogeny of the domain Archaea by reconstructing the most recent phylogenetic tree based on 8488 archaeal 16S rRNA sequences collected from all possible databases. The most recent findings on archeal ncRNA are explained along with the phylogeny of archaeal species, providing a deeper understanding of the functional and evolutionary roles of all types of archaeal RNA.

Keywords: Archaea, phylogenetic tree, non-coding RNA, RNA element.

* Institute for Advanced Biosciences, Keio University, Tsuruoka, Yamagata 997-0017, Japan. Tel: +81-235-29-0524, Fax: +81-235-29-0525, E-mail: akio@sfc.keio.ac.jp

1. Introduction

In 1977, Carl Woese and George Fox discovered that certain methanogens were phylogenetically distinct from both Bacteria and Eukarya based on their ribosomal RNA (rRNA) sequences, and proposed the classification of a third domain of life, termed Archaebacteria, currently known as Archaea (Woese and Fox 1977). In 1996, the first complete archaeal genome, from *Methanocaldococcus janaschii,* was sequenced by Craig Venter and colleagues (Bult et al. 1996), revealing a unique chimeric feature: an enzyme involved in central metabolism was similar to one found in bacteria, but the proteins involved in transcription and translation were more closely related to those in Eukarya. Furthermore, many archaeal species have been isolated from extreme environments such as deep-sea hyperthermal vents and hot springs, and since abundant evidence suggests that the last universal common ancestor (LUCA) was a thermophilic organism, it has been suggested that Archaea may still possess ancestral traits. Despite these interesting molecular and evolutionary features, most ongoing genome projects have focused on Bacteria (~1000 complete genome sequences) and Eukarya (114 genome sequences), while currently only about 80 archaeal genome sequences are available. This is also due to the fact that many Archaea are extremophiles, requiring complicated medium preparations and growth conditions, and generally possessing a long doubling time compared with that of Bacteria.

As more archaeal genome sequences have been revealed, it has become apparent that they contain genomic regions that transcribe non-coding RNAs (ncRNAs), which do not encode proteins. Until recently, it was believed that only a limited number of ncRNAs were found in archaeal genomes. These ncRNAs include rRNA, transfer RNA (tRNA), and RNase P RNA, which all play an important role in protein biosynthesis. In the past few years, the expansion of transcriptome research has made it possible to capture the expression of an entire genomic region with high resolution, using high-density genomic tiling arrays, next-generation sequencing, and many computational methods based on comparative genomics and RNA secondary structure analysis (Washietl et al. 2005; Hoheisel 2006; Metzker 2010; Weinberg et al. 2010). Such high-throughput technologies have revealed that vast numbers of intergenic and antisense coding regions are actually transcribed as ncRNAs in all domains of life. Interestingly, the chimeric feature of Archaea has also been observed among ncRNAs; for example, Archaea and Eukarya both possess enzymatically cleaved tRNA introns and snoRNAs, while both Archaea and Bacteria contain CRISPR RNAs that play an important role in viral defense, *cis*-antisense small RNAs that regulate sense-coding genes, and group II introns. In this review, we will discuss the distribution, biological function, and processing pathways of both classical and newly identified archaeal ncRNAs as revealed in recent studies (listed in Table 1), to provide an up-to-date summary of RNA diversity in Archaea.

2. Phylogeny of Archaea

Recent breakthroughs in sequencing technology have accelerated various genome projects, and complete genome sequences from over 80 archaeal species are currently available in the genome database at http://www.ebi.ac.uk/genomes/archaea.html. This is still a small number of sequences compared with those for Bacteria (~1000 genome sequences).

Metagenomic libraries from various environments have revealed a large number of genome fragments derived from uncultivated Archaea, reflecting the underlying archaeal diversity in the environmental samples (Wu et al. 2009). In addition, information about the sequence and function of ncRNAs has rapidly accumulated through transcriptomic and metatranscriptomic research (Shi, Tyson, and DeLong 2009). In this review, we discuss the types and distribution of archaeal ncRNAs. Currently, the archaeal species with complete genome sequences belong to five archaeal phyla: Euryarchaeota, Crenarchaeota (Woese, Kandler, and Wheelis 1990), Korarchaeota (Barns et al. 1996), Nanoarchaeota (Huber et al. 2002), and Thaumarchaeota (Brochier-Armanet et al. 2008) (Figure 1), although there is ongoing debate about the classification and nomenclature of archaeal subdivisions. Furthermore, most of these sequenced archaeal species were chosen either for their biophysical interest or cultivability, and may therefore represent only a small fraction of all Archaea.

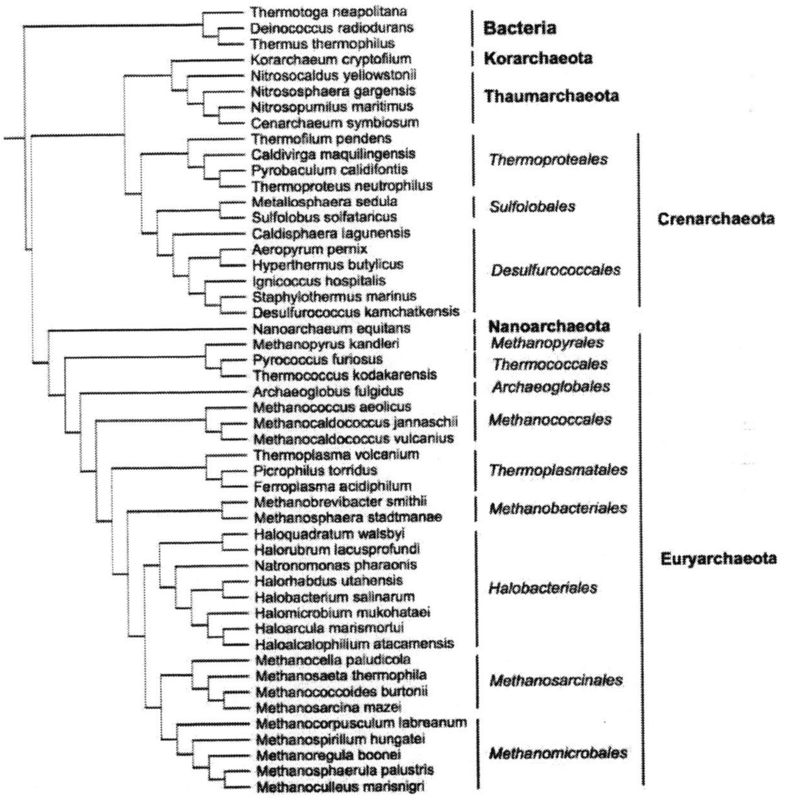

Figure 1. Phylogenetic tree of archaeal species with complete genome sequences. A 16S rRNA phylogenetic tree of 47 archaeal species with complete genome sequences was constructed using ClustalW 2.0 (Larkin et al. 2007) and interactive Tree Of Life (iToL) (Letunic and Bork 2007) using three deep-branching bacterial species as an outgroup. The nomenclature of five archaeal phyla (bold) and 12 archaeal orders (italic) is summarized.

To elucidate archeal diversity, we have collected a total of 8848 16S rRNA sequences from the Ribosomal Database Project, Release 10 (http://rdp.cme.msu.edu/) (Cole et al. 2009), which includes environmental sequences, and current complete and draft genome sequences.

Table 1. Unique ncRNAs found in Archaea

RNA type	Conservation across species	Method of prediction / identification	Experimental/ predicted	Reference
tRNA				
Multiple Intron-containing tRNA	Found in all Archaea Enriched in Thermoproteales	Predicted by SPLITS	0 / 88	Sugahara et al. (2008)
split tRNA	*Nanoarchaeum equitans*	Predicted by split-tRNA-search RT-PCR, DNA sequencing	6 / 6	Randau et al. (2005)
	Caldivirga maquilingensis	Predicted by SPLITS RT-PCR, DNA sequencing	4 / 4	Fujishima et al. (2009)
Tri-split tRNA	*Caldivirga maquilingensis*	RT-PCR, DNA sequencing	2 / 2	Fujishima et al. (2009)
Ribozyme				
RNase P RNA	Not found in *Nanoarchaeum equitans* nor in many Thermoproteales	-	-	Randau et al. (2008) RNase P database
Group II intron	*Methanosarcinales*	Computational prediction	0 / 15	Dai and Zimmerly (2003) Deppenmeier et al. (2002)
	Methanospirillum hungatei	Computational prediction	0 / 2	Dai and Zimmerly (2003)
	ANME-1	Computational prediction	0 / 5	Dai and Zimmerly (2003)
HEARO RNA	Environmental sample	Computational prediction	0 / 1	Weinberg et al. (2009)
snoRNA				

sR family (C/D box)	Found in many Crenarchaeota and Euryarchaeota	Computational prediction RT-PCR cDNA sequencing Northern blot analysis Deep sequencing	-	Omer et al. (2000) Gaspin et al. (2000) Tang et al. (2002) Tang et al. (2005) Zago et al. (2005) Grosjean et al. (2008) Wurtzel et al. (2010)
Pyrococcus C/D box	Thermococcales	Computational prediction	0 / 69	Omer et al. (2000) Gaspin et al. (2000) Rfam 10.0 database
H/ACA box	Thermococcales	Computational prediction Northern blot analysis RACE-PCR	10 / 31	Klein et al. (2002) Rozhdestvensky et al. (2003) Muller et al. (2008) Rfam 10.0 database
RNA type	Conservation across species	Method of prediction / identification	Experimental/ predicted	Reference
	Archaeoglobus fulgidus	Northern blot analysis	4 / 4	Tang et al. (2002)
	Sulfolobus solfataricus	Northern blot analysis	1 / 1	Zago et al. (2005)
Other small RNAs				
CRISPR RNA (crRNA)	Found in 90% of Archaea	-	-	CRISPRdb (2010-04-14 updated)
HgcC family	Found in Korarchaeota and many Crenarchaeota and Euryarchaeota	Northern blot analysis RACE-PCR	6 / 62	Klein et al. (2002) Rfam 10.0 database
SscA	Thermococcales	Northern blot analysis RACE-PCR	1 / 9	Klein et al. (2002) Rfam 10.0 database

Table 1 (Continued)

Pyrobac-1 motif	*Pyrobaculum* species	Comparative genomics	0 / 8	Weinberg et al. (2010)
putative small RNA (psRNA)	Thermoproteales, Marine group II euryarchaeota	cDNA pyrosequencing RT-qPCR	-	Shi et al. (2009)
Antisense RNA	Archaeoglobus fulgidus	cDNA cloning Northern blot analysis	33 / 33	Tang et al. (2002)
	Sulfolobus solfataricus	cDNA cloning Northern blot analysis	19 / 19	Tang et al. (2005)
		RT-PCR, cDNA cloning Northern blot analysis	7 / 7	Zago et al. (2005)
		Deep sequencing	185 / 185	Wurtzel et al. (2010)
	Haloferax volcanii	cDNA cloning Northern blot analysis	18 / 18	Straub et al. (2009)
	Methanosarcina mazei	Deep sequencing	42 / 42	Jager et al. (2009)
small RNAs expressed from the intergenic region	*Archaeoglobus fulgidus*	cDNA cloning Northern blot analysis	9 / 9	Tang et al. (2002)
	Sulfolobus solfataricus	cDNA cloning Northern blot analysis	11 / 11	Tang et al. (2005)
		RT-PCR, cDNA cloning Northern blot analysis	3 / 3	Zago et al. (2005)
		Deep sequencing	90 / 90	Wurtzel et al. (2010)
	Sulfolobus tokodaii	Computational prediction	0 / 2	Tang et al. (2005)
	Haloferax volcanii	cDNA cloning Northern blot analysis	20 / 20	Straub et al. (2009)
RNA type	Conservation across species	Method of prediction / identification	Experimental/ predicted	Reference
	Methanosarcina mazei	Deep sequencing	156 / 156	Jager et al. (2009)

RNA elements				
TPP riboswitch	Thermoplasmatales	Comparative genomics	0 / 11	Rodionov et al. (2002)
	Korarchaeum cryptofilum	-	0 / 1	Rfam 10.0 database
M-box riboswitch	*Methanosarcina mazei*	Computational prediction	0 / 1	Rfam 10.0 database
	Methanospirillum hungatei	Computational prediction	0 / 1	Rfam 10.0 database
SECIS element	*Methanocaldococcus jannaschii*	Predicted by archaeal SECISearch	0 / 8	Kryukov and Gladyshev (2004)
	Methanococcus maripaludis S2/C5/C6/C7	Predicted by archaeal SECISearch	0 / 4	Bock and Rother (2004) Rfam 10.0 database
	Methanopyrus kandleri	Predicted by archaeal SECISearch	0 / 7	Kryukov and Gladyshev (2004)
Potassium channel RNA editing signal	*Methanosphaerula palustris*	Computational prediction	0 / 1	Rfam 10.0 database
	Methanospirillum hungatei	Computational prediction	0 / 1	Rfam 10.0 database
Pyrobac-HINT motif	*Pyrobaculum* species	Comparative genomics	0 / 4	Weinberg et al. (2010)

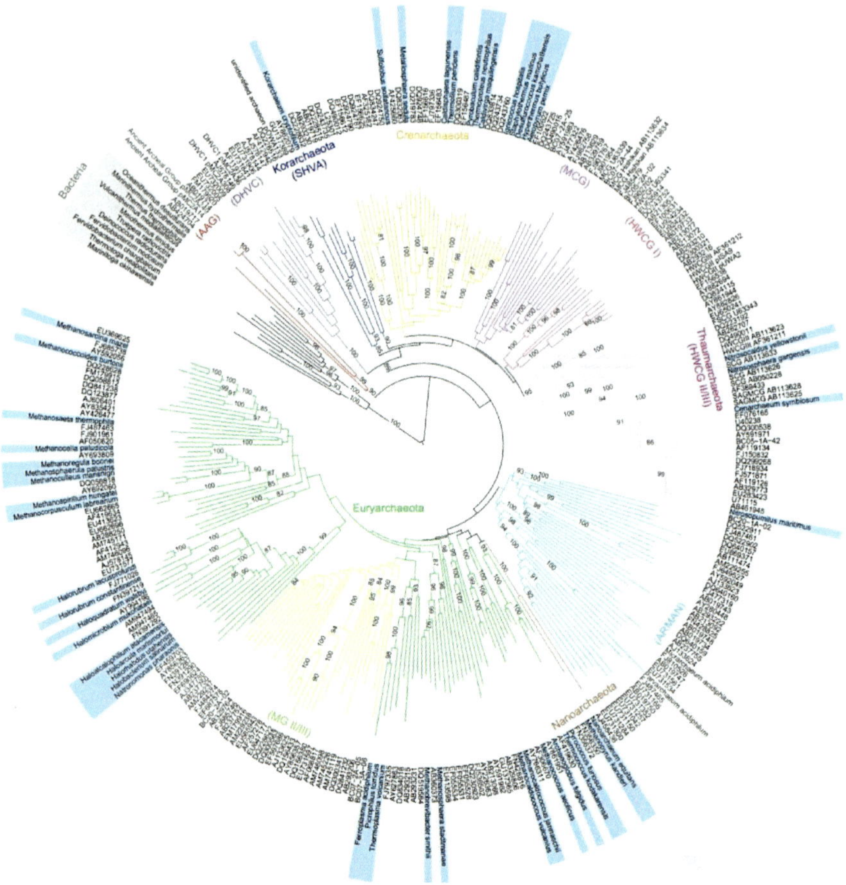

Figure 2. Phylogenetic tree representing archaeal diversity. A neighbor-joining tree was constructed and visualized using ClustalW 2.0 (Larkin et al. 2007) and interactive Tree Of Life (iToL) (Letunic and Bork 2007) based on 296 sequences selected from the 8488 archaeal 16S rRNA sequences registered in the Ribosomal Database Project, Release 10 (Cole et al. 2009). Reliable bootstrap values (> 80) are shown. 11 deep-branching bacterial species were used as an outgroup. Branch colors represent different phyla and archaeal groups that were characterized in the following previous studies: the Ancient Archaeal Group (AAG), the Deep-sea Hydrothermal Vent Crenarchaeotic Group (Takai and Horikoshi 1999), the Shallow marine Hydrothermal Vent Archaea (SHVA), the Miscellaneous Crenarchaeal Group (MCG), the Hot Water Crenarchaeal Group (HWCG) (Inagaki et al. 2003; Nunoura et al. 2005), Archaeal Richmond Mine Acidophilic Nanoorganisms (ARMAN) (Baker et al. 2006) and the Marine Group II and III Euryarchaeota (MG II/III) (Fuhrman and Davis 1997; Massana, DeLong, and Pedros-Alio 2000). Species with complete genome sequences are highlighted in blue.

The 8848 sequences were reduced to 296 sequences by filtering similar sequences with greater than 80% identity to maintain archaeal diversity as much as possible, and these 296 sequences were then used to construct an archeal phylogenetic tree. The distribution of archaeal species from complete and draft genome sequences shows a clear bias (Figure 2, highlighted in blue), indicating that much of the archaeal genome remains to be analyzed. The revised archaeal tree also includes several previously uncharacterized groups that have only been identified from environmental sequences. For example, samples from the Ancient Archaeal group (AAG) and the Deep-sea Hydrothermal Vent Crenarchaeotic Group (DHVC) were obtained from deep-sea hydrothermal fields (Takai and Horikoshi 1999), while those from the Miscellaneous Crenarchaeal Group (MCG) and Hot Water Crenarchaeal Groups I, II and III (HWCG I/II/III) came from subsurface gold mines (Inagaki et al. 2003; Nunoura et al. 2005), an Archaeal Richmond Mine Acidophilic Nanoorganisms (ARMAN) were obtained from from acidic mine drainage (Baker et al. 2006) and the Marine Group II and II (MG II/III) Euryarchaeota were detected in marine ecosystems (Fuhrman and Davis 1997; Massana, DeLong, and Pedros-Alio 2000). These uncharacterized archaeal groups cover one third of the archaeal diversity and thus, their complete genome sequences are anticipated to more fully reveal the biophysical properties and ecological impacts of these organisms.

3. Classical ncRNAs and Their Processing

3.1. Transfer RNA

Transfer RNA (tRNA) is a classical ncRNA conserved in all three domains of life (Bacteria, Archaea, and Eukarya). In Bacteria, tRNA is encoded by a single contiguous gene (general tRNA) except for the case of self-splicing group I introns located in the anticodon loop of the tRNA genes of α-, β-, and γ-Proteobacteria (Reinhold-Hurek and Shub 1992). In Eukarya, tRNA introns are also found within the anticodon loop at position 37/38 (the canonical position), but these introns are spliced by specific enzymatic reactions (Li, Trotta, and Abelson 1998). Archaeal tRNA genes also contain interspersed intronic sequences but their abundance is much greater than in Eukaryotes.

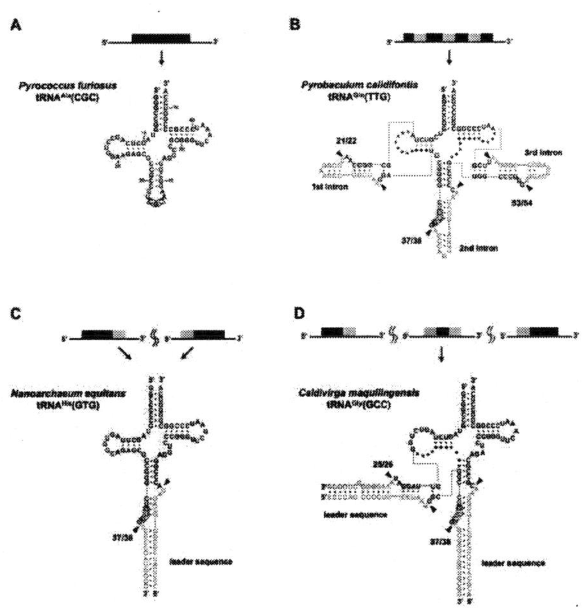

Figure 3. Schematic representation of four different types of archaeal tRNA. (A) contiguous tRNA (normal tRNA), (B) intron-containing tRNA, (C) split tRNA and (D) tri-split tRNA are shown. tRNA gene structures (exons shown in black and intron/leader sequences shown in gray) and their RNA secondary structures are represented. Nucleotide positions are marked on the contiguous tRNA. Anticodons are boxed. The locations of BHB splicing motifs and their cleavage positions are indicated by arrowheads.

Currently, 2662 tRNA genes from 58 archaeal species are registered in the archaeal tRNA database SPLITSdb (Sugahara et al. 2008), of which contiguous tRNAs (Figure 3A) comprise only 75%. While most euryarchaeal tRNA introns are located in canonical positions, crenarchaeal tRNA introns are frequently found in non-canonical positions (Marck and Grosjean 2002). Recent refinements of tRNA prediction software have revealed many multiple intron-containing tRNA genes in the crenarchaeal order Thermoproteales, including unique tRNA genes that possess three intron sequences (Sugahara et al. 2007) (Figure 3B).

Compared with group I introns, archaeal tRNA introns are short in length (11 to 50 nt) and contain a Bulge-Helix-Bulge (BHB) motif at the exon–intron junction, which is recognized and cleaved by a splicing endonuclease.

The origin of the tRNA intron is one of the unsolved mysteries in this field, although recent comprehensive analysis of hundreds of Thermoproteales tRNA introns revealed that certain groups of tRNA introns are translocated among different tRNA genes, which are likely derived from extracellular DNA sources (Fujishima et al. 2010). In 2005, a new type of disrupted tRNA, the so-called split tRNA, was found in the hyperthermophilic archaeon *Nanoarchaeum equitans*, in which the 5' and 3' tRNA ends are encoded by separate genes (Randau et al. 2005). These individually expressed 5' and 3' tRNA fragments possess long leader sequences with high complementarity, thus enabling the two halves to join and form a BHB splicing motif (Figure 3C). More recently, a novel type of split tRNA consisting of three fragments and termed "tri-split tRNA" was found in the hyperthermoacidophilic crenarchaeon *Caldivirga maquilingensis* (Fujishima et al. 2009). This tRNA was found to be encoded by three separate genes with distinct genomic loci. The second fragment encodes a tRNA exon only 12 nt in length, including a $tRNA^{Gly}$(GCC) antocodon with two leader sequences on either end that are able to base-pair with the first and third fragments (Figure 3D). Intriguingly, this second fragment can be replaced by a similar fragment encoding the $tRNA^{Gly}$(TCC) anticodon, and thus in *C. maquilingensis*, alternative trans-splicing produces different types of mature tRNAs from tRNA fragments. This substitution is also observed in *N. equitans*, representing a common feature of split tRNA processing. A sequence connecting the two types of tRNAs (split and intron-containing) has been found, in which the leader sequence of *N. equitans* $tRNA^{Lys}$ possesses 50% sequence identity with the *Pyrobaculum aerophilum* $tRNA^{Arg}$ intron sequence, and both are disrupted at the same nucleotide position (Fujishima et al. 2008). Significantly greater sequence similarity (90%) was observed between the leader sequence of *C. maquilingensis* $tRNA^{Ala}$ and the *Pyrobaculum islandicum* $tRNA^{Ala}$ intron, suggesting that these two $tRNA^{Ala}$ genes evolved from a common ancestor (Fujishima et al. 2009).

3.2. Ribosomal RNA and BHB motif

Ribosomal RNA is a classical ncRNA that is a core component of the protein biosynthesis organelle known as the ribosome. Microbial ribosomal RNAs are classified into three types: 16S rRNA, 23S rRNA, and 5S rRNA, and genes encoding these rRNAs are typically arranged into operons (rrn operons). Approximately half of all Archaea species possess a single rrn operon sequence, while methanophiles and halophiles typically have multiple copies of this sequence (Lee, Bussema, and Schmidt 2009), while the rRNA operon is disrupted in the genome of *N. equitans* (Waters et al. 2003). The archaeal rRNA operon is transcribed as a long multicistronic pre-rRNA transcript, and often includes one or two tRNA genes (Garrett et al. 1991). Different types of ribonucleases contribute to the processing of pre-rRNA, including RNase P and tRNase Z that excise pre-tRNA, and pre-5S rRNA sequences (Frank and Pace 1998; Holzle et al. 2008).

More important is the involvement of a splicing endonuclease that cleaves the BHB motif found in the primary transcript of 16S and 23S rRNA (Chant and Dennis 1986; Russell, Ebhardt, and Dennis 1999; Ciammaruconi and Londei 2001; Tang et al. 2002b). The unique consensus structure of the BHB motif found in both the tRNA and rRNA splicing pathways (Figure 4A and 4B) suggests a general role for splicing endonucleases in archaeal RNA processing. Indeed, the BHB motif has also been found at the exon–intron boundary in *cbf5* mRNA, encoding RNA pseudouridine synthase, which is found in many crenarchaeal species (Yoshinari et al. 2006; Yokobori et al. 2009) (Figure 4C). Currently, the *cbf5* genes are the only known example of protein-coding genes that have an enzymatically cleaved intron sequence. Likewise, long introns are interspersed throughout pre-16S/23S rRNA sequences in Crenarchaeota and HWCG III, which sometimes contain an open reading frame (ORF) encoding a LAGLIDADG homing endonuclease (Itoh, Nomura, and Sako 2003; Nomura et al. 2005; Nunoura et al. 2005).

3.3. RNA Processing Enzymes

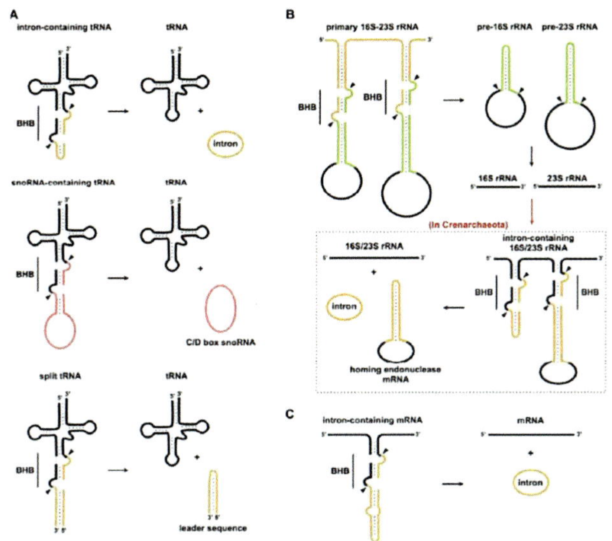

Figure 4. Non-coding RNAs requiring a BHB motif for processing. (A) Processing of intron-containing tRNA and split tRNA in Archaea requires splicing endonuclease cleavage at the exon–intron junction forming a BHB motif followed by a ligation reaction. This yields mature tRNA and a circularized intron and joined leader sequence. Archaeal $tRNA^{Trp}$ encodes C/D box snoRNA within its canonical intron. (B) Pre-rRNA processing is a two step reaction in which the BHB splicing of primary rRNA transcript produces circularized pre-16S and pre-23S rRNA, and is then further processed into mature rRNA sequence. Crenarchaeal 16S and 23S rRNA often includes one or two introns cleaved by the splicing endonuclease. The large intronic loop usually carries an ORF containing homing endonuclease (Dalgaard, Garrett, and Belfort 1993). (C) Intron-containing mRNA yields mature mRNA and a circularized intron. Currently, BHB-type mRNA introns are only known to be found in *cbf5* mRNA encoding pseudouridine synthase in the orders Desulfurococcales and Sulfolobales.

Archaeal species possess three types of splicing endonucleases: heterotetrameric, homodimeric, and homotetrameric, each with distinct structure and function (Tocchini-Valentini, Fruscoloni, and Tocchini-Valentini 2005). While the homotetrameric form can only cleave strict BHB motifs (a 4 bp helix flanking a 3 nt bulge on both ends), homodimeric endonucleases are capable of cleaving both strict and relaxed BHB motifs located at the canonical position (37/38) in the tRNA sequence (Tocchini-Valentini, Fruscoloni, and Tocchini-Valentini 2007). These two forms have been found in almost all Euryarchaeota and Korarchaeota (Sugahara et al. 2009).

In contrast, the heterotetrameric form can recognize relaxed BHB motifs at multiple nucleotide positions and has been found in all Crenarchaeota, Nanoarchaeota, Thaumarchaeota, and *Methanopyrus kandleri* (a member of Euryarchaeota). Indeed, these archaeal species tend to have many tRNA introns, especially in the order Thermoproteales, in which a high percentage of tRNA genes contain multiple introns (Sugahara et al. 2008).

It should be noted that archaeal RNA splicing yields 2'-3' cyclic phosphate and 5'-OH termini at both ends (Peebles, Gegenheimer, and Abelson 1983) and is followed by a ligation reaction that seals the broken RNA ends to produce two different products: a ligated exon and unique circularized intron (Kjems and Garrett 1988; Burggraf et al. 1993; Salgia et al. 2003).

In contrast to the diversity of DNA ligases, only a few RNA ligases have been studied so far. In yeast and wheat germ, ligation of spliced tRNA is catalyzed by two enzymes, RNA ligase Trl1 (Gegenheimer et al. 1983; Westaway, Phizicky, and Abelson 1988) and 2'-phosphotransferase Tpt1 (Spinelli et al. 1998). Trl1 is a multifunctional enzyme that contains an N-terminal ligase domain, kinase module, and C-terminal cyclic phosphodiesterase (CPDase) domain. In contrast, the molecular basis of archaeal RNA ligation remains to be characterized, as the responsible archaeal ligase has not been identified.

However, recently it has been shown that certain archaeal proteins have analogous functions to the Trl1 domains, such as the 2'-5' RNA ligase with CDPase activity (PF0027) found in *Pyrococcus furiosus* that opens 2'-3' cyclic phosphate, a hallmark of spliced RNA ends (Kanai et al. 2009); an archaeal polynucleotide kinase (Jain and Shuman 2009), which is a homolog of human Clp1 (hClp1) that functions in tRNA splicing (Weitzer and Martinez 2007); and a homodimeric T4-like RNA ligase characterized in both *Pyrococcus abyssi* (Pab1020) and *Methanobacterium thermoautotrophicum* (MthRn1), which catalyzes ligation of single-stranded RNA and DNA with 3'-OH and 5'-phosphate (Brooks et al. 2008; Torchia, Takagi, and Ho 2008). Since evidence suggests that ligation occurs coordinately with splicing of the BHB motif (Brooks et al. 2008; Torchia, Takagi, and Ho 2008), a dimeric conformation may enable the symmetrical ligation to produce the ligated exon and circularized intron, although several questions remain to be answered, including the mechanism of 5' and 3'-terminal recognition.

4. Small RNAs

4.1. Small Nucleolar RNA

Small RNA is a collective term for short ncRNAs (generally < 200 nt), either expressed individually or as part of long transcript that is later processed into short transcripts. One of the major findings in the field of archaeal ncRNA research was the discovery of small nucleolar RNA (snoRNA) in Archaea (Omer et al. 2000). snoRNAs were first discovered in the eukaryotic nucleolus, where they guide chemical modifications of other RNA molecules such as rRNA and tRNA in a type of ribonucleoprotein (RNP) (Balakin, Smith, and Fournier 1996), and hence it was thought that this small guide ncRNA was Eukarya specific. However, many ribose methylated sites have been found in the rRNA of *Sulfolobus solfataricus* (Noon, Bruenger, and McCloskey 1998), which later led to the discovery of C/D box snoRNAs in *Sulfolobus acidocaldarius* by coimmunoprecipitation with archaeal homologs of eukaryotic snoRNA-associated proteins (Omer et al. 2000).

Archaeal C/D box snoRNA is 50–60 nt in length, contains a C (UGAUGA) and D (CUGA) motif, and associates with three proteins (L7Ae, Fib, and Nop5) to guide ribose methylation (Rashid et al. 2003). Currently, a group of C/D box snoRNAs termed the sR family is found in diverse Crenarchaeota and Euryarchaeota, and appears frequently in hyperthermophiles such as Pyrococcus and Pyrobaculum species, which grow at temperatures approaching 100°C, and possess a highly enriched set of snoRNAs compared to other archaeal species (Dennis, Omer, and Lowe 2001). This is due to the requirement for thermodynamic structural stability of rRNA in such extreme environments.

Indeed, a unique type of C/D box snoRNAs, which includes snoPyro_CD and snoR9, with a structure significantly different from their eukaryotic homologs, has been identified from the order Thermococcales (Gaspin et al. 2000; Klein, Misulovin, and Eddy 2002). These snoRNAs are believed to play an important role in rRNA folding. Interestingly, unusual circular C/D box RNAs were discovered in *Pyrococcus furiosus,* which can form an RNP, suggesting that these RNAs play a role in ribose modification (Starostina et al. 2004). Unlike the snoRNAs found in the intron sequence of tRNA, which are processed by endonuclease splicing, the precise mechanism of the circularization of *P. furiosus* snoRNA remains unknown.

Another type of snoRNA, known as H/ACA box snoRNA, has also been found in diverse archaeal species, and guides pseudouridylation of specific rRNA nucleotides previously identified as novel small ncRNAs expressed from intergenic regions (Klein, Misulovin, and Eddy 2002; Tang et al. 2002a). H/ACA snoRNA is approximately 65–75 nucleotides in length, and constructs snoRNP with the four proteins Cbf5, Gar1, Nop10 and L7Ae (Baker et al. 2005). The variety and abundance of archaeal H/ACA box snoRNAs are minor compared to archaeal C/D box snoRNA, and indeed, only a few pseudouridine modifications have been observed in *Sulfolobus* species (Massenet et al. 1999). Currently, only about a dozen H/ACA snoRNAs have been experimentally verified in Archaea (Table 1).

4.2. CRISPR RNA

RNA interference (RNAi) is a major gene regulatory mechanism found in a diverse array of eukaryotes. In this process, tiny RNA fragments (20–25 nt), known as microRNA (miRNA) and small interfering RNA (siRNA) regulate expression of genes at the transcriptional or translational level (Ambros et al. 2003). A similar mechanism has also been shown to occur in Bacteria, in which a small RNA (< 200 nt) interacts with its target mRNA with the help of RNA chaperones to regulate target transcripts (Gottesman 2005; Liu and Camilli 2010).

Recently, a unique viral immune system called the Clustered Regularly Interspaced Short Palindromic Repeats (CRISPR) system has been characterized in a wide range of microorganisms, and functions in a manner analogous to that of eukaryotic RNAi (Sorek, Kunin, and Hugenholtz 2008). CRISPR comprises sequences that are found in 50% of bacterial genomes and 90% of archaeal genomes.

Only four archaeal species (*Archaeoglobus profundus, Methanococcus maripaludis, Candidatus Methanoregula boonei*, and *Halobacterium sp. NRC-1*) have been shown to lack CRISPR loci, each belonging to a distinct archaeal order (CRISPRdb, http://crispr.u-psud.fr/crispr/) (Grissa, Vergnaud, and Pourcel 2007). Three recently identified ARMAN species, *Candidus Micrarchaeum acidiphilum* (ARMAN-2), *Candidatus Parvarchaeum acidiphilum* (ARMAN-4) and *Candidatus Parvarchaeum acidophilus* (ARMAN-5), also lack CRISPR, and are hence often infected with one or two viruses (Baker et al. 2010).

Figure 5 represents the processing of CRIPSR loci, in which repeat/spacer arrays are expressed as long precursor transcripts that undergo endonucleolytic cleavage by CRISPR-associated (Cas) proteins to produce small RNAs (35–46 nt) termed CRISPR RNAs (crRNAs) or prokaryotic silencing RNAs. During the effector stage, crRNA forms a complex with additional Cas proteins to bind and cleave phage or viral mRNA or DNA. The distinction between RNA and DNA targeting may rely on the composition of the crRNA-protein complexes, since in *Pyrococcus furiosus,* six Repeat-associated mysterious proteins (RAMP) containing mature crRNA can recognize and cleave RNA targets (Hale et al. 2009).

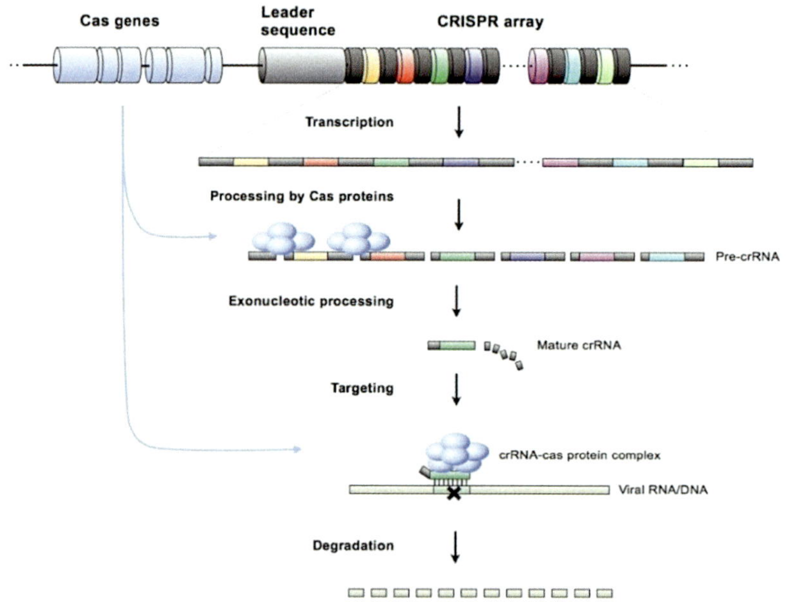

Figure 5. Model of the CRISPR system. The genome structure of a clustered, regularly interspaced short palindromic repeat (CRISPR) locus is represented. CRISPR sequences consist of phage or virus-derived spacer sequences (colored) interspersed with a repeat sequence (black) in an arrayed form. Leader sequence (gray) lies just upstream of the first repeat and is involved in the acquisition of new spacer sequence (Jernigan et al. 2007). CRISPR-associated (Cas) genes include a series of RNases that recognize precursor CRISPR RNA (crRNA) expressed from the repeat or spacer region to produce mature crRNA. Mature crRNA is then incorporated in to a complex with additional Cas proteins to base pair with target phage or viral RNA or DNA, leading to degradation.

4.3. Other Archaeal Small RNAs

Several unique small RNAs have been identified in hyperthermophiles, mainly through computational prediction and experimental verification, but many of their functions remain unknown.HgcC family RNA is a small ncRNA first predicted in *Methanocaldococcus jannaschii* by searching for genomic regions with high GC content (Klein, Misulovin, and Eddy 2002). Based on Northern blot analysis, this small RNA is detected at around 110 to 120 nt in length and, according to the Rfam database, homologous sequences are also found in three archaeal phyla: Korarchaeota, Crenarchaeota and Euryarchaeota. In *Pyrococcus furiosus,* hgcC isadjacent to transposase, and comparison with syntenic regions in other *Pyrococcus* species suggests that this locus (hgcC + transposase) was inserted in the genome later, suggesting a further link between hgcC family RNAs and transposition events. Similarly, a small RNA termed sscA was also identified by screening for high GC content in the *P. furiosus* genome, and is located downstream of C/D box snoRNA (Klein, Misulovin, and Eddy 2002). SscA is unique to the archaeal order Thermococcales, and Northern blot analysis has shown two distinct bands (48 and 98 nt), suggesting further processing of this small RNA. Novel small RNAs have also been predicted through comparative genomics. The *Pyrobac*-1 motif is found specifically in the genus *Pyrobaculum*, and has been proposed to function as a trans-acting RNA (Weinberg et al. 2010). It possesses neither sequence similarity nor common sequence features with any other known small RNA, representing a new class of ncRNA. Furthermore, recent metatranscriptomic analysis has also revealed a total of 66 known and putative small RNA (psRNA) groups that comprised at least 100 overlapping cDNA reads from different depth of ocean water samples (25m, 75m, 125m and 500m) (Shi, Tyson, and DeLong 2009). Eight psRNA groups are found as Archaea-specific and most of the reads were obtained from 500m depth reflecting a depth-dependent community. Archaeal psRNAs are 69–338 nt in length and located at intergenic region and UTR region of ORFs belonging to order Thermoproteales and Marine Group II Euryarchaeota, although their precise function remains unknown.

Small RNA that is expressed from the antisense strand of an ORF is called *cis*-antisense RNA, and numerous *cis*-antisense RNAs have been discovered in Bacteria, functioning to either suppress or up-regulate the expression of sense-coding genes through post-transcriptional modification (Thisted, Nielsen, and Gerdes 1994; Opdyke, Kang, and Storz 2004). Some small RNAs are mapped to the intergenic region and may encode short peptides or function as *trans*-acting RNA. Traditional transcriptome analyses (cDNA sequencing combined with Northern blotting) have revealed that both *cis*-antisense RNA and intergenic small RNAs exist in Archaea (Tang et al. 2002a; Tang et al. 2005; Zago, Dennis, and Omer 2005; Straub et al. 2009), and more recently, deep-sequencing analysis has revealed that hundreds of these small RNAs may be found in a single species (Table 1). Millions of mapped sequence reads have demonstrated that small RNAs in *Sulfolobus solfataricus* contain a TATA-like promoter sequence at positions –29 to –23, similar to that found in protein-coding genes (Wurtzel et al. 2010). In *Methanosarcina mazei*, the expression of most small RNAs is sensitive to nitrogen; indeed, a new motif (box I), which seems to be associated with nitrogen-responsive regulation, was found near the TATA box and IIB recognition element (BRE) in the core promoter (Jager et al. 2009).

5. Ribozymes

5.1. RNase P RNA

RNase P RNA is an essential component of RNase P ribonucleoprotein, a universally conserved holoenzyme comprised of RNase P RNA and several related proteins, and is involved in the processing of the 5' end of pre-tRNA (Lai et al. 2010). In all three domains of life, RNase P RNA functions as a ribozyme which can act alone as a catalyst (Guerrier-Takada et al. 1983). In Archaea, RNase P RNA interacts with four RNase P proteins: RPP21, RPP29, RPP30 and POP5, which form two binary complexes, RPP21-RPP29 and RPP30-POP5, that have contrasting affinities for pre-tRNA molecules (Tsai et al. 2006; Pulukkunat and Gopalan 2008). The most recent study suggests that archaeal RNase P RNA may interact with L7Ae, which is known to be a key component of snoRNP (Lai et al. 2010). Despite the essentiality of RNase P RNA, this molecule has not been detected in *N. equitans* or Thermoproteales to date, except in *Thermofilum pendens*. Interestingly, in *N. equitans*, a promoter sequence located upstream of all tRNA genes produces 5'-leaderless tRNA to compensate for the loss of RNase P function (Randau, Schroder, and Soll 2008). This feature is not observed in Thermoproteales species, suggesting that these species have obtained either an alternative mechanism to deal with the loss of RNase P RNA, or that they utilize an irregular form of RNase P RNA which cannot be detected by current computational analysis.

5.2. Group II Introns

Self-splicing group II introns are a major class of ribozymes 300–3000 nt in length, which have been found in diverse Bacteria, in Eukaryotic organelles (mitochondria and plastids) and most recently, in Archaea. Group II introns catalyze a self-splicing reaction forming a 2'-5' lariat intermediate resembling a structure formed by nuclear spliceosomal introns (Peebles et al. 1986).

In Bacteria, group II introns usually encode a reverse transcriptase, which is related to a non-LTR-retrotransposon (Curcio and Belfort 1996). Thus, group II introns are thought to be ancestors of spliceosomal introns and retrotransposons in Eukarya.

However, in Archaea, a small number of group II introns have been computationally predicted from the genome sequences of Methanosarcinales species, *Methanospirillum hungatei,* and the uncultured archaeon ANME-1 (Deppenmeier et al. 2002; Dai and Zimmerly 2003). Sequence evidence suggests that these group II introns were likely acquired from Bacterial domains via horizontal transfer (Rest and Mindell 2003). More recently, a novel type of RNA involved in mobile genetic elements, termed HNH Endonuclease-Associated RNA and ORF (HEARO) RNA, was discovered by comparative sequence analysis of bacterial metagenome data (Weinberg et al. 2009). These ncRNAs are approximately 300 nt in length and often contain an embedded ORF encoding an HNH Endonuclease, a trait common to group I and group II introns. Only one HEARO RNA has been predicted in Archaea to date.

6. RNA Elements

The term 'RNA element' generally indicates a *cis*-acting RNA motif placed either upstream, downstream, or within an mRNA.

One major RNA element is known as a riboswitch, an RNA structural motif that monitors cellular conditions by interacting with small ligands such as metal ions and metabolites and regulates gene expression. Currently, more than a dozen classes of riboswitches have been discovered or predicted in the three domains of life, although most are found in Bacteria (Barrick and Breaker 2007).

In Archaea, two types of riboswitches, the thiamin pyrophosphate (TPP) riboswitch and M-box riboswitch, have been predicted based on computational analysis, although no experimental studies of these riboswitches have been reported thus far. The TPP riboswitch is one of the most abundant classes in all three domains, and is usually located near operons containing genes for TPP metabolism.

Archaeal TPP riboswitches are found in the 5' UTR of thiamine-related transporter genes in Thermoplasmatales (Rodionov et al. 2002) and in a CytX gene encoding a putative hydroxymethylpyrimidine transporter in *CandidatusKorarchaeum cryptofilum* (Rfam database 10.0).

The secondary structure of the archaeal TPP riboswitch is well conserved with those of Bacteria and Eukarya, and consists of five major stems (P1–P5) and one variable stem-loop (Figure 6A). The M-box riboswitch is a metal-sensing RNA element that binds to magnesium ions (Mg^{2+}) and regulates the expression of downstream genes encoding the Mg^{2+} transporter. The M-box riboswitch was originally detected as a conserved RNA element: the ykoK leader, found upstream of the ykoK gene, encodes a divalent metal transporter in *Bacillus subtilis* (Barrick et al. 2004). According to the Rfam database, an archaeal M-box riboswitch is predicted upstream of the mgtE gene known to function as an Mg^{2+} transporter in *Methanosarcina mazei,* and downstream of the Mhun_2363 gene encoding 3-isopropylmalate dehydrogenase in *Methanospirillum hungatei.*

RNA elements are also used as landmarks for recruiting proteins to mRNA. A Selenocysteine insertion sequence (SECIS) element has been found adjacent to and within mRNAs encoding selenoproteins and is known to associate with SECIS binding protein 2 (SBP2) to recruit the selenocysteyl-$tRNA^{[Ser]Sec}$ complex to ribosomes (Tujebajeva et al. 2000). The SECIS element has a simple stem-loop structure 60 nucleotides in length, and this structure is conserved across the three domains of life (Figure 6B).

In Bacteria, SECIS elements appear just after the UGA codon, while in Eukarya, they are located in the 3' UTR of mRNAs. In Archaea, SECIS elements are located in the 5' or 3' UTR of mRNAs found in various methanogens belonging to the archaeal orders Methanopyrales and Methanococcales (Kryukov and Gladyshev 2004; Bock and Rother 2005) (Table 1). Another type of RNA element, termed a potassium (K^+) channel RNA editing signal, has been predicted in two Methanomicrobales species (Rfam database 10.0), and was originally identified as a small mRNA hairpin structure (approximately 114 nt) in an intronless human mRNA encoding a K^+ channel (Bhalla et al. 2004). This motif guides A to I editing mediated by adenosine deaminase acting on RNA (ADAR), and causes an isoleucine/valine recoding event in humans and flies. Previously, A to I editing has been observed in archaeal tRNA (Grosjean et al. 1995), but not in archeal mRNA. Since the K^+ channel RNA editing signalis found in mRNA encoding an ion transport protein in *Methanospirillum hungatei,* this may represent the first example of mRNA A to I editing in Archaea.

Also comparative genomics approach has predicted novel RNA element termed *Pyrobac*-HINT motif conserved in the genus *Pyrobaculum* (Weinberg et al. 2010). This motif is found just upstream of a gene encodinghistidine triad nucleotide binding protein (HINT) protein.

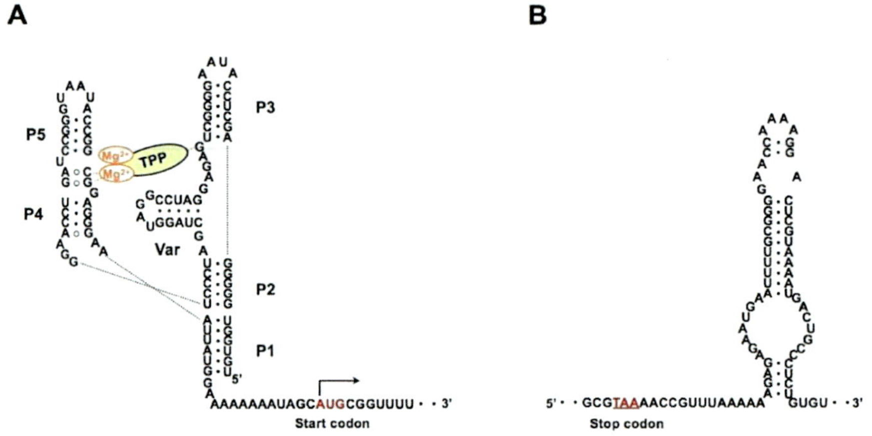

Figure 6. Examples of RNA elements predicted in Archaea. A schematic overview of (A) A Thiamine Pyrophosphate (TPP)-sensing riboswitch predicted in the 5'UTR of the Ta0442 gene in *Thermoplasma acidophilum* and (B) a Selenocysteine insertion sequence (SECIS) element downstream of the *vhuD* gene in Methanococcus maripaludis. The crystal structure of the *E. coli* TPP riboswitch (Serganov et al. 2006) was used as a reference to create a secondary structural model of the *T. acidophilum* TPP riboswitch and its ligands. Secondary structure of the *M. maripaludis* SECIS element was illustrated based on the consensus structure registered in the Rfam database (Gardner et al. 2009). See the text for details.

Conclusion

In this review, we have summarized the diversity and function of both classical and recently annotated archaeal ncRNAs, and described their processing. Most archaeal ncRNAs have been predicted or identified from the small set of archaeal genome sequences (~80 species), which only cover a small portion of the phylogenetic diversity of Archaea (Figure 2). A prospective study has suggested that the currently identified metagenomic samples only comprise 50% of actual sequence diversity (Wu et al. 2009), and the variety and number of archaeal ncRNAs is expected to increase with the expansion of genome and transcriptome research. While this review has mainly focused on ncRNAs, there is also a vast number of ncRNA-related proteins that play important roles in various cellular processes (e.g., splicing endonucleases, ligases, and modification enzymes for RNA processing, snoRNA-associated proteins for snoRNA, Cas proteins for CRISPR RNA). Thus, it is also important to discover these counterparts of the newly identified ncRNAs. Our group has worked on predicting and identifying novel DNA- and RNA-binding proteins from the hyperthermophilic archaeon *Pyrococcus furiosus* using both computational and molecular biology approaches on a proteome level (Kanai et al. 2003; Kanai et al. 2006; Fujishima et al. 2007). We hope to detect new RNA processing mechanisms or protein-RNA complexes involved in unknown biological activities. Archaea is an intriguing model due to the unique phylogenies and living environments of these organisms, and understanding the roles of archaeal ncRNA will provide new insights into the molecular evolution of functional RNAs in all three domains of life as well as their possible origins, which may lie in the "RNA world" (Joyce 2002).

Acknowledgments

We thank all the members of the RNA group (Institute for Advanced Biosciences, Keio University) for insightful discussions. This work was supported in part by a Grant-in-Aid for Scientific Research (B) #22370066 from the Ministry of Education, Culture, Sports, Science and Technology of Japan, by a research fund at the Institute for Fermentation, Osaka, Japan, and by research funds at the Yamagata Prefectural Government and Tsuruoka City in Japan.

References

Ambros, V., R. C. Lee, A. Lavanway, P. T. Williams, and D. Jewell. 2003. MicroRNAs and other tiny endogenous RNAs in C. elegans. *Curr. Biol.* 13:807-818.

Baker, B. J., L. R. Comolli, G. J. Dick, L. J. Hauser, D. Hyatt, B. D. Dill, M. L. Land, N. C. Verberkmoes, R. L. Hettich, and J. F. Banfield. 2010. Enigmatic, ultrasmall, uncultivated Archaea. *Proc. Natl. Acad. Sci. U S A* 107:8806-8811.

Baker, B. J., G. W. Tyson, R. I. Webb, J. Flanagan, P. Hugenholtz, E. E. Allen, and J. F. Banfield. 2006. Lineages of acidophilic archaea revealed by community genomic analysis. *Science* 314:1933-1935.

Baker, D. L., O. A. Youssef, M. I. Chastkofsky, D. A. Dy, R. M. Terns, and M. P. Terns. 2005. RNA-guided RNA modification: functional organization of the archaeal H/ACA RNP. *Genes. Dev.* 19:1238-1248.

Balakin, A. G., L. Smith, and M. J. Fournier. 1996. The RNA world of the nucleolus: two major families of small RNAs defined by different box elements with related functions. *Cell* 86:823-834.

Barns, S. M., C. F. Delwiche, J. D. Palmer, and N. R. Pace. 1996. Perspectives on archaeal diversity, thermophily and monophyly from environmental rRNA sequences. *Proc. Natl. Acad. Sci. U S A* 93:9188-9193.

Barrick, J. E., and R. R. Breaker. 2007. The distributions, mechanisms, and structures of metabolite-binding riboswitches. *Genome. Biol.* 8:R239.

Barrick, J. E., K. A. Corbino, W. C. Winkler, A. Nahvi, M. Mandal, J. Collins, M. Lee, A. Roth, N. Sudarsan, I. Jona, J. K. Wickiser, and R. R. Breaker. 2004. New RNA motifs suggest an expanded scope for riboswitches in bacterial genetic control. *Proc. Natl. Acad. Sci. U S A* 101:6421-6426.

Bhalla, T., J. J. Rosenthal, M. Holmgren, and R. Reenan. 2004. Control of human potassium channel inactivation by editing of a small mRNA hairpin. *Nat. Struct. Mol. Biol.* 11:950-956.

Bock, A., and M. Rother. 2005. A pseudo-SECIS element in Methanococcus voltae documents evolution of a selenoprotein into a sulphur-containing homologue. *Arch Microbiol.* 183:148-150.

Brochier-Armanet, C., B. Boussau, S. Gribaldo, and P. Forterre. 2008. Mesophilic Crenarchaeota: proposal for a third archaeal phylum, the Thaumarchaeota. *Nat. Rev. Microbiol.* 6:245-252.

Brooks, M. A., L. Meslet-Cladiere, M. Graille, J. Kuhn, K. Blondeau, H. Myllykallio, and H. van Tilbeurgh. 2008. The structure of an archaeal homodimeric ligase which has RNA circularization activity. *Protein. Sci.* 17:1336-1345.

Bult, C. J., O. White, G. J. Olsen, L. Zhou, R. D. Fleischmann, G. G. Sutton, J. A. Blake, L. M. FitzGerald, R. A. Clayton, J. D. Gocayne, A. R. Kerlavage, B. A. Dougherty, J. F. Tomb, M. D. Adams, C. I. Reich, R. Overbeek, E. F. Kirkness, K. G. Weinstock, J. M. Merrick, A. Glodek, J. L. Scott, N. S. Geoghagen, and J. C. Venter. 1996. Complete genome sequence of the methanogenic archaeon, Methanococcus jannaschii. *Science* 273:1058-1073.

Burggraf, S., N. Larsen, C. R. Woese, and K. O. Stetter. 1993. An intron within the 16S ribosomal RNA gene of the archaeon Pyrobaculum aerophilum. *Proc. Natl. Acad. Sci. U S A* 90:2547-2550.

Chant, J., and P. Dennis. 1986. Archaebacteria: transcription and processing of ribosomal RNA sequences in Halobacterium cutirubrum. *EMBO J.* 5:1091-1097.

Ciammaruconi, A., and P. Londei. 2001. In vitro processing of the 16S rRNA of the thermophilic archaeon Sulfolobus solfataricus. *J. Bacteriol.* 183:3866-3874.

Cole, J. R., Q. Wang, E. Cardenas, J. Fish, B. Chai, R. J. Farris, A. S. Kulam-Syed-Mohideen, D. M. McGarrell, T. Marsh, G. M. Garrity, and J. M. Tiedje. 2009. The Ribosomal Database Project: improved alignments and new tools for rRNA analysis. *Nucleic Acids Res.* 37:D141-145.

Curcio, M. J., and M. Belfort. 1996. Retrohoming: cDNA-mediated mobility of group II introns requires a catalytic RNA. *Cell* 84:9-12.

Dai, L., and S. Zimmerly. 2003. ORF-less and reverse-transcriptase-encoding group II introns in archaebacteria, with a pattern of homing into related group II intron ORFs. *RNA* 9:14-19.

Dalgaard, J. Z., R. A. Garrett, and M. Belfort. 1993. A site-specific endonuclease encoded by a typical archaeal intron. *Proc. Natl. Acad. Sci. U S A* 90:5414-5417.

Dennis, P. P., A. Omer, and T. Lowe. 2001. A guided tour: small RNA function in Archaea. *Mol. Microbiol.* 40:509-519.

Deppenmeier, U., A. Johann, T. Hartsch, R. Merkl, R. A. Schmitz, R. Martinez-Arias, A. Henne, A. Wiezer, S. Baumer, C. Jacobi, H. Bruggemann, T. Lienard, A. Christmann, M. Bomeke, S. Steckel, A. Bhattacharyya, A. Lykidis, R. Overbeek, H. P. Klenk, R. P. Gunsalus, H. J. Fritz, and G. Gottschalk. 2002. The genome of Methanosarcina mazei: evidence for lateral gene transfer between bacteria and archaea. *J. Mol. Microbiol. Biotechnol. 4*:453-461.

Frank, D. N., and N. R. Pace. 1998. Ribonuclease P: unity and diversity in a tRNA processing ribozyme. *Annu. Rev. Biochem.* 67:153-180.

Fuhrman, J. A., and A. A. Davis. 1997. Widespread Archaea and novel Bacteria from the deep sea as shown by 16S rRNA gene sequences. *Mar. Ecol. Prog. Ser.* 150:275-285.

Fujishima, K., M. Komasa, S. Kitamura, H. Suzuki, M. Tomita, and A. Kanai. 2007. Proteome-wide prediction of novel DNA/RNA-binding proteins using amino acid composition and periodicity in the hyperthermophilic archaeon Pyrococcus furiosus. *DNA Res.* 14:91-102.

Fujishima, K., J. Sugahara, K. Kikuta, R. Hirano, A. Sato, M. Tomita, and A. Kanai. 2009. Tri-split tRNA is a transfer RNA made from 3 transcripts that provides insight into the evolution of fragmented tRNAs in archaea. *Proc. Natl. Acad. Sci. U S A* 106:2683-2687.

Fujishima, K., J. Sugahara, M. Tomita, and A. Kanai. 2010. Large-scale tRNA intron transposition in the archaeal order Thermoproteales represents a novel mechanism of intron gain. *Mol. Biol. Evol.* 10:2233-2243.

Fujishima, K., J. Sugahara, M. Tomita, and A. Kanai. 2008. Sequence evidence in the archaeal genomes that tRNAs emerged through the combination of ancestral genes as 5' and 3' tRNA halves. *PLoS One* 3:e1622.

Gardner, P. P., J. Daub, J. G. Tate, E. P. Nawrocki, D. L. Kolbe, S. Lindgreen, A. C. Wilkinson, R. D. Finn, S. Griffiths-Jones, S. R. Eddy, and A. Bateman. 2009. Rfam: updates to the RNA families database. *NucleicAcids Res.* 37:D136-140.

Garrett, R. A., J. Dalgaard, N. Larsen, J. Kjems, and A. S. Mankin. 1991. Archaeal rRNA operons. *Trends Biochem. Sci.* 16:22-26.

Gaspin, C., J. Cavaille, G. Erauso, and J. P. Bachellerie. 2000. Archaeal homologs of eukaryotic methylation guide small nucleolar RNAs: lessons from the Pyrococcus genomes. *J. Mol. Biol.* 297:895-906.

Gegenheimer, P., H. J. Gabius, C. L. Peebles, and J. Abelson. 1983. An RNA ligase from wheat germ which participates in transfer RNA splicing in vitro. *J. Biol. Chem.* 258:8365-8373.

Gottesman, S. 2005. Micros for microbes: non-coding regulatory RNAs in bacteria. *Trends Genet.* 21:399-404.

Grissa, I., G. Vergnaud, and C. Pourcel. 2007. The CRISPRdb database and tools to display CRISPRs and to generate dictionaries of spacers and repeats. *BMCBioinformatics* 8:172.

Grosjean, H., F. Constantinesco, D. Foiret, and N. Benachenhou. 1995. A novel enzymatic pathway leading to 1-methylinosine modification in Haloferax volcanii tRNA. *Nucleic Acids Res.* 23:4312-4319.

Guerrier-Takada, C., K. Gardiner, T. Marsh, N. Pace, and S. Altman. 1983. The RNA moiety of ribonuclease P is the catalytic subunit of the enzyme. *Cell* 35:849-857.

Hale, C. R., P. Zhao, S. Olson, M. O. Duff, B. R. Graveley, L. Wells, R. M. Terns, and M. P. Terns. 2009. RNA-guided RNA cleavage by a CRISPR RNA-Cas protein complex. *Cell* 139:945-956.

Hoheisel, J. D. 2006. Microarray technology: beyond transcript profiling and genotype analysis. *Nat. Rev. Genet.* 7:200-210.

Holzle, A., S. Fischer, R. Heyer, S. Schutz, M. Zacharias, P. Walther, T. Allers, and A. Marchfelder. 2008. Maturation of the 5S rRNA 5' end is catalyzed in vitro by the endonuclease tRNase Z in the archaeon H. volcanii. *RNA* 14:928-937.

Huber, H., M. J. Hohn, R. Rachel, T. Fuchs, V. C. Wimmer, and K. O. Stetter. 2002. A new phylum of Archaea represented by a nanosized hyperthermophilic symbiont. *Nature* 417:63-67.

Inagaki, F., K. Takai, H. Hirayama, Y. Yamato, K. H. Nealson, and K. Horikoshi. 2003. Distribution and phylogenetic diversity of the subsurface microbial community in a Japanese epithermal gold mine. *Extremophiles* 7:307-317.

Itoh, T., N. Nomura, and Y. Sako. 2003. Distribution of 16S rRNA introns among the family Thermoproteaceae and their evolutionary implications. *Extremophiles* 7:229-233.

Jager, D., C. M. Sharma, J. Thomsen, C. Ehlers, J. Vogel, and R. A. Schmitz. 2009. Deep sequencing analysis of the Methanosarcina mazei Go1 transcriptome in response to nitrogen availability. *Proc. Natl. Acad. Sci. U S A* 106:21878-21882.

Jain, R., and S. Shuman. 2009. Characterization of a thermostable archaeal polynucleotide kinase homologous to human Clp1. *RNA* 15:923-931.

Jernigan, F. E., 3rd, N. A. Sieracki, M. T. Taylor, A. S. Jenkins, S. E. Engel, B. W. Rowe, F. A. Jove, G. P. Yap, E. T. Papish, and G. M. Ferrence. 2007. Sterically bulky tris(triazolyl)borate ligands as water-soluble analogues of tris(pyrazolyl)borate. *Inorg. Chem.* 46:360-362.

Joyce, G. F. 2002. The antiquity of RNA-based evolution. *Nature* 418:214-221.

Kanai, A., H. Oida, N. Matsuura, and H. Doi. 2003. Expression cloning and characterization of a novel gene that encodes the RNA-binding protein FAU-1 from Pyrococcus furiosus. *Biochem. J.* 372:253-261.

Kanai, A., A. Sato, Y. Fukuda, K. Okada, T. Matsuda, T. Sakamoto, Y. Muto, S. Yokoyama, G. Kawai, and M. Tomita. 2009. Characterization of a heat-stable enzyme possessing GTP-dependent RNA ligase activity from a hyperthermophilic archaeon, Pyrococcus furiosus. *RNA* 15:420-431.

Kanai, A., A. Sato, J. Imoto, and M. Tomita. 2006. Archaeal Pyrococcus furiosus thymidylate synthase 1 is an RNA-binding protein. *Biochem. J.* 393:373-379.

Kjems, J., and R. A. Garrett. 1988. Novel splicing mechanism for the ribosomal RNA intron in the archaebacterium Desulfurococcus mobilis. *Cell* 54:693-703.

Klein, R. J., Z. Misulovin, and S. R. Eddy. 2002. Noncoding RNA genes identified in AT-rich hyperthermophiles. *Proc. Natl. Acad. Sci. U S A* 99:7542-7547.

Kryukov, G. V., and V. N. Gladyshev. 2004. The prokaryotic selenoproteome. *EMBO Rep.* 5:538-543.

Lai, L. B., A. Vioque, L. A. Kirsebom, and V. Gopalan. 2010. Unexpected diversity of RNase P, an ancient tRNA processing enzyme: challenges and prospects. *FEBS Lett.* 584:287-296.

Larkin, M. A., G. Blackshields, N. P. Brown, R. Chenna, P. A. McGettigan, H. McWilliam, F. Valentin, I. M. Wallace, A. Wilm, R. Lopez, J. D. Thompson, T. J. Gibson, and D. G. Higgins. 2007. Clustal W and Clustal X version 2.0. *Bioinformatics* 23:2947-2948.

Lee, Z. M., C. Bussema, 3rd, and T. M. Schmidt. 2009. rrnDB: documenting the number of rRNA and tRNA genes in bacteria and archaea. *Nucleic Acids Res.* 37:D489-493.

Letunic, I., and P. Bork. 2007. Interactive Tree Of Life (iTOL): an online tool for phylogenetic tree display and annotation. Bioinformatics 23:127-128.

Li, H., C. R. Trotta, and J. Abelson. 1998. Crystal structure and evolution of a transfer RNA splicing enzyme. *Science* 280:279-284.

Liu, J. M., and A. Camilli. 2010. A broadening world of bacterial small RNAs. *Curr. Opin. Microbiol.* 13:18-23.

Marck, C., and H. Grosjean. 2002. tRNomics: analysis of tRNA genes from 50 genomes of Eukarya, Archaea, and Bacteria reveals anticodon-sparing strategies and domain-specific features. *RNA* 8:1189-1232.

Massana, R., E. F. DeLong, and C. Pedros-Alio. 2000. A few cosmopolitan phylotypes dominate planktonic archaeal assemblages in widely different oceanic provinces. *Appl. Environ. Microbiol.* 66:1777-1787.

Massenet, S., I. Ansmant, Y. Motorin, and C. Branlant. 1999. The first determination of pseudouridine residues in 23S ribosomal RNA from hyperthermophilic Archaea Sulfolobus acidocaldarius. *FEBSLett* 462:94-100.

Metzker, M. L. 2010. Sequencing technologies - the next generation. *Nat. Rev.Genet.* 11:31-46.

Nomura, N., Y. Morinaga, N. Shirai, and Y. Sako. 2005. I-ApeKI [corrected]: a novel intron-encoded LAGLIDADG homing endonuclease from the archaeon, Aeropyrum pernix K1. *Nucleic Acids Res.* 33:e116.

Noon, K. R., E. Bruenger, and J. A. McCloskey. 1998. Posttranscriptional modifications in 16S and 23S rRNAs of the archaeal hyperthermophile Sulfolobus solfataricus. *J. Bacteriol.* 180:2883-2888.

Nunoura, T., H. Hirayama, H. Takami, H. Oida, S. Nishi, S. Shimamura, Y. Suzuki, F. Inagaki, K. Takai, K. H. Nealson, and K. Horikoshi. 2005. Genetic and functional properties of uncultivated thermophilic crenarchaeotes from a subsurface gold mine as revealed by analysis of genome fragments. *Environ. Microbiol.* 7:1967-1984.

Omer, A. D., T. M. Lowe, A. G. Russell, H. Ebhardt, S. R. Eddy, and P. P. Dennis. 2000. Homologs of small nucleolar RNAs in Archaea. *Science* 288:517-522.

Opdyke, J. A., J. G. Kang, and G. Storz. 2004. GadY, a small-RNA regulator of acid response genes in Escherichia coli. *J. Bacteriol.* 186:6698-6705.

Peebles, C. L., P. Gegenheimer, and J. Abelson. 1983. Precise excision of intervening sequences from precursor tRNAs by a membrane-associated yeast endonuclease. *Cell* 32:525-536.

Peebles, C. L., P. S. Perlman, K. L. Mecklenburg, M. L. Petrillo, J. H. Tabor, K. A. Jarrell, and H. L. Cheng. 1986. A self-splicing RNA excises an intron lariat. *Cell* 44:213-223.

Pulukkunat, D. K., and V. Gopalan. 2008. Studies on Methanocaldococcus jannaschii RNase P reveal insights into the roles of RNA and protein cofactors in RNase P catalysis. *Nucleic Acids Res.* 36:4172-4180.

Randau, L., R. Munch, M. J. Hohn, D. Jahn, and D. Soll. 2005. Nanoarchaeum equitans creates functional tRNAs from separate genes for their 5'- and 3'-halves. *Nature* 433:537-541.

Randau, L., I. Schroder, and D. Soll. 2008. Life without RNase P. *Nature* 453:120-123.

Rashid, R., M. Aittaleb, Q. Chen, K. Spiegel, B. Demeler, and H. Li. 2003. Functional requirement for symmetric assembly of archaeal box C/D small ribonucleoprotein particles. *J. Mol. Biol.* 333:295-306.

Reinhold-Hurek, B., and D. A. Shub. 1992. Self-splicing introns in tRNA genes of widely divergent bacteria. *Nature* 357:173-176.

Rest, J. S., and D. P. Mindell. 2003. Retroids in archaea: phylogeny and lateral origins. *Mol. Biol. Evol.* 20:1134-1142.

Rodionov, D. A., A. G. Vitreschak, A. A. Mironov, and M. S. Gelfand. 2002. Comparative genomics of thiamin biosynthesis in procaryotes. New genes and regulatory mechanisms. *J. Biol. Chem.* 277:48949-48959.

Russell, A. G., H. Ebhardt, and P. P. Dennis. 1999. Substrate requirements for a novel archaeal endonuclease that cleaves within the 5' external transcribed spacer of Sulfolobus acidocaldarius precursor rRNA. *Genetics* 152:1373-1385.

Salgia, S. R., S. K. Singh, P. Gurha, and R. Gupta. 2003. Two reactions of Haloferax volcanii RNA splicing enzymes: joining of exons and circularization of introns. *RNA* 9:319-330.

Serganov, A., A. Polonskaia, A. T. Phan, R. R. Breaker, and D. J. Patel. 2006. Structural basis for gene regulation by a thiamine pyrophosphate-sensing riboswitch. *Nature* 441:1167-1171.

Shi, Y., G. W. Tyson, and E. F. DeLong. 2009. Metatranscriptomics reveals unique microbial small RNAs in the ocean's water column. *Nature* 459:266-269.

Sorek, R., V. Kunin, and P. Hugenholtz. 2008. CRISPR--a widespread system that provides acquired resistance against phages in bacteria and archaea. *Nat. Rev. Microbiol.* 6:181-186.

Spinelli, S. L., H. S. Malik, S. A. Consaul, and E. M. Phizicky. 1998. A functional homolog of a yeast tRNA splicing enzyme is conserved in higher eukaryotes and in Escherichia coli. *Proc. Natl. Acad. Sci. U S A* 95:14136-14141.

Starostina, N. G., S. Marshburn, L. S. Johnson, S. R. Eddy, R. M. Terns, and M. P. Terns. 2004. Circular box C/D RNAs in Pyrococcus furiosus. *Proc.Natl. Acad. Sci. U S A* 101:14097-14101.

Straub, J., M. Brenneis, A. Jellen-Ritter, R. Heyer, J. Soppa, and A. Marchfelder. 2009. Small RNAs in haloarchaea: identification, differential expression and biological function. *RNA Biol.* 6:281-292.

Sugahara, J., K. Fujishima, K. Morita, M. Tomita, and A. Kanai. 2009. Disrupted tRNA gene diversity and possible evolutionary scenarios. *J. Mol.Evol.* 69:497-504.

Sugahara, J., K. Kikuta, K. Fujishima, N. Yachie, M. Tomita, and A. Kanai. 2008. Comprehensive analysis of archaeal tRNA genes reveals rapid increase of tRNA introns in the order thermoproteales. *Mol. Biol. Evol.* 25:2709-2716.

Sugahara, J., N. Yachie, K. Arakawa, and M. Tomita. 2007. In silico screening of archaeal tRNA-encoding genes having multiple introns with bulge-helix-bulge splicing motifs. *RNA* 13:671-681.

Takai, K., and K. Horikoshi. 1999. Genetic diversity of archaea in deep-sea hydrothermal vent environments. *Genetics* 152:1285-1297.

Tang, T. H., J. P. Bachellerie, T. Rozhdestvensky, M. L. Bortolin, H. Huber, M. Drungowski, T. Elge, J. Brosius, and A. Huttenhofer. 2002a. Identification of 86 candidates for small non-messenger RNAs from the archaeon Archaeoglobus fulgidus. *Proc. Natl. Acad. Sci. U S A* 99:7536-7541.

Tang, T. H., N. Polacek, M. Zywicki, H. Huber, K. Brugger, R. Garrett, J. P. Bachellerie, and A. Huttenhofer. 2005. Identification of novel non-coding RNAs as potential antisense regulators in the archaeon Sulfolobus solfataricus. *Mol. Microbiol.* 55:469-481.

Tang, T. H., T. S. Rozhdestvensky, B. C. d'Orval, M. L. Bortolin, H. Huber, B. Charpentier, C. Branlant, J. P. Bachellerie, J. Brosius, and A. Huttenhofer. 2002b. RNomics in Archaea reveals a further link between splicing of archaeal introns and rRNA processing. *Nucleic Acids Res.* 30:921-930.

Thisted, T., A. K. Nielsen, and K. Gerdes. 1994. Mechanism of post-segregational killing: translation of Hok, SrnB and Pnd mRNAs of plasmids R1, F and R483 is activated by 3'-end processing. *EMBO J.* 13:1950-1959.

Tocchini-Valentini, G. D., P. Fruscoloni, and G. P. Tocchini-Valentini. 2005. Structure, function, and evolution of the tRNA endonucleases of Archaea: an example of subfunctionalization. *Proc. Natl. Acad. Sci. U S A* 102:8933-8938.

Tocchini-Valentini, G. D., P. Fruscoloni, and G. P. Tocchini-Valentini. 2007. The dawn of dominance by the mature domain in tRNA splicing. *Proc.Natl. Acad. Sci. U S A* 104:12300-12305.

Torchia, C., Y. Takagi, and C. K. Ho. 2008. Archaeal RNA ligase is a homodimeric protein that catalyzes intramolecular ligation of single-stranded RNA and DNA. *Nucleic Acids Res.* 36:6218-6227.

Tsai, H. Y., D. K. Pulukkunat, W. K. Woznick, and V. Gopalan. 2006. Functional reconstitution and characterization of Pyrococcus furiosus RNase P. *Proc. Natl. Acad. Sci. U S A* 103:16147-16152.

Tujebajeva, R. M., P. R. Copeland, X. M. Xu, B. A. Carlson, J. W. Harney, D. M. Driscoll, D. L. Hatfield, and M. J. Berry. 2000. Decoding apparatus for eukaryotic selenocysteine insertion. *EMBO Rep.* 1:158-163.

Washietl, S., I. L. Hofacker, M. Lukasser, A. Huttenhofer, and P. F. Stadler. 2005. Mapping of conserved RNA secondary structures predicts thousands of functional noncoding RNAs in the human genome. *Nat.Biotechnol.* 23:1383-1390.

Waters, E., M. J. Hohn, I. Ahel, D. E. Graham, M. D. Adams, M. Barnstead, K. Y. Beeson, L. Bibbs, R. Bolanos, M. Keller, K. Kretz, X. Lin, E. Mathur, J. Ni, M. Podar, T. Richardson, G. G. Sutton, M. Simon, D. Soll, K. O. Stetter, J. M. Short, and M. Noordewier. 2003. The genome of Nanoarchaeum equitans: insights into early archaeal evolution and derived parasitism. *Proc. Natl. Acad. Sci. U S A* 100:12984-12988.

Weinberg, Z., J. Perreault, M. M. Meyer, and R. R. Breaker. 2009. Exceptional structured noncoding RNAs revealed by bacterial metagenome analysis. *Nature* 462:656-659.

Weinberg, Z., J. X. Wang, J. Bogue, J. Yang, K. Corbino, R. H. Moy, and R. R. Breaker. 2010. Comparative genomics reveals 104 candidate structured RNAs from bacteria, archaea, and their metagenomes. *Genome Biol.* 11:R31.

Weitzer, S., and J. Martinez. 2007. The human RNA kinase hClp1 is active on 3' transfer RNA exons and short interfering RNAs. *Nature* 447:222-226.

Westaway, S. K., E. M. Phizicky, and J. Abelson. 1988. Structure and function of the yeast tRNA ligase gene. *J. Biol. Chem.* 263:3171-3176.

Woese, C. R., and G. E. Fox. 1977. Phylogenetic structure of the prokaryotic domain: the primary kingdoms. Proc Natl Acad Sci U S A 74:5088-5090.

Woese, C. R., O. Kandler, and M. L. Wheelis. 1990. Towards a natural system of organisms: proposal for the domains Archaea, Bacteria, and Eucarya. *Proc. Natl. Acad. Sci. U S A* 87:4576-4579.

Wu, D., P. Hugenholtz, K. Mavromatis, R. Pukall, E. Dalin, N. N. Ivanova, V. Kunin, L. Goodwin, M. Wu, B. J. Tindall, S. D. Hooper, A. Pati, A. Lykidis, S. Spring, I. J. Anderson, P. D'Haeseleer, A. Zemla, M. Singer, A. Lapidus, M. Nolan, A. Copeland, C. Han, F. Chen, J. F. Cheng, S. Lucas, C. Kerfeld, E. Lang, S. Gronow, P. Chain, D. Bruce, E. M. Rubin, N. C. Kyrpides, H. P. Klenk, and J. A. Eisen. 2009. A phylogeny-driven genomic encyclopaedia of Bacteria and Archaea. *Nature* 462:1056-1060.

Wurtzel, O., R. Sapra, F. Chen, Y. Zhu, B. A. Simmons, and R. Sorek. 2010. A single-base resolution map of an archaeal transcriptome. *Genome Res.* 20:133-141.

Yokobori, S., T. Itoh, S. Yoshinari, N. Nomura, Y. Sako, A. Yamagishi, T. Oshima, K. Kita, and Y. Watanabe. 2009. Gain and loss of an intron in a protein-coding gene in Archaea: the case of an archaeal RNA pseudouridine synthase gene. *BMC Evol. Biol.* 9:198.

Yoshinari, S., T. Itoh, S. J. Hallam, E. F. DeLong, S. Yokobori, A. Yamagishi, T. Oshima, K. Kita, and Y. Watanabe. 2006. Archaeal pre-mRNA splicing: a connection to hetero-oligomeric splicing endonuclease. *Biochem. Biophys. Res. Commun.* 346:1024-1032.

Zago, M. A., P. P. Dennis, and A. D. Omer. 2005. The expanding world of small RNAs in the hyperthermophilic archaeon Sulfolobus solfataricus. *Mol. Microbiol.* 55:1812-1828.

In: Archaea: Structure, Habitats and Ecological...
Editor: Sakura Y. Katō, pp. 95-120

ISBN 978-1-61761-932-8

Chapter 4

Archaea Diversity in Brazilian Aquatic Ecosystems

Flávia Talarico Saia[1], *Ana Carolina Araujo*[2], *Cristina Rossi Nakayama*[2], *Diego Assis das Graças*[3], *Juliana Perez Chaparro*[3], *Fernando Andreote*[4], *Rodrigo Taketani*[5], *Fernanda Piza*[6], *Artur Silva*[3], *Vivian Helena Pellizari*[2], *Itamar Soares*[5], *Gilson Manfio*[7] *and Rosana Filomena Vazoller*[8]

[1] Laboratory of Biological Process, Engineering School of São Carlos, University of São Paulo, São Carlos, SP, Brazil,
[2]Institute of Oceanography, University of São Paulo. Praça do Oceanográfico, Brazil
[3]Laboratory of Polimorphism of DNA, Federal University of Pará, Guamá, Belem, PA,Brazil
[4]Departament of Soil Science, Superior School of Agriculture "Luiz de Queiroz", University of São Paulo, Piracicaba, SP, Brazil
[5] Laboratory of Environmental Microbiology, Embrapa Environment, Jaguariúna, SP, Brazil
[6] Instituto Vita Nova, Hortolândia, SP, Brazil
[7] Natura Inovação e Tecnologia de Produtos Ltda, Cajamar, SP, Brazil
[8] Instituto Samuel Murgel Branco, São Paulo, SP, Brazil

Abstract

Comparative sequence analysis of 16S RNA molecules carried out by Carl Woese in the 1970s revealed that life on Earth can be divided into three primary lineages (referred to as domains): *Eukarya*, *Bacteria* and *Archaea*. Presently, most known archaeal representatives are included into two kingdoms: *Crenarchaeota* and *Euryarchaeota*, with remarkable diversity of organisms. *Archaea* can be found in some of the most extreme environments on the planet, but are not restricted to them. Although the important role that *Archaea* play in biogeochemical cycles, such as carbon, sulfur and nitrogen cycles, physiology and ecological significance of archaeal populations are still poorly known.

This lack of knowledge is, in large part, due to the limitations of cultivation procedures. In addition, information about the archaeal community in tropical ecosystems is still scarce. In this chapter we present a survey of the archaeal community present in aquatic ecosystems of Brazil: Madeira River and Tucuruí Power Plant Reservoir in Amazon; Santos and São Vicente Estuary and mangroves in the Southeast coast. The study of archaeal groups in freshwater and saline tropical aquatic ecosystems in Brazil revealed highly diverse communities and suggested the presence of new archaeal groups. In Amazonian environments, *Crenarchaeota* representatives such as Marine Group I, 1.1.band *Thermoplasmata* were frequent, suggesting a possible role of *Archaea* in nitrogen and carbon metabolism. Methanogenic groups were also detected in sediment and water where anaerobic conditions prevailed from both regions studied. In Santos and São Vicente Estuarine System, higher abundance of methanogenic *Archaea* was found in a highly impacted site when compared to a less contaminated area. In Mangrove sediments, the detection of MCG, MBG-B and MBG-D as an important component of the microbial community indicates the possible role of these groups in carbon cycle. This work emerges as a valuable contribution to the understanding of the biogeography and diversity of *Archaea* Domain in these rich ecosystems, besides providing hints as to help characterizing the role and dynamics of archaeal communities in major processes such methane production in tropical environments, subjected or not to the influence of human activities.

1. Introduction

The pioneering work of Carl Woese in the 1970s, utilizing sequence data from small subunit ribosomal RNA (SSU rRNA) to classify all life forms, led to the surprising discovery of the Domain *Archaea* (Woese and Fox 1977), a third line of evolutionary descent, distinct from the domain *Bacteria* as well as from eukaryotic cells from the domain *Eukarya*(Figure 1). The domain *Archaea* tree, based on SSU rRNA, consists of two major phyla, *Crenarchaeota* and *Euryarchaeota*, although two other phyla have been proposed: *Korarchaeota* (Stein *et al.* 1996) and *Nanoarchaeota* (Huber *et al.* 2002) with no isolated representative to date. Phylogenetic studies based on ribosomal proteins sequences have led to the suggestion of the *Archaea* domain to be divided in 3 phyla: *Euryarchaeota, Crenarchaeota* and the newly proposed, *Thaumarchaeota*, composed by non-thermophilic species related to *Crenarchaeota* members (Brochier-Armanet *et al.* 2008).

Microorganisms belonging to *Crenarchaeota* phylum are thermophilic, both aerobes and anaerobes, and a wide diversity of physiologies are represented, including chemoheterotrophs, chemolithotrophs and fermenters.

The cultured organisms within the phylum*Euryarchaeota* are more diverse, comprising methanogens, extreme halophiles (*Halobacteriales*), sulfate reducers (*Archaeoglobales*), and various thermophiles (*Thermococcales* and *Thermoplasmatales*) (Chaban *et al.* 2006).

Cenarchaeum symbiosum (Könneke *et al.* 2005) is the first cultured representative from the *Thaumarchaeota* phyla, and presents a nitrifying metabolism. These phyla groups are non-thermophilic microorganisms previously classified under the *Crenarchaeota* division. In this chapter, *Thaumarchaeota* will be referred to as non-thermophilic *Crenarchaeota*, as original works presented here were elaborated before the changes in classification of this group.

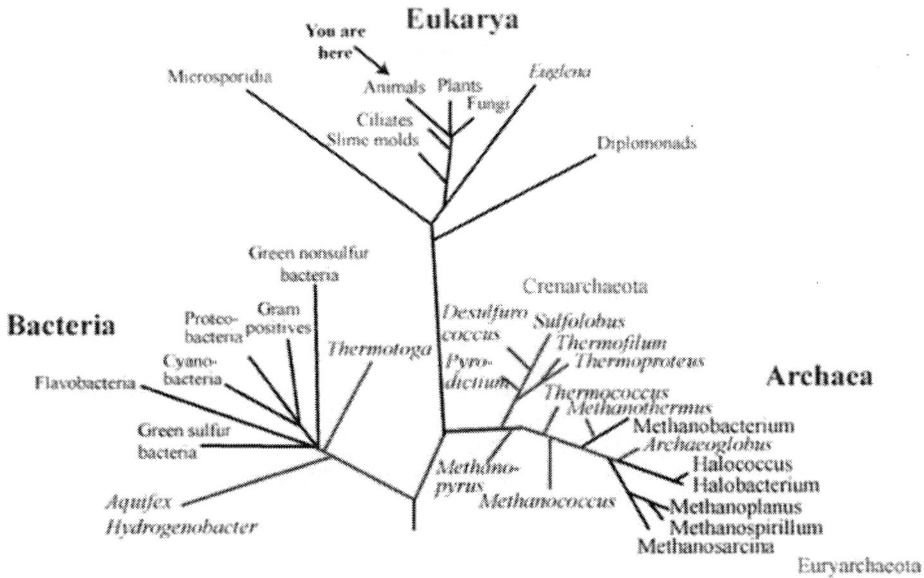

Figure 1. Phylogentic tree of life based on the ribosomal RNA gene. This tree is divided into three major Domains - the Bacteria, Archaea and Eukarya. Archaea Domain is divided in Euryarchaeota and Crenarchaeota Phyla (http://oceanservice.noaa.gov adapted from Woese et al., 1990).

Archaea organisms occur in diverse environments and commonly occupy a significant fraction of the total microbiota, typically 10% of total rRNA phylotypes.The remainder is composed mainly of bacteria (Robertson *et al.* 2005). Environmental *Archaea*sequences have been found in extreme (De Long et *al.* 1994; Füttere*ret al.* 2004) and non–extreme environments, such as freshwater lake sediments (Vissers *et al.*2009), continental shelf sediments (Vetriani *et al.* 1999), estuaries (Vieira *et al.* 2007, Clementino *et al.*2007, 2008) and mangroves (Yan *et al.*2006,Taketani *et al.* 2010a, 2010b). Although the application of molecular surveys conducted during the last two decades have unveiled the ubiquity and abundance of *Archaea* in both aquatic and terrestrial environments, showing a great diversity and activity not restricted to "extreme" conditions, the mere application ofmolecular analyses is not sufficient to link microbial identity and function, especially when the target microorganism belongs to a poorly known phylogenetic group as occur with *Archaea* (Robertson *et al.* 2005, Plasencia *et al.* 2010).This is mainly due to the lack of isolated representatives from which to resolve their metabolic capabilities and infer their contribution to ecosystem functioning. There is a great discrepancy between the molecular diversity of *Archaea* (with more than 40.000 16S rRNA gene sequences deposited in public databases) and their representativeness in culture collections. From all 445 archaeal isolates available at the DSMZ collection (http://www.dsmz.de), 52% are methanogens, 23% halophiles, 23% thermophiles, and 2% acidophiles (Plasencia *et al.* 2010). Thus, to obtain insights into diversity of *Archaea* in the environment and the ecosystem services that they provide, multiple, complementary approaches, including molecular tools, metagenomics, cultivation and (bio)geochemistry are necessary (Robertson *et al.* 2005).

Tropical aquatic environments, such as mangroves, estuaries, rivers and reservoirs are of prime ecological and economical relevance. Estuaries and mangroves represent the interface between freshwater and marine water systems and are subject to periodic flooding by high tides. They are regions of high biological activity and host great biological diversity. Rivers are of primary importance to the establishment of urban areas, since they provide transportation, water and food resources. In Brazil, rivers are of special interest, since around 80% of electric power is generated in hydroelectric power stations. Hydroelectric Power Reservoirs are man-made ecosystems to energy generation purposes. They are characterized by permanent flooding of areas previously covered by vegetation or human occupation. The main characteristic of these ecosystems is that sediments become anoxic and with a highly reduced redox potential. The lack of oxygen coupled with the abundance of organic matter creates an optimal environment for several groups of anaerobic organisms, such as methanogenic Archaea, which rely on simple substrates (e.g., methanol, mono-, di-, trimethylamine, acetate, hydrogen) as carbon and energy sources. Methanogens are thus of prime ecological relevance, once they are responsible for biogenic methane production during the mineralization of organic matter in anaerobic environments. Methane is the second most important greenhouse gas contributing to around 15% to the greenhouse effect.

Despite the importance of *Archaea* for the functioning and maintenance of these tropical ecosystems, information about the archaeal community is still scarce. Among the studies that have been carried out to assess the diversity of *Archaea* in tropical estuaries there are the researches of Vieira *et al.* (2007) and Clementino *et al.*(2007, 2008) that found Methanogens, Halophiles, and uncultured members of the phyla *Crenarchaeota* and *Euryarchaeota* in Guanabara Bay (Brazil). In mangrove ecosystems, only a few valuable studies have described archaeal communities in tropical and subtropical climates as composed primarily by*Crenarchaeota* and *Euryarchaeota* phyla (Yan *et al.* 2006, Lyimo *et al.* 2009,Taketani *et al.* 2010b). Taketani *et al.* (2010b) also detected the *mcrA* gene in mangrove samples, which encodes for the methyl coenzyme-M reductase that is involved in the methane cycle. To date, there has been no study on archaeal diversity in aquatic ecosystems of Brazilian Amazonia. The characterization of the archaeal community in flooded environments in Amazonia will contribute to the understanding of the ecophysiological processes involved in the decay of organic matter and green house gas (GHG) emissions.

1.1. Freshwater Ecosystems in Brazilian Amazon: Madeira River and Tucuruí Hydroelectric Power Station Reservoir

Brazil harbours 13% of total freshwater available in the world (www.ana.gov.br) Most of this water is distributed in the two biggest Brazilian hydrographic basins – the Amazon and the Tocantins-Araguaia – both located in the Amazon biome. They are nationally important due to food resources, transportation, hydroelectricity generation and maintenance of the Amazon biome which hosts approximately a quarter of all global biodiversity (Betts *et al.* 2008) as well as mineral and water resources.

The Amazon basin with 6,000,000 km^2 of drainage has a tropical warm climate and a discharge of 3,767.8 km^3 a year. The average humidity is 80%. One of the important characteristics of this region is the high water level fluctuation between the rainy and dry seasons promoting a corresponding hydro social cycle (fisheries, floodplain exploitation, local navigation) (Junk *et al.* 1990).

The hydrographic basin of Madeira River ranges 1.52 million km^2, the equivalent to 24% of the Amazon Basin. Madeira River, 1459 km long, is the most important tributary to the

Amazon River. It presents an average flow rateof 23000 m^3s^{-1} with a high seasonal variance and is characterized by the huge amounts of wood and soil material it carries, equivalent to 35% of total solid material (40 x 10^6 t) discharged annually in the Atlantic Ocean by the Amazon River (Bernardi *et al.* 2009).

The huge hydric potential of the Amazon region has been harnessed with the construction of hydroelectric power stations (HPS). One of the main consequences to the environment which emerge from building HPS reservoirs is the permanent flooding of areas previously occupied by forests. With the creation of a reservoir, not only large amounts of organic matter are trapped when land is flooded but more is flushed in after from upstream of the dam. In HPS reservoirs located in the tropics, the decomposition of the submerged organic matter results in the formation of methane (CH4) and carbon dioxide (CO2), both well known greenhouse gases (GHG) (Giles2006).

TucuruíHydroelectric Power Reservoir is located in the Brazilian State of Pará on the Tocantins River, main river from the Tocatins-Araguaia Basin. Tucuruí HPS Reservoir has a surface area of approximately 2430 km^2 and a storage capacity of about 45 km^3.The reservoir inundated a vast area of native forest that was not removed before being flooded. Several regions of the reservoir display interesting characteristics, such as Caraipé, located about 20 km upstream of the dam, and known to present anoxic zones (Evangelista 1993) and hydrographical characteristics of a typical Amazonian *igarapé*, with low volumetric flow rates and calm waters. In this zone, vegetation is only temporarily flooded during the rainy season.

1.2. Brackish Ecosystems in São Paulo State: Santos and São Vicente Estuarine System and Cananeia And Bertioga Mangroves

The Santos-São Vicente Estuarine System and Bertioga Mangrove are located in the central coast of São Paulo State, Brazil, in a metropolitan region known as Baixada Santista. This region has a permanent population of 1.47 million and an estimated fluctuating population of 3 million people and comprehends the municipalities of Bertioga, Guarujá, Santos, São Vicente, Cubatão, Praia Grande, Mongaguá, Itanhaém and Peruíbe. Climate is hot and wet, with a yearly average temperature of 22°C and relative humidity rates that reach more than 80% throughout the year. Rainfall rates range between 2,000 to 2,500mm, with a more intense regime in summer months in the south hemisphere - January to March (Cantagalo *et al.* 2008). Both ecosystems are confined between the Serra do Mar cliff and the Atlantic Ocean. Rivers coming down from the Serra do Mar flow fast and intensely, losing energy when they reach the plains, which have little or no declivity. As a result, a complex pattern of streams and creeks are formed, transforming vast regions in wetlands covered by mangroves.

The São Vicente Estuary is located at the Western outlet of the system and the Santos Estuary, where the Santos Port is situated, lies at the eastern outlet. Both estuaries are interconnected at the upper part of the system (Hortellani et al. 2005, Rodrigues et al. 2009), where another port can be found, the Cubatão Maritime Terminal, a private port to the Cubatão unit of USIMINAS company (former Companhia Siderurgica Paulista – Steelworks Company of São Paulo, COSIPA).The mangroves studied were mangrove of Bertioga, in the Bertioga Channel at the lower part of the Santos Estuary and Cananeia Mangrove. The mangrove in Cananeia is located at the Cardoso Island in the estuary, and it belongs to a protected wildlife reserve and is thus considered pristine in terms of direct human impact. The mangroves from Bertioga have been affected either by oil or by sewage discharge and solid residues deposition. As for the oil contamination, the last spill was in 1983, when 35 million liters of oil were released in the area. Native vegetation in the mangrove is still undergoing regeneration. Bertioga and Cananeiamangroves also differ in their flooding systems. While the first receive mixed water from the sea and the Iriri River, the mangrove located in Cananeia is located in an estuary and is mostly under tidal influence from the Atlantic Ocean.

2. Synthesis of the Study

The peculiar characteristics of Amazonian rivers and dams and of the estuary and mangroves at the southeast of São Paulo State make them interesting for the research on microbial diversity. A number of Brazilian research programs and projects have been created in the last years that included the study of the microbial diversity in these Brazilian ecosystems. This chapter brings information generated in three of these programs: (1) The Virtual Institute of Biodiversity (BIOTA-FAPESP), an initiative of the São Paulo State Research Foundation (FAPESP) to collect, organize and disseminate information about the biodiversity in the State (Joly et al. 2010); (2) The Program for Supporting Centers of Excellence (PRONEX-Rondônia), financed by the National Council for Scientific and Technological Development (CNPq) with the aim of promoting the scientific development of the Amazonian region and the study of tropical diseases; and (3) The Research and Development Programme of the Electrical Energy Sector, regulated by the Brazilian Electricity Regulatory Agency (ANEEL) and supported by Centrais Elétricas do Norte do Brasil (ELETRONORTE), CNPq and the Consejo Superior de Investigaciones Cientificas (Superior Council of Scientific Research, CSIC, Spain).

In order to gain new insights into the archaeal communities of aquatic freshwater and coastal ecosystems of Brazil and to provide reference information for future research, a 16S rRNA gene clone library approach was used. For archaeal molecular diversity assessments in freshwater ecosystems, the river Madeira and the Tucuruí hydroelectrical plant reservoir in the Brazilian Amazon were selected. Brackish ecosystems were represented by the Santos and São Vicente Estuarine System and Bertioga and Cananeiamangroves in São Paulo State coast (southern coast of Brazil).

3. Experimental Setup and Methods

3.1. Sampling and Environmental Parameters

Freshwater Ecosystems

Madeira River. Archaeal diversity studies were carried out with sediment samples collected with sterile spatula in the margins of the river at Rondônia State, near Porto Velho city (Figure 2). In this area, the river is used for transportation of people and agricultural products, being of crucial importance to the maintenance of the state. The water at the sampling site exhibited a neutral pH, warm temperature and was well oxygenated, typical characteristics of white water rivers (Patel *et al.* 1999) (Table 1). The sediment showed a sandy silt composition according (Shepard 1954), and was poorly selected (Folk and Ward 1957). Organic matter content was low (Table 1) and similar to other non-organic river sediments (Jia and Peng 2003, Kamaruzzaman *et al.* 2009).

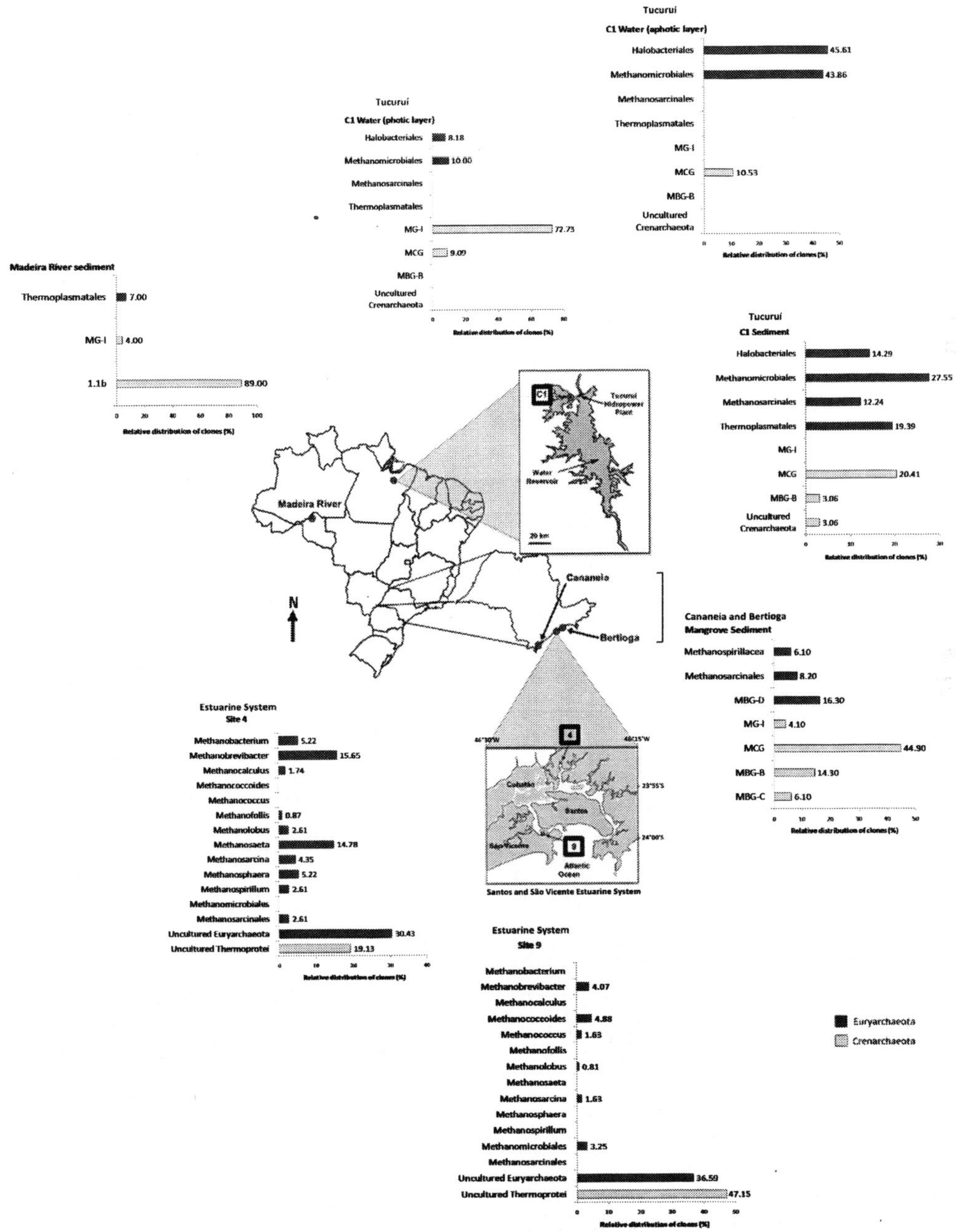

Figure 2. Map of Brazil showing the area studied and archaeal diversity of each area. Santos São Vicente Estuary, in São Paulo State,and location of sampling sites 4 and 9; Cananeia and Bertioga Mangroves in São Paulo State; Madeira River in Porto Velho city, Rondônia State; Tucuruí Hidropower Plant in Pará State and the location of sampling C1.

Tucuruí Hydroelectric Power Reservoir. Assessment of archaeal diversity at Tucuruí reservoir was performed with water and sediment samples collected at the Caraipé (C1) region (Figure 2). The water samples were collected at the photic (surface) and at the aphotic (40 m deep) layers. Sediment was sampled with a van Veen grab (Table 1).

Table 1. Environmental parameters of sampling sites at studied freshwater Amazonian ecosystems.

	Madeira River	Tucuruí Reservoir
GPS	S 08° 33' 17.0" W 063° 33' 14.6"	S 03°43' to 5°15' W 49°12' to 50°00'
pH (water)	6.5	ND[1]
Dissolved oxygen ($mg.L^{-1}$)	6.6	ND
Temperature (°C)	31.2±2.9	ND
Organic matter (%)	5.03	ND

[1]Not determined

Table 2. Environmental parameters of sampling sites at studied brackish ecosystems in São Paulo State coast.

	Santos and São Vicente Estuarine System				Mangroves			
	Site 4		Site 9		Bertioga		Cananeia	
	Water	Sediment	Water	Sediment	Water	Sediment	Water	Sediment
Location	S 23°52′31.1" W 46°22'35.9"		S 23°58′28.4" W' 46°24'48.7"		(A) S 23° 53' 49"; W 46° 12' 28" (B) S 23° 54' 06"; W 45° 15' 03"		S 25° 05' 02" W 47° 57' 42"	
pH	7.7	8.2	6.6	7.4	7.09	6.56-6.67	6.5	5.9-6.35
Dissolved oxygen ($mg.L^{-1}$)	ND[1]	ND	ND	ND	4.0	<DL[2]	7.11	<DL
Temperature (°C)	25.3	ND	25.0	ND	22.4	22	20.6	19.0
Eh (mV)	200	-371	64	-330	ND	ND	ND	ND
Salinity (‰)		16.9		23.2	17.4	4.7	15.5	1.7 - 12.7
	Highly contaminated - hydrocarbons of anthropogenic origin		Presence of hydrocarbons of biogenic origin		Presence of contaminated areas: (A) Oil spill or (B) Wastewater and solid residues		Preserved area	

[1] Not determined; [2]Below detection limit

Brackish Ecosystems

Santos and São Vicente Estuarine System. Two sites were sampled at the estuarine system (Figure 2). The first, site 4, was near USIMINAS port, an area at the brackish end of the estuary receiving wastewater from steel works, petrochemical and fertilizer industries. The sediment in the site is highly contaminated with hydrocarbons and other organic and inorganic compounds (Rodrigues *et al.* 2009), is anaerobic, slightly alkaline, and presents mild temperatures (Table 2). The second area, site 9, is less contaminated and exhibited neutral pH, mild temperatures and low oxygen content (Table 2). As site 9 is nearer the marine end of the estuary, it presents higher salinity values. This site receives effluents from domestic sewage and hydrocarbons in the area are mostly from biogenic origin and in concentrations much lower than those found in site 4 (Table 3). Sediment was sampled with a van Veen grab at a maximum depth of 1.3m below the water surface.

Table 3. Hydrocarbons concentrations in sites 4 and 9 of the Santos and São Vicente Estuarine System.

Site	Total n-alkanes ($\mu g.g^{-1}$)[1]	Total aliphatics ($\mu g.g^{-1}$)	ΣPAHs ($\mu g.g^{-1}$)[2]
4	113.0	2.508	407.560
9	4.1	4.1	14

[1]Dry weight sediments; [2] ΣPAHs = total polycyclic aromatic hydrocarbons. Source: Piza 2004.

Bertioga and Cananeia mangroves. Sediment samples were collected from three distinct areas (Figure 2). In Bertioga mangrove, two areas were sampled, one affected by oil and the other by sewage discharge and solid residues deposition. The third area was located in the pristine Cananeia mangrove. Samples were collected manually with cores (7 cm diameter x 30 cm deep). Water and sediment in both areas exhibited neutral pH and temperatures around 20°C (Table 2). Salinity gradients were more evident at sediment samples, ranging from 1.7 to 12.7‰. Dissolved oxygen measurements showed that surficial waters were well oxygenated, while sediments were anoxic, which is consistent to mangrove sediments, where oxygen is rapidly consumed by organic matter degradation.

Table 4. Procedures for culture-independent characterization of archaeal communities in sediment and water samples.

	Madeira River	Tucuruí reservoir			Santos and São Vicente Estuarine System		Bertioga and Cananeia mangroves	
DNA Extraction	Fast DNA Spin kit for soil (Q-Biogene)	Ultra clean soil DNA kit (MoBio)			Van Elsas & Smalla, 1995 (modified)		Power Soil DNA kit (MoBio)	
Gene	Archaeal 16S rDNA	Archaeal 16S rDNA			Archaeal 16S rDNA		Archaeal 16S rDNA[2]	mcrA
Primers	109F-915R (Grosskopf et al. 1998)	341F-1492R			344F-1400R (Casamayor et al. 2002; Kudo et al. 1997)		21F-958R (Ovreas et al. 1997)	MLF-MFR (Luton et al. 2002)
Samples	Sediment	Water (photic layer)	Water (aphotic layer)	Sediment	Sediment (site 4)	Sediment (site 9)	Sediment	
Sample size	0.9g in three aliquots	5 L	5 L	2.0g in five aliquots	2.5g in five aliquots (per site)		Pool: 27 samples of 0.3g	
No. clones[1]	81	110	57	98	121	123	84	59
Reference	Araujo, 2010	Silva et al. (unpublished)			Piza, 2004		Taketani et al.(2010a; b)	

[1]After removal of chimeric and bad quality sequences. [2]DGGE analysis of mangrove sediments showed no significant difference between communities in Bertioga and Cananeia sites. For this reason, DNA from all samples was pooled and used to compose one 16S rDNA and one *mcr*A library.

3.2. Assessment of Archaeal Diversity Using Culture-Independent Methods

The characterization of archaeal diversity in water and sediment samples was carried out through the construction of 16S rRNA clone libraries. In Cananeia sediments, the methanogenic specific gene, *mcr*A, was also used to assess the diversity of the group. Table 4 summarizes the procedures adopted for each site.

4. Characterization of *Archaea* Communities in Brazilian Ecosystems

4.1. Madeira River

Studies on archaeal diversity generally focus on marine rather than riverine ecosystems and, consequently, there is a lack in the knowledge about the physiology and ecology of riverine *Archaea* (Herfort *et al.* 2009). Considering tropical freshwater ecosystems, the lack is even bigger since, in our revision, no published studies were found about *Archaea* in this kind of environment. In our work, from 96 sequenced clones, 16 were of poor quality or chimera suspected. 93% (74) belonged to the kingdom *Crenarchaeota* and only 7% (6) were related to species belonging to the *Euryarchaeota* kingdom (Figure 2). The prevalence of *Crenarchaeota* over *Euryarchaeota* representatives is widely known in marine ecosystems, but has also been detected in freshwater (Herfort *et al.* 2009) and sediments (Rastogi *et al.* 2009). None of the clones was related to previously cultured representatives. Only three sequences (4%) were related to group 1.1a or Marine Group-1, typical of marine environments, but very diverse and with members already found in terrestrial habitats(Takai *et al.* 2001).96% of the clones belonging to the kingdom *Crenarchaeota* were related to group 1.1b, characteristic of soils (Auguet *et al.* 2010), but also found in other terrestrial environments. The sequences belonging to group 1.1b were similar to sequences previously found in soils with different physical-chemical characteristics (Simon *et al.*2000, Takai *et al.*2001, Quaiser *et al.* 2002, Sliwinski and Goodman 2004, Cadillo-Quiroz *et al.*2008, Nelson *et al.* 2009). The predominance of typical soil members of the *Crenarchaeota* kingdom may be due to a lack of more studies focusing *Archaea* in river sediments, mainly in tropical environments, but also to the huge amount of soil material that Madeira River carries when it is flooding. The alochtonous hypothesis was used to explain the abundance of lacustrine phylotypes found in the waters of Mackenzie River (Galand *et al.* 2006), that might have been carried by the river from the plains it washed. In surface waters of the Rhine River, a high proportion of *Crenarchaeota* sequences belonging to group 1.1b during the winter was explained by precipitation and runoff of members of this group typical of soils (Herfort *et al.* 2009). Eight sequences from the Madeira 16S rRNA gene clone library belonging to group 1.1b were related to the species *Nitrososphaera gargensis* (Hatzenpichler *et al.* 2008), the first ammonium oxidizing *Archaea* isolated from a terrestrial thermal spring. This group of *Archaea* was only recently discovered and its importance in the nitrogen cycle is still under debate, especially in terrestrial environments, where few studies are available. Sequences related to nitrifying *Archaea* were also found the plankton of Rhine River, but they were related to group 1.1a (Herfort *et al.* 2009).

Six clones from the Madeira sediment belonged to the euryarchaeotal class*Thermoplasmata*, but had less than 90% of similarity with sequences found in the NCBI database. The uncultured members of *Thermoplasmata* class are divided in two clades, E2 – including TMEG ("Terrestrial Miscellaneous Euryarchaeotic Group") and species from hydrothermal vents; and E3 – including Marine Groups II and III. Both groups were described in numerous environments, as marine sediments (Kendall *et al.* 2007); marine hot (Dhillon *et al.* 2005) and cold seeps (Knittel *et al.* 2005); hypersaline biofilms (Sørensen *et al.* 2005, Robertson *et al.* 2009). These groups were also detected in biodigestors and in an eutrophic lake (Briée *et al.*2007), but there were not more information about these works in the databases. Sequences belonging to clusters RC-V (Grosskopf *et al.* 1998) and LDS (Glissman 2004) are frequently encountered in freshwater environments (Herfort *et al.* 2009, Galand *et al.* 2006) and are distantly related to *Thermoplasma.* Although the sequences encountered in this work could not be related to LDS or RC-V clusters, they might represent another cluster of non-extremophile *Thermoplasma*. Yan *et al.* (2006) also observed the predominance of the *Thermoplasmatales* order (93.6% of the euryarchaeal clones analyzed) in mangrove sediment samples in China. Thus, the finding of sequences related to cultured extremophile species in non extreme environments reflects a lack of cultured species rather than an incongruence between the physiology and the phylogeny of the group. The detection of clones with such a low similarity to sequences occurring in databases also reveals the need for more studies involving archaea in the Amazonian environment.

Table 5. Diversity and richness estimators of clone libraries from Amazonian freshwater and São Paulo coastal brackish samples.

	Madeira River	Tucuruí reservoir			Santos and São Vicente Estuarine System		Bertioga and Cananeia mangroves	
Libraries	Sediment	Water (photic layer)	Water (aphotic layer)	Sediment	Sediment (site 4)	Sediment (site 9)	Archaeal 16S rDNA	mcrA
OTUs	69	35	33	64	55	63	50	42
Coverage (%)	28	57	60	46	ND	ND	ND	ND
Chao1	76.08	77	83	114	100.2	163.6	141.0	88.2
ACE	88.8	79	106	153	96.46	162.5	148.0	98.7

Finally, the absence of methanogenic groups in Madeira River library is also noteworthy, since methane production has been detected in the area and enrichment cultures showed methane production in relatively short incubation times (Araujo 2010). The low coverage value obtained for the Madeira River library (Table 5) indicates that much of the archaeal diversity in the site needs to be revealed. Further studies using different primers or larger libraries are required to better characterize the archaeal community and possibly assess methanogenic diversity in this site. Most knowledge on methanogenic metabolism is based in cultures obtained fromengineered environments, such as bioreactors used for wastewater treatment, temperate regions or environments considered to be extreme (Chaban *et al.* 2006 for a review). Tropical habitats have not yet been extensively studied (Saia *et al.* 2010), what turns them into a hot spot for the discovery of new methanogenic strains. Due the importance of the Amazonian wetland to biological methane emissions worldwide (Melack *et al.*

2004)attempts to culture methanogens are crucial to better understand their capabilities and their role in this environment.

4.2. Tucuruí Reservoir

The relative abundance of *Crenarchaeota* and *Euryarchaeota* sequences in Tucuruí reservoir varied with samples analyzed. Even though coverage estimates suggest that libraries did not explain all the diversity present in the reservoir (Table 5), it was possible to identify interesting patterns and observe clear differences between water layers and the sediment. In the photic layer, *Crenarchaeota* represented over 82% of the clone library (Figure 2).A profusion of *Crenarchaeota* in the oxic layer of Lake Michigan was reported by Macgregor *et al.* (1997). In Tucuruí reservoir, two *Crenarchaeota* groups were found: Marine Group I (73%) and Miscellaneous Crenarchaeotic Group (9%). The Marine Group I (MG-I) is a major component in marine picoplankton and is also present in subsurficial marine sediments (Teske and Sørensen 2008). The group has been related to oxidized poor sediment surface layers and some cultivated representatives, as *Nitrosopumilus maritimus* (Könneke *et al.* 2005) have been associated to autotrophic aerobic ammonia oxidation. Organic carbon assimilation has also been detected in the group, suggesting that autotrophy is not exclusive and that mixotrophic metabolism may occur (Teske and Sørensen 2008). The Miscellaneous Crenarchaeotic Group (MCG) is an ubiquitous group that has been detected in habitats ranging from terrestrial to marine environments (Teske and Sørensen 2008). It has been associated to organic rich or poor sediments, and carbon isotope studies suggested that the MCG may be a heterotrophic anaerobic group able to use complex organic compounds (Teske and Sørensen 2008). Detection of this group in oxygenated layers of the water column is intriguing, as it is also the presence of sequences related to strict anaerobic methanogenic *Archaea* belonging to the Order *Methanomicrobiales* (10% of clone library). One reason for that behavior may be the transport of cells due to sediment ressuspension and water column mixing processes. Although not detected in the study described in this chapter, total anoxia can develop in water column at Caraipé site at the limits of the euphotic zone (Espíndola *et al.* 2000), favouring the growth of anaerobic microorganisms that can be transported into the photic zone. The area is also highly productive, presenting the highest fish capture values and zooplankton abundance (Espíndola *et al.* 2000). Fecal pellets could then be another source of methanogenic cells, as detected in other works (Reeburgh 2007), since members of *Methanomicrobiales* family has already been detected inhabiting gastrointestinal tracts of invertebrates (Chaban *et al.* 2006). Members of *Methanomicrobiales* have also been detected in the surfaces of rice roots (Grosskopf *et al.* 1998) and some tolerance to oxic environments might be expected for this family. However, further studies would be necessary to determine if those cells are metabolically active or if there has been only a fortuitous transport of the cells, since they were abundant in the aphotic zone, as presented below. Another euryarchaeotal group found in the photic layer was the order *Halobacteriales* (8% of clone library). This order consists of a large group of chemoautotrophic or photosynthetic aerobic microbes that live and grow in hypersaline environments, but are not restrict to them (Elshaned *et al.* 2004). Sequences belonging to DHVEG-6 uncultured division were recovered in the clone library. This division is affiliated to *Halobacteriales* class and has been found in both marine and terrestrial environments (Takai and Horikoshi 1999, Takai *et al.* 2001), but its metabolism and physiology are still unknown.

In the aphotic layer, *Euryarchaeota* group was found to be significantly more present (90% prevalence) than in the photic layer (Figure 2). This phylum was represented by two orders: *Halobacteriales* (46%) and *Methanomicrobiales* (44%). Within the *Halobacteriales* order, 81% of sequences belonged to DHVEG-6 uncultured division and 16% of the sequences belonged to Miscellaneous Euryarchaeotic Group (MEG). The MEG is a phylogenetically diverse euryarchaeotic group found in several environments as geothermal springs (Kormas *et al.* 2009), microbial mats (Robertson *et al.* 2009) and waters from gold mines (Takai *et al.* 2001). Members of *Methanomicrobiales* are hydrogenotrophic methanogens, characterized to use H_2 and CO_2as substrates for methanogenesis, which suggests that this can be an important methane producing pathway taking place at this layer at C1 site. Although not measured in the study described in this chapter, the occurrence of anoxic conditions in the aphotic layer can be an explanation for the abundance of *Methanomicrobiales* members.

In the sediment, as observed in the aphotic layer, *Euryarchaeota* group was found to bepredominant (73% of the clone library). Also, a larger diversity of *Archaea* was observed in the sediment (Figure 2) with methanogens being the predominant group (40%). Main methanogenic groups found included *Methanosarcinales* (12%) and *Methanomicrobiales* (28%). Members of *Methanosarcinales* are characterized to be the most versatile group within methanogens as they use H_2, acetate, methanol and C1-compounds for methane production. The presence of sequences related to this Order suggests that not only hydrogenotrophic methanogens, but also acetoclastic and methylotrophic methanogens may be important to methane production at the sediment layer. The percentage of *Euryarchaeota* belonging to the *Thermoplasmatales* order was also considerable (19%). Although the cultured members of this order are both thermo- and acidophilic (Huber and Stetter 2007, for a review), a vast amount of uncultured groups are affiliated to this order and have been found in non-extreme environments, such as the TMEG (Terrestrial Miscellaneous Euryarchaeotic Group) (Takai *et al.* 2001) and SAGMEG (South African Gold Mine Euryarchaeotic Group). The TMEG have been found in soils and both marine and freshwater sediments (Teske and Sørensen 2008) and the SAGMEG division (Takai *et al.* 2001) was initially found in terrestrial environments but harbors marine representatives. In the sediment, sequences assigned to these groups were detected as well as sequences assigned to the groups DHVEG-1 and 6. DHVEG-1 and 6 are often detected in deep sea hydrothermal vents and black smoker chimney (Takai and Horikoshi 1999, Yakimov *et al.* 2007), hypersaline microbial mats (Robertson *et al.* 2009) and lakes (Elsaied andMaruyamal 2008), but their metabolism is not clear yet. Other sequences encountered in this work do not present enough similarity be assigned to these divisions, but might represent a group of uncultured non-extremophile *Thermoplasmatales* (ANT06-05,CCA47 and VC2.1 Arc6). Although the *Crenarchaeota* group was less abundant in sediment samples, crenarchaeotal diversity was higher if compared to water samples. Some of the Crenarchaeota groups detected included Miscellaneous Crenarchaeotic Group (MCG) and MBG-B (Marine Benthic Group B). The MCG, as discussed above, is an ubiquitous group that has been detected in habitats ranging from terrestrial to marine environments, associated to organic rich or poor sediments, and carbon isotope studies suggested that the MCG may be a heterotrophic anaerobic group able to use complex organic compounds (Teske and Sørensen 2008). The MBG-B has been associated to methane rich environments (Teske and Sørensen 2008, Wegener *et al.* 2008). The large diversity found in the sediment of Tucuruí reservoir sample (Figure 2) is probably related to the amount of available organic matter content from flooded vegetation and soil. In this sense, Lozupone and Knight (2007) reported that the sediment hosts a larger phylogenetic diversity than aquatic environments. Similarly, Torsvik *et al.* (2002) speculate that prokaryotic diversity may be influenced by the soil matrix complexity but also by physicochemical gradients. It is also speculated that freshwater environments harbour higher diversity due to its late colonization by the *Archaea* domain, especially in freshwater sediments where no indicator *Archaea* group could be established (Auguet *et al.* 2010). However, more work on tropical freshwater sediments is necessary to understand the ecological properties of *Archaea* in this environment.

Finally, it is also interesting to note the presence of sequences related to Miscellaneous Crenarchaeotic Group (MCG), in all layers, with higher abundance at the sediment. According to Biddle *et al.* (2006), the MCG *Archaea* are found globally distributed in both surface and subsurface environments, indicating a high ecophysiological flexibility of this group. To date, no cultivated MCG *Archaea* are available and almost no metabolic or physiological properties of this group of *Archaea* are known, except that it was suggested to have a heterotrophic lifestyle based on stable isotope analysis (Biddle *et al.* 2006).

4.3. Estuary inthe Southeast Coast

The site 4 clone library, constructed from hydrocarbon contaminated sediment collected at the brackish end of the estuary, revealed that most archaeal clones were closely related to known methanogenic *Archaea*(53%). The others were related to uncultured representatives of the crenarchaeotal class*Thermoprotei* (18%) and uncultured *Euryarchaeota* (29%)(Figure 2).

In contrast, the clone library of site 9, less contaminated and located at the marine end of the estuary, revealed *Thermoprotei* as the dominant group (48%), followed by non methanogenic uncultured *Euryarchaeota* (36%). Only 16% of the clones were related to known methanogenic groups (Figure 2). Although the sites differ in environmental parameters such as salinity and pH (Table 2), the most remarkable difference is the hydrocarbon concentrations found in both sediments (Table 3). Site 4 is highly contaminated with hydrocarbons of anthropogenic origin, while site 9 presents lower amounts of predominately biogenic hydrocarbons. The presence of higher amounts of organic compounds may lead to the rapid consumption of oxygen and impose an anaerobic metabolism to active microbes in this environment, creating a selective pressure that could favor methanogenic *Archaea*metabolic activity. The strict anaerobic conditions generated, as well as the availability of higher amounts of methanogenic substrates, reduce competition with sulphate reducing bacteria and may be responsible for the observed increase in the abundance of methanogenic *Archaea*in site 4 sediment.

The methanogenic groups found in the site 4 library revealed great functional diversity including hydrogenotrophic, acetoclastic, and methylotrophic groups (Table 6). The proximity of this site to diverse contamination sources may result in a varied range of compounds available for anaerobic degradation leading to the enrichment of the different methanogenic groups observed. In a highly polluted area of Guanabara Bay, Vieira *et al.*(2007) also recovered several phylotypes related to organisms from methanogenic environments. In site 4, a predominance of hydrogenotrophic taxa was observed (Table 6). This group was also the most abundant (31% of total clones analyzed, and 56% of methanogenic taxa), indicating that hydrogenotrophic methanogenesis may be the main process of methane formation at the contaminated brackish end of the estuary. Clones related to genus *Methanosaeta*, a highly specialized obligate acetate user(Table 6), accounted for 26.5% of methanogenic *Archaea* sequences found in site 4. Similar results were observed in sediments of different origins (Dojka *et al.*1998, Grosskopf *et al.* 1998, Von Wintzingerode *et al.* 1999, Purdy *et al.*2002, 2003), indicating that this group is an active member of the microbial community of sediments and are prime candidates for a globally important acetoclastic methanogenic group. The detection of these microorganisms as an important component of the microbial community in a site contaminated with hydrocarbons corroborates data obtained by other authors (Dojka *et al.* 1998, Ficker *et al.* 1999, Watanabe *et al.* 2002) and highlights the idea that acetoclastic methanogenesis can be the main methanogenic process associated with the degradation of hydrocarbons. The remaining methanogenic sequences were related to the C1 user *Methanolobus* (5%) and to the versatile *Methanosarcina* (9%), that can use hydrogen, acetate, methanol or C1-compunds for methane production (Garcia *et al.* 2000).

Table 6. Methanogenic groups and respective preferential methanogenic substrates in sediment samples of Santos and São Vicente Estuarine System.

Methanogenic Group	Substrate for methanogenesis[1]	Location
Order Methanosarcinales		
Methanosaeta	Acetate	Site 4
Methanosarcina	H_2+CO_2, acetate, methanol, methylamines	Site 4 and 9
Methanococcoides	Methanol, methylamines	Site 9
Methanolobus	Methanol, methylamines	Site 4 and 9
Order Methanobacteriales		
Methanobrevibacter	H_2+CO_2	Site 4 and 9
Methanobacterium	H_2+CO_2	Site 4
Methanosphaera	H_2+CO_2	Site 4
Order Methanomicrobiales		
Methanospirillum	H_2+CO_2	Site 4
Methanocalculus	H_2+CO_2	Site 4
Methanofollis	H_2+CO_2	Site 4
Order Methanococcales		
Methanococcus	H_2+CO_2	Site 9

[1]Madigan *et al.* (2003)

Another important euryarchaeotal group detected in site 4 library was the ANME-2. This group is known to be involved in anaerobic methane oxidation in association with sulphate reducing bacteria and it is believed that this is an important process regulating methane emissions from oceans to the atmosphere (Knittel and Boetius 2009). Anaerobic oxidation of methane is usually associated to high methane and sulphate concentrations, which suggests that methanogenic activity at site 4 is high. Other environmental conditions reported to be related to anaerobic methane oxidation include pH between 7 and 8 and strict anaerobiosis (Knittel and Boetius 2009), all occurring at site 4.

Uncultured archaeal groups in site 4 accounted for the remaining 47% of the clones analyzed. All *Crenarchaeota* clones (18%) belonged to uncultured groups in the *Thermoprotei* class. *Euryarchaeota* clones showed very low similarity indices to sequences in the database, suggesting that new groups may be present in the sediment at this site. A very different situation was encountered in the site 9 library, constructed from the less impacted sediment from the marine end of the Estuary. Most clones in this library (84%) were related to uncultured *Archaea*, and methanogenic groups were less abundant. This situation is most commonly found in environmental and uncontaminated samples. Clementino *et al.* (2007) observed uncultured environmental *Euryarchaeota* in a tropical lagoon and Vieira *et al.* (2007) also reported that the majority of clones in Guanabara Bay were related to uncultured archaeal species. In site 9, most uncultured clones were related to *Crenarchaeota*, belonging to the class *Thermoprotei*. According Robertson *et al.*(2005) non thermophilic *Crenarchaeota* is one of the most abundant and well distributed types of organisms on Earth. However, due to the difficulty of cultivating these organisms, there is a lack of information on their metabolism (Robertson *et al.* 2005). As for the *Euryarchaeota* related clones, crenarchaeotal sequences bore low similarity with database, indicating that new groups may be present at the site.

Methanogenic groups in site 9 were mostly represented by hydrogenotrophic groups (*Methanobrevibacter*, *Methanococcus* and *Methanomicrobiales*, summing 55% of methanogenic clones), followed by groups that use non competitive substrates for methanogenesis (*Methanolobus* and *Methanococcoides*, representing 35% of the group). *Methanosarcina* was also present and accounted for 10% of the methanogenic archaeal clones. Despite the low number of representatives of methanogenic *Archaea*, their distribution in site 9 library suggests that methylotrophic and hydrogenotrophic methanogenesis are the main processes at the marine end of the Estuary. No obligatory acetotrophic methanogens were detected in this site.

Among the obligatory C1-compounds users, representatives of *Methanolobus*, recovered also in site 4 libraryhave also been detected in different locations at estuarine systems, subjected to high and low sulphate concentrations(Muson *et al.* 1997, Purdy *et al.* 2002), suggesting a widespread distribution of these organisms along estuarine gradients. In contrast, clones related to *Methanococcoides*, recovered only in the site 9 library, have been reported in sediments at the marine end of estuaries (Muson *et al.* 1997, Purdy *et al.* 2002) and as active member of the archaeal community from coastal marine sediment (Purdy *et al.* 2003). These findings suggest that these methanogens may be important components of archaeal communities in marine sediments.

4.4. Mangrove in the Southeast Coast

In the clone library of Bertioga and Cananeia mangroves, most archaeal clones belonged to the *Crenarchaeota* (69.4% corresponding to 34 OTUs) whereas *Euryarchaeota* phyla represented 30.6% of clone the library, corresponding to 15 OTUs (Figure 2). The prevalence of *Crenarchaeota* sequences was also detected in a Chinese mangrove (Yan *et al.* 2006) but in distinct values, *Crenarchaeota* (80.4%) and *Euryarchaeota* (19.6%) and with distinct groups of microorganisms. Among the *Crenarchaeota*, 45% of the clones were related to Miscellaneous Crenarchaeotic Group (MCG), 14.3% to Marine Benthic Group (MBG-B),

6.1% to MBG-C, a sub-group of MCGand 4% to Marine Group I(MG I). Interesting to note that clones related to MCG and to MBG-B were also found in Tucuruí Reservoir, as discussed above. The MCG *Archaea* are globally distributed in terrestrial and marine, hot and cold, surface and subsurface environments, indicating a high ecophysiological flexibility (Teske and Sørensen 2008). Clones related to MBG-B are synonymous with the Deep-Sea Archaeal Group (DSAG) (Inagaki *et al.* 2003) but are not limited to deep sea marine sediments. Members of this group have been detected in coastal and intertidal sediments (Kim *et al.* 2005) in methane–sulfate transition zones and also anywhere in methanogenic, anaerobic sediments (Teske and Sørensen 2008). To date, no cultivated MCG *Archaea*and MBG-B are available and almost no metabolic or physiological properties of this group of *Archaea*are known, except that it was suggested to have a heterotrophic lifestyle based on stable isotope analysis (Biddle *et al.* 2006). The detection of MCG and MBG-B as an important component of the microbial community indicates the possible role of these groups in heterotrophic carbon degradation in mangrove sediments. Clones related to Marine Group I accounted for 4% of the clone library. MG-I *Archaea* were originally identified by sequencing of environmental 16S rRNA genes from seawater (DeLong 1992, Fuhrman *et al.* 1992) and further studiesdetected MG-I *Archaea* in the marine sedimentary subsurface (Sørensen *et al.* 2004). Some members of the MG-I *Archaea* are aerobic, autotrophic ammonia oxidizers, as indicated by functional gene surveys (Francis *et al.* 2005) and pure-culture study (Könneke *et al.* 2005). However, the ability to assimilate amino acids (Ouverney and Fuhrman 2000) shows that autotrophy is not obligate. Besides, carbon-isotopic composition of MG-I archaeal lipids suggests some degree of organic carbon assimilation, perhaps in a mixotrophic metabolism (Ingalls *et al.* 2006). The detection of MG I in our study suggest the possible role in nitrogen cycle in mangrove sediments.

For *Euryarchaeota*, the sequences were better affiliated with described genera of the methanogenic groups *Methanosarcinales* and *Methanospirillaceae* and such affiliation was the first indicative of the possible role of the archaeal community in the methanogenic process in mangrove sediments.Clones related to MBG-D were also found (Figure 2), corresponding to 16.3% of the clone library. This group is commonly abundant in marine environments and predominated in cold sediments from where *Methanosarcinales* and *Methanomicrobiales* strains were cultivated (Kendall *et al.* 2007). The co-occurrence of active methanogenic groups and MBG-D representatives has also been detected in estuarine sediments (Webster *et al.* 2010) and in a freshwater lake in both the water column and sediment (Jiang *et al.* 2008). There is not yet a cultured representative of the MBG-D but its metabolism seems to be related to methanogenic conditions. In a SIP study, the addition of ^{13}C-acetate led to the enrichment of a strain belonging to MBG-D from estuarine sediment as verified by the DGGE analysis (Webster *et al.* 2009). Furthermore, this group was prevalent in the *Euryarchaeota* community of a Tanzanian mangrove sediment (Lyimo *et al.* 2009) indicating that they can have an important role in these environments.An important characteristic of the mangrove sediments is the absence of oxygen at a few millimetres below the surface (Lyimo *et al.* 2002). As mentioned before, the lack of oxygen, coupled with the abundance of organic matter, creates an optimal environment for anaerobic organisms such as methanogenic *Archaea*(Dar *et al.* 2008). The presence of these organisms have been observed in previous studies based on 16S rRNA gene clone libraries (Yan *et al.* 2006, Lyimo *et al.* 2009) and functional approachesin mangroves in China (Yan *et al.* 2006), Tanzania (Lyimo *et al.* 2002) and Brazil(Taketani *et al.* 2010b). These studies have shown the presence of 16S rRNA gene belonging to *Euryarchaeota* and the *mcrA* gene, which encodes for the methyl coenzyme-M reductase that is involved in the methanogenesis.

On molecular based surveys of the archaeal community in mangrove soils, *Euryarchaeota* group was less prevalent, corresponding to 19.6% of total clones (Yan *et al.* 2006).Methanogens seems to be minor players in the Euryarchaeotal communities (Yan *et al.* 2006; Lyimo *et al.*2009) and prevalently shows methylotrophic metabolism (Lyimo *et al.* 2009). Similar result was obtained by Dias et al (*submitted*), who found that sequences belonging to the *Euryarchaeota* accounting for 29% of the archaeal clones observed in Brazilian mangrove. However, Mendes *et al.* (personal communication) observed a different community profile. In this study, the *Euryarchaeota* were dominant over the *Crenarchaeota* on shallow sediment depths (0-20cm). This later study indicates that *Euryarchaea* correspond to about a third of the total amount of *Archaea*found in mangrove sediments.

On the functional level, Taketani *et al.* (2010b) has observed that sediment depth selects for specific functional groups, corroborating the observation of Mendes *et al.* (*personal communication*). As discussed by the authors, this selectivity over methanogenic community by the sediment depth is commonly observed in marine environments (Wilms *et al.* 2006b, Wilms *et al.* 2006a, Parkes *et al.* 2007, Wilms *et al.* 2007, Bagwell *et al.* 2009). Taketani *et al.* (2010b) observed *mcr*A OTUs similar to the *mcr*A of *Methanopyrus kandlery* and *Methanothermococcus thermolithotrophicus.* The only exception was the OTU represented by the sequence M40E11-09, which clustered closer to the *Methanosarcina* cluster. Although these sequences shared similarities to the genera *Methanopyrus* and *Methanothermococcus*, these archaeal groups are described as hydrogenotrophic and thermophilic, what makes difficult its occurrence in mangrove sediments. However, considering the low amount of sequences of *mcrA* gene in databases obtained from isolates, and the low average similarity (0.25 phylogenetic distance) between the sequences, the presence of these sequences seen justified. In addition, the detection of genes indicative of methanogens and sulphate-reducers along the sediment profile reveal the co-occurrence of these two functional groups in mangrove sediments (Taketani *et al.* 2010b, Lyimo *et al.* 2002). This finding indicates that either the abundance of methanogenic substrates allows the presence of these organisms by reducing substrate competition(Oremland *et al.* 1982) or the organisms present in these sediments are methylotrophic methanogens (Wilms *et al.* 2007), which are capable of using substrates that are not used by sulphate-reducing bacteria(Madigan *et al.* 2003).

In previous studies, the existence of sulphate reducers and methanogens suggested the occurrence of anaerobic methane oxidation (ANME) carried out syntrophically by these two groups (Knittel and Boetius2009, Orphan *et al.* 2002). This process is expected to also take place in mangrove sediments, however none of these studies were able to detect sequences related to known archaeal ANME groups. Although molecular clues concerning the presence of ANME have been found in a few sediments (Thomsen *et al.* 2001, Goffredi *et al.* 2008,Harrison*et al.* 2009), many studies have failed to detect them (Parkes *et al.* 2005, Wilms *et al.* 2007). This difficulty in the direct detection of this group might explain the lack of ANME clones in all mangrove clone libraries published so far.

Conclusion

The overall observation is that mangroves, estuaries and freshwater ecosystems in Brazil bear niches for several groups of *Archaea*. The archaeal community in these ecosystems is abundant, even under varying environmental conditions, such as oxygen and salinity gradients, and contamination by hydrocarbons and other compounds of industrial and domestic origin. In general, sequences associated to *Crenarchaeota* phylum, mostly related to uncultured clones, were predominant. The exceptions were the highly contaminated site 4 in Santos and São Vicente Estuary and sediment and aphotic layer of the Tucuruí Reservoir. In Amazonian ecosystems (Madeira River and Tucuruí reservoir), the crenarchaeotal Marine Group-I, group 1.1b and *Thermoplasmata* were important groups, suggesting a possible role of archaea in the metabolism of nitrogen and in heterotrophic carbon degradation. However, relatively low coverage values and the presence of several clones bearing low similarity with GenBank sequences suggest that there is still much of the archaeal diversity to be unveiled. In Mangrove sediments, the detection of MCG, MBG-B and MBG-D as an important component of the microbial community indicates the possible role of these groups in carbon cycle.

Except for Madeira River,all studied ecosystems showed sequences related to known methanogenic *Archaea*, with higher abundance in impacted environments and/or where anaerobic conditions prevailed, such as in site 4 in Santos and São Vicente Estuary and at the sediment of Tucuruí Reservoir. The absence of methanogenic groups in Madeira River may be related to methodological constraints and research is needed to further characterize archaeal - and methanogenic - communities in this ecosystem. In the highly contaminated site 4, acetoclastic methanogens represented more than 25% of the clones of methanogens, indicating that this is one of the major methanogenic pathways. This behavior was not observed in site 9, less contaminated, where hydrogenotrophic and methylotrophic groups predominated. To our knowledge, culture-dependent studies on methanogenic metabolism, such as methanogenic activity from different substrates, have scarcely been carried applying samples from tropical natural environmentsand, therefore, attempts to culture methanogenic strains from unexplored environments might reveal new species and new metabolic pathways contributing to the better understanding of the function of methanogens to methane cycle. Site 4 also calls attention as the only area that showed the presence of the ANME-2 group, related to anaerobic methane oxidation, a process that usually requires high concentrations of methane and sulphate, strict anaerobiosis and slightly alkaline pH.

These findings contribute to the understanding of the ecophysiology of *Archaea*in saline and freshwater ecosystems in Brazil, and the influence of environmental conditions and changes on methanogenic and methanotrophic archaeal activity, with obvious consequences to methane emission rates to the atmosphere. They also reinforce the need for the application of second and third generation molecular tools in order to cover a wider range of diversity and for research to improve cultivation methods for archaeal groups so that we will be able to increase our knowledge on the physiological and ecological capabilities of the *Archaea* domain.

Acknowledgments

The study on Tucuruí Hydropower is part of the Research and Development Programme of the Electrical Energy Sector regulated by ANEEL and Centrais Elétricas do Norte do Brasil. The authors are grateful to the Conselho Nacional de Desenvolvimento Científico e Tecnológico (National Council for Scientific and Technological Development – Brazil) and the Consejo Superior de Investigaciones Cientificas (Spain).

The study on Madeira River was supported by PRONEX-Rondônia (grant SP_CV_013)

Estuary and mangrove studies were supported by grants from the State of São Paulo Research Foundation (FAPESP/BIOTA grants 98/05068-0, 2004/13910-6). The authors also would like to thank FAPESP for the fellowships: F.D. Andreote (postdoctoral 2007/56360-4), A.C.F. Dias (graduate 2008/54013-8), Fernanda Piza (doctoral 99/10739-4). R.G. Taketani (post-doctoralgrant from CNPq) also participated of the mangrove research. We also thank the support from CETESB and the Oceanographic Institute (IO, USP, São Paulo), especially Dr. Ricardo P. Menghini and João L. Silva for their support in mangrove and estuary expeditions.

References

Agência Nacional de Águas. Brasil: Ministério do Meio Ambiente. Available at: http://www.ana.gov.br[2010 Feb 20].

Araujo, ACV. *Diversidade Molecular de Arqueias em Sedimentos de Rios da Amazônia e Caracterização de Espécies Metanogênicas Cultivadas* [PhD Thesis]. São Paulo: University of São Paulo at São Paulo; 2010.

Auguet, JC; Barberan, A; Casamayor, EO. Global ecological patterns in uncultured Archaea. *ISME J,* 2010 4, 182-190.

Bagwell, CE; Formolo, M; Ye, Q; Yeager, CM; Lyons, TW; Zhang, CL. Direct analysis of sulfate reducing bacterial communities in gas hydrate-impacted marine sediments by PCR-DGGE. *J Basic Microbiol,* 2009 49 Suppl1, S87-92.

Bernardi, JVE; Lacerda, LD; Dórea, JG; Landim, PMB; Gomes, JPO; Almeida, R; et al. Aplicação da análise das componentes principais na ordenação dos parâmetros físico-químicos no alto Rio Madeira e afluentes, Amazônia Ocidental. *Geochim Bras,*2009 23(1), 79-90.

Betts, RA; Malhi, Y; Roberts, JT. The future of the Amazon: new perspectives from climate*Philos Trans R Soc Lond B Biol Sci,*2008 363(1498), 1729-35.

Biddle, JF; Lipp, JS; Lever, MA; Lloyd, KG; Sørensen, KB; Anderson, R; et al. Heterotrophic Archaea dominate sedimentary subsurface ecosystems off Peru. *P Natl Acad Sci U S A,* 2006 103(10), 3846-51.

Briée, C; Moreira, D; Lopez-Garcia, P. Archaeal and bacterial community composition of sediment and plankton from a suboxic freshwater pond. *Res Microbiol*, 2007 158, 213-227.

Brochier-Armanet C; Boussau B; Gribaldo S; Forterre P. Mesophilic Crenarchaeota: proposal for a third archaeal phylum, the Thaumarchaeota. *Nat Rev Microbiol, 2008 6, 245-252*

Cadillo-Quiroz, H; Yashiro, E; Yavitt, JB; Zinder, SH. Characterization of the archaeal community in a minerotrophic fen and terminal restriction fragment length polymorphism-directed isolation of a novel hydrogenotrophic methanogen. *Appl Environ Microb*, 2008 74(7), 2059-68.

Casamayor, EO; Massana, R; Benlloch, S; Øvreås, L; Díez, B; Goddard, VJ; et al. Changes in archaeal, bacterial and eukaryal assemblages along a salinity gradient by comparison of genetic fingerprinting methods in a multipond solar saltern. *Environ Microbiol*, 2002 4(6), 338-48.

Chaban, B; Ng, SY; Jarrell, KF. Archaeal habitats - from the extreme to the ordinary. *Can J Microbiol,* 2006 52, 73–116.

Clementino, MM; Fernandes, CC; Vieira, RP; Cardoso, AM; Polycarpo, CR; Martins, OB; et al. Archaeal diversity in naturally occurring and impacted environments from a tropical region. *J Appl Microbiol*, 2007 103(1), 141-51.

Clementino, MM; Vieira, RP; Cardoso, AM; Nascimento, AP; Silveira, CB; Riva, TC; et al. Prokaryotic diversity in one of the largest hypersaline coastal lagoons in the world. *Extremophiles*, 2008 12(4), 595-604.

Dar, SA; Kleerebezem, R; Stams, AJM; Kuenen, JG; Muyzer, G. Competition and coexistence of sulfate-reducing bacteria, acetogens and methanogens in a lab-scale anaerobic bioreactor as affected by changing substrate to sulfate ratio. *Appl Microbiol Biot*, 2008 78, 1045-1055.

DeLong, EF. Archaea in coastal marine environment*Proc Natl Acad Sci U S A*. 1992 89(12), 5685-9.

Deutsche Sammlung von Mikroorganismen und Zellkulturen. Available at: http://www.dsmz.de [2010 Jun 16].

Dhillon, A; Lever, M; Lloyd, KG; Albert, DB; Sogin, ML; Teske, A.Methanogen diversity evidenced by molecular characterization of methyl coenzyme M reductase A (*mcrA*) genes in hydrothermal sediments of the Guaymas Basin. *Appl Environ Microb*, 2005 71(8), 4592–4601.

Dojka, MA; Hugenholtz, P; Haack, SK; Pace, NR. Microbial diversity in a hydrocarbon- and chlorinated-solvent-contaminated aquifer undergoing intrinsic bioremediation. *Appl Environ Microb*, 1998 64(10), 3869–3877. Verificar se é 1998

Elsaied, HE; Maruyama, A. Biodiversity of archaea in Manzala lake, Egypt, based on 16S rRNA gene. *Egypt J Genet Cytol*, 2008 37, 57-72.

Espíndola, EL; Matsumura-Tundisi, T; Rietzler, AC; Tundisi, JG. Spatial heterogeneity of the Tucuruí Reservoir (State of Pará, Amazonia, Brazil) and the distribution of zooplanktonic species. *Rev Bras Biol,* 2000 60(2), 179-194.

Evangelista, RAO. *Estudo limnoquímico do reservatório da Usina HidrelétricaTucuruí.* [Masters Thesis]. Belém: Universidade Federal do Pará, Centro de Geociências at Belém; 1993.

Ficker, M; Krastel, K; Orlicky, S; Edwards, E. Molecular characterization of a toluene*Appl Environ Microbiol*, 1999 65(12), 5576-85.

Folk, RL; Ward, WC. Brazos River Bar: a study in the significance of grain size parameters. *J Sediment Petrol*, 1957 27(1), 3-26.

Francis, CA; Roberts, KJ; Beman, JM; Santoro, AE; Oakley, BB. Ubiquity and diversity of ammonia- oxidizing archaea in water columns and sediments of the ocean. *Proc Natl Acad Sci U S A*, 2005 102, 14683-88.

Fütterer, O; Angelov, A; Liesegang, H; Gottschalk, G; Schleper, C; Schepers, B; et al. Genome sequence of *Picrophilus torridus* and its implications for life around pH 0. *Proc Natl Acad Sci U S A*, 2004 101, 9091-96.

Fuhrman, JA; McCallum, K; Davis AA. Novel major archaebacterial group from marine plankton. *Nature,* 1992 356(6365), 148-9.

Galand, PE; Lovejoy, C; Vincent, WF.Remarkably diverse and contrasting archaeal communities in a large arctic river and the coastal arctic ocean. *Aquat Microb Ecol*, 2006 44, 115-26.

Garcia, JL; Patel, BKC; Ollivier, B. Taxonomic, phylogenetic and ecological diversity of methanogenic archaea. *Anaerobe*, 2000 6, 205-26.

Giles, J. Methane quashes green credentials of hydropower. *Nature* 2006 444(7119), 524-5.

Goffredi, SK; Wilpiszeski, R; Lee, R; Orphan, VJ. Temporal evolution of methane cycling and phylogenetic diversity of archaea in sediments from a deep-sea whale-fall in Monterey Canyon, California. *ISME J*, 2008 2, 204-20.

Grosskopf, R; Stubner, S; Liesack, W. Novel euryarchaeotal lineages detected on rice roots and in the anoxic bulk soil of flooded rice microcosms. *Appl Environ Microb*, 1998 64 (12), 4983-9.

Harrison, BK; Zhang, H; Berelson, W; Orphan, VJ. Variations in archaeal and bacterial diversity associated with the sulfate-methane transition zone in continental margin sediments (Santa Barbara Basin, California). *Appl Environ Microbiol*, 2009 75, 1487-99. Confirmer se tem ANME

Hatzenpichler, R; Lebedeva, EV; Spieck, E; Stoecker, K; Richter, A; Daims, H; Wagner, M.A moderately thermophilic ammonia-oxidizing crenarchaeote from a hot spring. *Proc Natl Acad Sci U S A*, 2008 105(6), 2134-9.

Herfort, L; Jung-Hyun, K; Coolen, MJL; Abbas, B; Schouten, S; Herndl, GJ; Damsté, JSS. Diversity of archaeaand detection of crenarchaeotal *amoA* genes in the Rivers Rhine and Têt. *Aquat Microb Ecol*, 2009 55, 189-201.

Hortellani, MA; Sarkis, JES; Bonetti, J; Bonetti, C.Evaluation of mercury contamination in sediments from Santos - São Vicente Estuarine System, São Paulo State, Brazil. *J Braz Chem Soc*, 200516, 1140-9.

Huber, H; Hohn, MJ; Rachel, R; Fuchs, T; Wimmer, VC; Stetter, KO. A new phylum of Archaea represented by a nanosized hyperthermophilic symbiont. *Nature*, 2002 417(6884), 63-7.

Huber, H; Stetter, K. Thermoplasmatales. In: Dworkin, M. editor. *The Prokaryotes.* New York: Springer; 2007; 101-112.

Inagaki, F; Suzuki, M; Takai, K; Oida, H; Sakamoto, T; Aoki, K; et al. Microbial communities associated with geological horizons in coastal subseafloor sediments from the Sea of Okhotsk. *Appl Environ Microbiol*, 2003 69(12), 7224-35.

Ingalls, AE; Shah, SR; Hansman, RL; Aluwihare, LI; Santos, GM; Druffel, ER; Pearson, A. Quantifying archaeal community autotrophy in the mesopelagic ocean using natural radiocarbon. *Proc Natl Acad Sci U S A*, 2006 103(17), 6442–7.

Jia, GD; Peng, PA. Temporal and spatial variations in signatures of sedimented organic matter in Lingding Bay (Pearl estuary), southern China. *Mar Chem*, 2003 82, 47-54.

Jiang, H; Dong, H; Yu, B; Ye, Q; Shen, J; Rowe, H; Zhang, C. Dominance of putative marine benthic Archaea in Qinghai Lake, North-Western China. *Environ Microbiol*, 2008 10(9), 2355-67.

Joly, CA; Rodrigues, RR; Metzger, JP; Haddad, CF; Verdade, LM; Oliveira, MC; Bolzani, VS. Ecology. Biodiversity conservation research, training, and policy in São Paulo. *Science*, 2010 328(5984), 1358-9.

Junk, WJ; Mello, JASN. Impactos ecológicos das represas hidrelétricas na bacia amazônica brasileira. *Estudos Avançados*, 1990 4, 126-43.

Kamaruzzaman, BY;Siti Waznah,A; Ong, MC; Shahbudin, S; Jalal, KCA. Variability of organic carbon content in bottom sediment of Pahang River Estuary, Pahang, Malaysia. *J Applied Sciences*, 2009 9(24), 4253-7.

Kendall, MM; Wardlaw, GD; Tang, CF; Bonin, AS; Liu, Y; Valentine, DL. Diversity of Archaeain marine sediments from Skan Bay, Alaska, including cultivated methanogens, and description of *Methanogenium boonei* sp. nov. *Appl Environ Microb*, 2007 73(2), 407–14.

Kim, BS; Oh, HM; Kang, H; Chun, J. Archaeal diversity in tidal flat sediment as revealed by 16S rDNA analysis. *J Microbiol*, 2005 43(2), 144-51.

Knittel, K; Boetius, A. Anaerobic oxidation of methane: progress with an unknown process. *Annu Rev Microbiol*, 2009 63, 311-34.

Knittel, K; Lösekann, T; Boetius, A; Kort, R; Amann, R. Diversity and distribution of methanotrophic archaea at cold seeps. *Appl Environ Microb*, 2005 71(1), 467–79.

Könneke, M; Bernhard, AE; de la Torre, JR; Walker, CB; Waterbury, JB; Stahl, DA. Isolation of an autotrophic ammonia-oxidizing marine archaeon. *Nature*, 2005 437(7058), 543-6.

Kormas, KA; Tamaki, H; Hanada, S; Kamagata, Y. Apparent richness and community composition of Bacteria and Archaea in geothermal springs. *Aquat Microb Ecol*, 2009 57, 113-22.

Kudo, Y; Shibata, S; Miyaki, T; Aono, T; Oyaizu, H. Peculiar archaea found in Japanese paddy soils. *Biosci Biotechnol Biochem,* 1997 61(5), 917-20.

Lozupone, CA; Knight, R. Global patterns in bacterial diversity. *Proc Natl Acad Sci U S A*, 2007 104(27), 11436-40.

Luton, PE; Wayne, JM; Sharp, RJ; Riley, PW. The *mcrA* gene as an alternative to 16S rRNA in the phylogenetic analysis of methanogen populations in landfill. *Microbiology*. 2002 148(Pt 11), 3521-30.

Lyimo, TJ; Pol, A; Jetten, MS; den Camp, HJ. Diversity of methanogenic archaea in a mangrove sediment and isolation of a new *Methanococcoides* strain. *FEMS Microbiol Lett*, 2009 291, 247-53.

Lyimo, TJ; Pol, A; Op den Camp, HJ. Sulfate reduction and methanogenesis in sediments of Mtoni mangrove forest, Tanzania. *Ambio*, 2002 31, 614-16.

MacGregor, BJ; Moser, DP; Alm, EW; Nealson, KH; Stahl, DA. Crenarchaeota in Lake Michigan sediment. *Microbiology*, 1997 63(3), 1178-81.

Madigan, MT; Martinko, JM; Parker, J. *Brock Biology of Microorganisms*. 9th edition. Saddle River, NJ: Pearson Education Inc; 2003.

Melack, JM; Hess, LL; Gastil, M; Forsberg, BR; Hamilton, SK; Lima, IBT; Novo, EMLM. Regionalization of methane emissions in the Amazon Basin with microwave remote sensing. *Global Change Biol*, 2004 10, 530-44.

Munson, MA; Nedwell, DB; Embley, TM. Phylogenetic diversity of Archaea in sediment samples from a coastal salt marsh. *Appl Environ Microbiol,* 1997 63(12), 4729-33.

Nelson, KA; Moin, NS; Bernhard, AE.Archaeal diversity and the prevalence of *Crenarchaeota* in salt marsh sediments. *Appl Environ Microbiol*, 2009 75(12), 4211-15.

Oremland, RS; Marsh, LM; Polcin, S. Methane production and simultaneous sulfate reduction in anoxic, salt-marsh sediments. *Nature*, 1982 296, 143-145.

Orphan, VJ; House, CH; Hinrichs, KU; McKeegan, KD; DeLong, EF. Multiple archaeal groups mediate methane oxidation in anoxic cold seep sediments. *Proc Natl Acad Sci U S A*, 2002 99, 7663-8.

Ouverney, CC; Fuhrman, JA. Marine planktonic archaea take up amino acids. *Appl Environ Microbiol*. 2000 66(11), 4829-33.

Øvreas, L; Forney, L; Daae, FL; Torsvik, V. Distribution of bacterioplankton in meromictic Lake Saelenvannet, as determined by denaturing gradient gel electrophoresis of PCR-amplified gene fragments coding for 16S rRNA. *Appl Environ Microbiol*, 1997 63, 3367–73.

Parkes, RJ; Cragg, BA; Banning, N; Brock, F; Webster, G; Fry, JC; et al. Biogeochemistry and biodiversity of methane cycling in subsurface marine sediments (Skagerrak, Denmark). *Environ Microbiol*, 2007 9, 1146-61.

Parkes, RJ; Webster, G; Cragg, BA; Weightman, AJ; Newberry, CJ; Ferdelman, TG; et al. Deep sub-seafloor prokaryotes stimulated at interfaces over geological time. *Nature*, 2005 436, 390-4.

Patel, N; Mounier, S; Guyot, JL; Benamou, C; Benaim, JY. Fluxes of dissolved and colloidal organic carbon along the Purus and Amazonas rivers, Brazil.*Sci Total Environ*, 1999 229, 53-64.

Piza, F. *Ecologia molecular microbiana associada a sedimentos do Estuário de Santos-São Vicente (SP,Brasil)*. [PhD Thesis]. Campinas: Universidade Estadual de Campinas; 2004.

Plasencia, A; Bañeras, L; Llirós, M; Casamayor, EO; Borrego, C. *Maintenance of previously uncultured freshwater archaea from anoxic waters under laboratory conditions*. A Van Leeuw, 2010. In press.

Purdy, KJ; Munson, MA; Nedwell, DB; Embley, TM. Comparison of the molecular diversity*FEMS Microbiol Ecol*, 2002 39(1), 17-21.

Purdy, KJ; Nedwell, DB; Embley, TM. Analysis of the sulfate-reducing bacterial and methanogenic archaeal populations in contrasting Antarctic sediments*Appl Environ Microbiol*, 2003 69(6), 3181-91.

Quaiser, A; Ochsenreiter, T; Klenk, HP; Kletzin, A; Treusch, AH; Meurer, G; et al. First insight into the genome of an uncultivated crenarchaeote from soil. *Environ Microbiol*, 2002 10, 603-611.

Rastogi, G; Sani, RK; Peyton, BM; Moberly, JG; Ginn, TR. Molecular studies on the microbial diversity associated with mining-impacted Couer d'Alene river sediments. *Microb Ecol*, 2009 58, 129-39.

Reeburgh, WS. Oceanic methane biogeochemistry. *Chem Rev,* 2007 107(2), 486-513.

Robertson, CE; Harris, JK; Spear, JR; Pace, NR. Phylogenetic diversity and ecology of environmental Archaea. *Curr Opin Microbiol*, 2005 8, 638–642.

Robertson, CE; Spear, JR; Harris, K; Pace, NR.Diversity and stratification of archaea in a hypersaline microbial mat. *Appl Environ Microb*, 2009 75(7), 1801-10.

Rodrigues, DF; Sakata, SK; Comasseto, JV; Bícego, MC; Pellizari, VH. Diversity of hydrocarbon-degrading *Klebsiella* strains isolated from hydrocarbon-contaminated estuaries. *JAppl Microbiol*, 2009 106, 1304-14.

Shepard, F.P. Nomenclature based on sand-silt-clay ratios. *J Sediment Petrol*, 1954 24, 151-8.

Saia, F; Domingues, M; Pellizari, VH; Vazoller R. Occurrence of Methanogenic *Archaea* in Highly Polluted Sediments of Tropical Santos–São Vicente Estuary (São Paulo, Brazil). Current Microbiology (Print), 2010, 60, 66-70.

Silva, A: Graças, DA; Miranda, PR; Baraúna, RA; McCulloch, JA;Ghilardi Jr. R; Lorenzo, V; Schneider, MPC. Metagenomic analysis of anoxic zone from hydropower plant reservoir in Brazilian Amazon. *unpublished*

Simon, HM; Dodsworth, JA; Goodman, RM.Crenarchaeota colonize terrestrial plant roots. *Environ Microbiol*, 2000 2(5), 495-505.

Sliwinski, MK; Goodman, RM. Spatial heterogeneity of crenarchaeal assemblages within mesophilic soil ecosystems as revealed by PCR–single-stranded conformation polymorphism profiling. *Appl Environ Microb*, 2004 70(3), 1811-20.

Sørensen, KB; Canfield, DE; Teske, AP; Oren, A. Community composition of a hypersaline endoevaporitic microbial mat. *Appl Environ Microb*, 2005 71(11), 7352-65.

Sørensen, KB; Lauer, A; Teske, A. Archaeal phylo-types in a metal-rich, low-activity deep subsurface sediment of the Peru Basin, ODP Leg 201, Site 1231. *Geobiology*, 2004 2, 151-61.

Stein, JL; Marsh, TL; Wu, KY; Shizuya, H; DeLong, EF. Characterization of uncultivated prokaryotes: isolation and analysis of a 40-kilobase-pair genome fragment from a planktonic marine archaeon. *J Bacteriol*, 1996 178(3), 591-9.

Takai, K;Moser, DP; DeFlaun, M; Onstott, TC; Fredrickson, JK. Archaeal diversity in waters form deep South African gold mines. *Appl Environ Microb*, 2001 67(12), 5750-60.

Takai, K; Horikoshi, K. Molecular phylogenetic analysis of archaeal intron-containing genes coding for rRNA obtained from a deep-subsurface geothermal water pool. *Appl Environ Microb*, 1999 65(12), p. 5586-89.

Taketani, RG; Franco, NO; Rosado, AS; van Elsas, JD. Microbial community response to a simulated hydrocarbon spill in mangrove sediments. *J Microbiol*, 2010a 48(1), 7-15.

Taketani, RG; Yoshiura, CA; Dias, ACF; Andreote, FD; Tsai, SM. Diversity and identification of methanogenic archaea and sulphate-reducing bacteria in sediments from a pristine tropical mangrove. *Anton Leeuw Int J G*, 2010b. In press.

Teske, A; Sørensen, KB. Uncultured Archaea in deep marine subsurface sediments: have we caught them all? *ISME J*, 2008 2, 3-18.

Thomsen, TR; Finster, K; Ramsing, NB. Biogeochemical and molecular signatures of anaerobic methane oxidation in a marine sediment. *Appl Environ Microbiol*, 2001 67, 1646-56.

Torsvik, V; Øvreås, L; Thingstad, TF. Prokaryotic diversity - magnitude, dynamics, and controlling factors. *Science*, 2002 296(5570), 1064-6.

Vetriani, C; Jannasch, HW; MacGregor, BJ; Stahl, DA; Reysenbach, A-L. Population structure and phylogenetic characterization of marine benthic archaea in deep-sea sediments. *Appl Environ Microb*, 1999 65(10), 4375-84.

Vieira, RP; Clementino, MM; Cardoso, AM; Oliveira, DN; Albano, RM; Gonzalez, AM; et al. Archaeal communities in a tropical estuarine ecosystem: Guanabara Bay, Brazil. *Microb Ecol*, 2007 54(3), 460-8.

Vissers, EW; Bodelier, PL; Muyzer, G; Laanbroek, HJ. A nested PCR approach for improved recovery of archaeal 16S rRNA gene fragments from freshwater samples. *FEMS Microbiol Lett*, 2009 298(2), 193-8.

von Wintzingerode, F; Selent, B; Hegemann, W; Göbel, UB. Phylogenetic analysis of an anaerobic, trichlorobenzene-transforming microbial consortium. *Appl Environ Microbiol.* 1999 65(1), 283-6.

Watanabe, K; Kodama, Y; Hamamura, N; Kadu, N. Diversity abundance and activity archaeal population in oil-contaminated graundwater accumulated at the botton of an underground crude oil storage cavity.*Appl Environ Microbiol*, 2002 62, 4299-4301.

Webster, G; Rinna, J; Roussel, EG; Fry, JC; Weightman, AJ; Parkes, RJ. Prokaryotic functional diversity in different biogeochemical depth zones in tidal sediments of the Severn Estuary, UK, revealed by stable-isotope probing. *FEMS Microbiol Ecol*, 2010 72(2), 179-97.

Wegener, G; Niemann, H; Elvert, M; Hinrichs, KU; Boetius A. Assimilation of methane and inorganic carbon by microbial communities mediating the anaerobic oxidation of methane. *Environ Microbiol*,2008 10(9), 2287-98.

Wilms, R; Sass, H; Kopke, B; Koster, J; Cypionka, H; Engelen, B. Specific bacterial, archaeal, and eukaryotic communities in tidal-flat sediments along a vertical profile of several meters. *Appl Environ Microbiol*, 2006a 72, 2756-2764.

Wilms, R; Kopke, B; Sass, H; Chang, TS; Cypionka, H; Engelen, B. Deep biosphere-related bacteria within the subsurface of tidal flat sediments. *Environ Microbiol*, 2006b 8, 709-719.

Wilms, R; Sass, H; Kopke, B; Cypionka, H; Engelen, B. Methane and sulfate profiles within the subsurface of a tidal flat are reflected by the distribution of sulfate-reducing bacteria and methanogenic archaea. *FEMS Microbiol Ecol*, 2007 59, 611-621.

Woese, CR; Kandler, O; Wheelis, ML.Towards a natural system of organisms: Proposal of the domains Archaea, Bacteria, and Eucarya. *Proc. Natl. Acad. Sci.* USA, 1990 87, 4576-4579.

Woese, CR; Fox, GE. Phylogenetic structure of the prokaryotic domain: the primary kingdoms. *Proc Natl Acad Sci U S A*, 1977 74(11), 5088-90.

Yakimov, MM; Giuliano, L; Cappello, S; Denaro, R; Golyshin, PN. Microbial community of a hydrothermal mud vent underneath the deep-sea anoxic brine lake Urania (eastern Mediterranean). *Orig Life Evol Biosph*, 2007 37(2), 177-88.

Yan, B; Kui, H; Zi-Niu, Y.Archaeal communities in mangrove soil characterized by 16S rRNA gene clones.*J Microbiol*, 2006 44(5), 566-71.

In: Archaea: Structure, Habitats and Ecological...
Editor: Sakura Y. Katō, pp. 121-141
ISBN 978-1-61761-932-8

Chapter 5

The Archaea Ribonuclease P: Molecular Mechanisms of Thermal Adaptation

Francisco Miralles*
Département de Génétique et Développement,
INSERM U567-CNRS UMR8104, Institut Cochin, 24,
rue du Faubourg Saint-Jacques, 75014 PARIS, France

Abstract

Temperature is an essential and variable environmental factor, which impacts the phenotype, habitat distribution and evolution of organisms. The pervasive influence of temperature on organisms requieres an ensemble of compensatory adaptations that span all levels of biological organization, from behaviour to fine-scale molecular structure. The Archaea stand out as the only group of organisms that have species capable of growing at temperatures ranging from 0°C to 110°C. Thus, the recent availability of complete genome sequences for a large number of archaea species provides a powerful tool to analyze the mechanisms of thermal adaptation at the molecular level. In this study we have used comparative genomics to analyze thermal adaptation in the universally conserved ribonucleoprotein: Ribonuclease P (RNase P). We have analyzed compositional and structural features of both, the RNA and protein components of the archaeal RNase P, to identify molecular trends related to thermal adaptation. Our results show a robust correlation between the stems GC content of the RNase P RNA and the optimal growth temperature (OGT) of the archaea species. Moreover, we have detected significant differencesin the composition and distribution of several amino acids in the RNase P protein components of hyperthermophilic and mesophilic archaea. Thus, this study reveals that natural selection has adapted the RNase P holoenzyme to the OGT of the archaea species by modifying both, its RNA and proteins subunits.

Keywords: Archaea, RNase P, secondary structure, GC content, amino acid composition, thermal adaptation.

---

* Dépt. de Génétique et Développement, Institut Cochin, 24, rue du Faubourg Saint-Jacques, 75014 PARIS, France, Tel:01 44 41 23 08, Fax:01 44 41 23 02, E-mail: miralles.francisco@orange.fr

Introduction

Since the origins of life on earth, environmental parameters have driven the evolution of organisms. Among these, variation in environmental temperature appears to be particularly important. Because of its pervasiveness, temperature penetrates physical barriers and thus can have dramatic effects on the organisms. As a consequence, adaptative responses to temperature are observed at all levels of biological organization. Thus, "variations in environmental temperature can explain the evolution of biological phenomena as diverse as the behaviour of animals or the density of hydrogen bonds in nucleic acids" [1]. The recent availability of large number of genomes from organisms that occupy different ecological niches provides new ways for the analysis of thermal adaptation processes at the molecular level. Among the tree kingdoms of life, the Archaea characterize by the greatest diversity of optimal growth temperatures (OGT). The Archaea kingdom contain species embracing the whole spectrum of growing temperatures: psicrophiles (OGT < 25°C), mesophiles (OGT = 25°C-40°C), thermophiles (OGT = 40°C-75°C) and hyperthermophiles (OGT > 75°C). Therefore, the archaea species appear particularly well suited for the study of thermal adaptation at the molecular level. Several studies have addressed the question of the thermal adaptation at the transcriptomic and proteomic level. Transcriptomic studies analyzing thermal adaptation have focused on structural RNAs. These studies have shown that the GC content of structural RNAs correlates positively with the optimal growth temperature (OGT) in prokaryotes [2-5]. Moreover, these studies have established that the increase in GC content concentrates in the double-stranded stem regions of the molecule. Compared to AT pairs, GC pairs are more thermostable because they have an additional hydrogen bond. Thus, it is likely that the increase in the GC content of the stems of a nucleic acid will help to strength the stability of its secondary structures. This has led to the proposal of the thermal adaptation hypothesis. This hypothesis suggests that the increase in the GC content on the stem regions of the structural RNAs results from natural selection acting to favour the thermostability of the molecule. The studies analyzing the thermal adaptation at the proteome level have shown that the major effect of thermophily at the proteomic level is a significant reduction in the frequency of thermolabile amino acids histidine, glutamine and threonine [6-8]. The concomitant increase among thermophiles and hyperthermophilic species of both positively charged residues (arginine and lysine) and negatively charged residues (glutamic acid) suggests that ionic bonds between oppositely charged proteins may help to stabilize multimeric proteins at higher temperatures [1, 9].

To the date the studies supporting the thermal adaptation hypothesis in structural RNAs have mainly focused into ribosomal and transfer RNAs, rRNA and tRNA. However, the current availability of large genomic datasets offers the possibility to test more soundly this hypothesis by extending the study of thermal adaptation to other structural RNAs. Therefore, in the present study we addressed the question of the relationship between GC content and OGT in a different and universally conserved structural RNA, the RNA component of the archaeal RNase P ribonucleoprotein. Moreover, to have an integrated view of the thermal adaptation of the RNase P holoenzyme we have completed our study by analyzing the putative thermal adaptations of the protein components of this complex ribonucleoprotein.

The Ribonuclease P (RNase P) is an essential endoribonuclease, universally conserved throughout evolution, which is responsible for the maturation of the 5′ end of pre-tRNA. The

RNase P processes the 5′ end of pre-tRNAs, precisely removing the 5′ leader sequence (50 nucleotides long), resulting in a 5′ phosphate on the mature tRNA and a 3′ hydroxyl on the leader sequence. However, the RNase P is also able to cleave other substrates, including some viral and phage RNAs [10, 11], pre-4.5S RNA [12], pre-tmRNA [13], a few mRNAs [14] and riboswitches [15]. The enzymatic activity of the RNase P is dependent upon divalent cations, preferably magnesium, which are likely to fulfill both structural and functional roles.

In the three kingdoms of life (Bacteria, Archaea and Eukarya) the RNAse P holoenzyme is composed of both an essential RNA and a variable number of protein subunits.The bacterial RNase P holoenzyme is composed of a large RNA (typically 350–400 nt) and a single small protein (15 kDa). Both the RNA and protein components are essential *in vivo*, however the bacterial RNA has been shown to function independently *in vitro* and can accurately and efficiently catalyze pre-tRNA cleavage under high salt conditions [16]. Thus, the bacterial RNA is considered as a true ribozyme. Phylogenetic comparative analysis has provided most of the secondary structure data on the RNA and has led to the definition of two major subsets, the type A (ancestral) and type B (bacillus). The bacterial RNase P RNAs can be divided in two independently folding domains involved in both substrate cleavage, the C-domain (catalytic) and the substrate binding, specificity domain (S-domain). Sequence analysis has identified five conserved regions within the structures that are found in all RNase P RNAs, CRI-CRV [17].The archaeal and eukaryal RNAse P enzymes have considerably more complex subunit compositions, retaining a structurally related RNA (300–350 nt), but with an increased number of protein subunits compared to the bacterial enzymes. At least four proteins have been found in archaeal enzymes, with more found in the eukaryotic enzymes: nine proteins in yeast and ten proteins in human [18].

The archaeal RNase P holenzyme is composed of an RNA (300-400 nucleotides) and at least four proteins: Rpp21, Rpp29, Rpp30, and Rpp14 (Pop5). Unlike bacterial RNase P RNAs, the archaeal RNAs are not generally catalytically active *in vitro* in the absence of proteins. However, a small number of archaeal RNase P RNAs have shown trace amounts of catalytic activity*in vitro*, under conditions of extreme ionic strength [19]. The ability of these RNAs to display even trace activity indicates that they possess all of the structures necessary for substrate recognition and catalysis. The secondary structures of archaeal RNase P RNAs (Figure 1) conforms into at least two classes, those displaying similarity to the bacterial RNAs, type A, and others with an apparently less complex structure, type M [20, 21]. The archaeal type A RNAs lacks the P13/14 stems and the P18 stem found in the bacterial type A RNAs, and many show extensions and additional complexity in their P12 stem. The type M archaeal RNAs appear to be more diverged from the bacterial RNAs, also lacking the P8 helix from the cruciform region and the extended P15 region, including the pre-tRNA 5'-CCA recognition loop and the P6 pseudoknot. This structural divergence is also reflected in the physical properties of these RNAs. Those archaeal RNAs that display trace activity in the absence of proteins are all found to possess an archaeal type A structure. The type M RNAs do not show catalytic activity by themselves, and thus appear to be more dependent on their proteins to achieve catalysis.

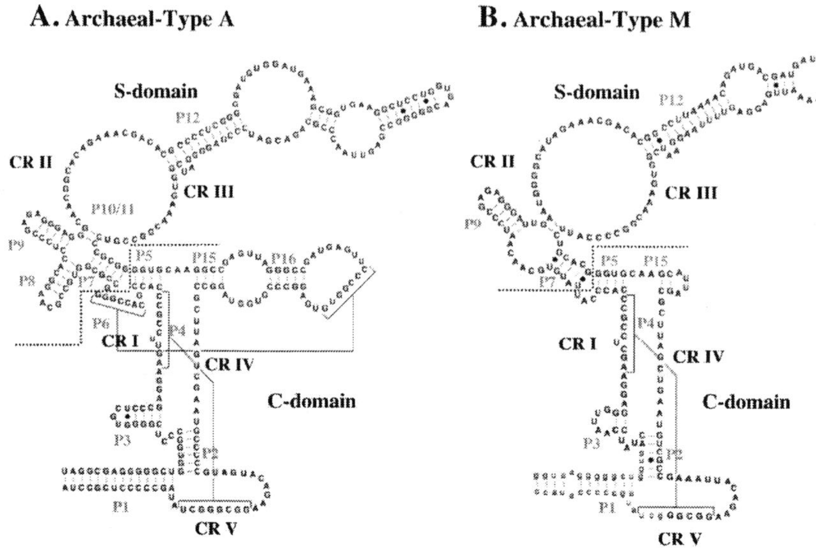

Figure 1. Archaea RNase-P-RNA secondary structures. The diagram shows the secondary structures of characteristic type A and type M archaeal RNase P RNAs as obtained by comparative sequence analysis (25). (A) Secondary structure of Pyrococcus horikoshii, Type A RNase P RNA. (B) Secondary structure of Methanococcus jannaschii, Type M RNase P RNA. The dots line separates the specification domain involved in substrate binding (S-domain) and the catalytic domain (C-domain). The helices are numbered as paired regions (P). The pseudoknots are represented by lines joining the nucleotides involved in the tertiary interaction. The five universally conserved regions are labelled as CRI to CRV.

Experiments of *in vitro* reconstitution, using recombinant proteins, have shown that the four archaeal Rpps function as two binary complexes (Rpp21-Rpp29 and Rpp14-Rpp30). Although each binary complex participates in the RNase P catalysis, kinetic analysis suggests that they fullfil different roles: Rpp21-Rpp29 increases affinity of the RNase P for the pre-tRNA substrate (18), while Rpp30-Rpp14 enhances the rate of chemical cleavage [22]. Recently, it has been proposed that the archaeal putative ribosomal protein L7Ae (homologous to human Rpp38) could be a fifth protein component of the archaeal RNase P. *In vitro* experiments have shown that this protein binds near the catalytic center and induces a conformational change which elevates the optimun temperature for the catalytic activity of the *Pyrococcus Horikoshii* RNase P from 50°C to 70°C [23]. However further studies will be need to determine if this protein is a real component of the archaeal RNase P holoenzyme.

The presence of the RNase P in all living organisms and its crucial role in the processing of the pre-tRNAs, suggests that the preservation of the functional and structural features of this ribonucleoprotein could be essential. Therefore, we reasoned that in face of a variable environmental parameter such as temperature, natural selection would have acted on the RNA and protein components of this molecule to preserve the stability of its secondary and tertiary structures. Thus, the RNAse P holoenzyme appeared as a good candidate to test the thermal adaptation hypothesis at both the RNA and protein level. Since the broadest range of growing temperatures is found among the Archaea we focused our study on the RNase P of these organisms. To this end, we generated the secondary structures of the RNase P RNAs of hyperthermophilic, thermophilic and mesophilic archaea species, and analyzed the GC content as a function of their OGT. Our study shows that there is a clear correlation between the GC content in the stems of the RNase P RNA and the OGT of the archaea species. This correlation is particularly strong in the S-domain of the molecule. The analysis of the four protein components of the RNase P holoenzyme suggests that the hyperthermophilic and thermophilic species have adapted these proteins to their growing temperatures by increasing the proportion of the amino acids Lys, Leu, Arg, Val and Trp. By analyzing the structures of these proteins we have observed that in the hyperthermophilic and thermophilic species, the proportion of charged basic amino acids is clearly increased in the α-helices, while the proportion of tiny and small amino acids is decreased, in particular in coiled regions. Thus, our study reveals that on the course of evolution, natural selection has adapted the RNase P holoenzyme to the OGT of the archaea species by modifying both its RNA and protein components.

Methods

Optimal Growth Temperature

Optimal growth temperatures for the archaea species analyzed in this study were retrieved from the Prokaryotic Growth Temperature Database, http://pgtdb.csie.ncu.edu.tw [24] and from the German Collection of Microorganisms and Cell Cultures, http://www.dsmz.de. The optimal growth temperature data were carefully checked in order to ensure that they referred unambiguously to the optimal growth temperature rather than to a permissive range. Some species are listed in these databases with slightly different optimal growth temperatures, in such cases we calculated and used for our study the average of the optimal growth temperatures reported.

Sequence Data

The sequences of the RNase P RNA component were downloaded from the RNase P database, http://www.mbio.ncsu.edu/RNaseP/threed.html (25, 26), from the Integrated Microbial Genomes database, http://img.jgi.doe. gov/cgi-bin/pub/main.cgi [27] and from the NCBI database, http://www.ncbi. nlm.nih.gov. A total of 65 archaea RNase P RNA sequences were retrieved. For each RNase P RNA sequence, the global GC content was calculated. In addition, we analyzed the secondary structure of these RNAs and calculated the GC content for the stem and loop regions. We also calculated the GC content on the pseudoknots.

RNA Secondary Structure Analysis

The secondary structures of the archaeal RNase P RNAs analyzed in this study were generated using the RNAalifold and RNAfold software's from the Vienna RNA package, http://rna.tbi.univie.ac.at [28, 29]. The RNAalifold software predicts the consensus structure of a set of aligned RNA sequences. It extends standard dynamic programming algorithms for RNA secondary structure prediction by averaging the energy contributions over all sequences and incorporating covariation terms into the energy model to reward compensatory mutations and penalize non-compatible base pairs. The output is a consensus structure for the aligned sequences. Then, this consensus structure is used as a constraint to fold each individual sequence using RNAfold. The RNase P RNA sequences were aligned with CLUSTAL W. The pseudoknots in the secondary structures of the archaeal RNase P were predicted by comparative sequence analysis and with the help of the software DotKnot, http://dotknot.csse.uwa.edu.au [30].

Proteins Composition Analysis

The sequences of the protein components (Rpp30, Rpp29, Rpp21 and Rpp14) of the archaeal RNase P holoenzyme were retrieved from the Integrated Microbial Genomes database, http://img.jgi.doe.gov/cgi-bin/pub/main.cgi [31] and from the NCBI database, http://www.ncbi.nlm.nih.gov. The amino acid composition of these sequences was analyzed using the ProtParam software, http://expasy.org/ tools/protparam.html. In addition, we calculated the theoretical isoelectric point (iP) of each sequence, as well as the percentage ratios of: tiny (A+C+G+S+T), small (A+B+C+D+G+N+P+S+T+V), aliphatic (A+I+L+V), aromatic (F+H+W+Y), polar (D+E+H+K+N+Q+R+S+T+Z), non-polar (A+C+F+G+I+-L+M+P+V+W+Y), charged (B+D+E+H+K+R+Z), hydrophilic (D, E, K, N, Q, and R), hydrophobic (A, C, F, I, L, M, F, P, V, W and Y), basic (H+K+R) and acid amino acids(B+D+E+Z). The sequences of the different Rpps were aligned using the EXPRESSO (3DCoffee) software, http://tcoffee.org. This software computes structure based multiple alignments. The PDB files used as guides for the alignments were: 2AV5 (Rpp14), 2K3R (Rpp21), 1X0T (Rpp21), 2K17 (Rpp29-Rpp21), and 1V77 (Rpp30): The secondary structure elements of the protein components of the archaeal RNase P were predicted using the EXPRESSO alignments and the protein secondary structure determination software Jpred3, http://www.compbio.dundee.ac.uk/ www-jpred. We used these predictions to analyze the distribution of amino acids in three major secondary structural regions: helix (H), beta-strands (B), and coiled (C).

Statistical Analyses

The statistical analyses were performed using the statistics software package MATLAB. The non-parametric Kruskal-Wallis rank sum test was used to compare data among three temperature groups: mesophilic, thermophilic and hyperthermophilic microorganisms. The Mann-Whitney-Wilcoxon rank sum test (*U*-test) was used to compare the average GC content between two temperature groups such as mesophiles and thermophiles. For the correlation analysis, the Pearson correlation coefficient (r) was used to evaluate the strength of correlation.

A strong correlation is defined as an absolute value of $r > 0.9$; a value between 0.5 and 0.9 represents a good correlation; and a value of $r < 0.5$ is defined as a weak or no correlation [32]. We used the method of comparative analysis by independent contrasts to control for phylogenetic non-independence for temperature and GC content in stems [33].

The contrasts were calculated using the PDAP:PDTREE module of the MESQUITE software for phylogenetic analysis [34, 35]. The independent contrasts method assumes a Brownian motion model of evolution where variance in a trait increases through evolutionary time and calculates the amount of change between sister taxa at each node of the phylogeny.

Independent contrasts are generated by substracting trait values between all pairs of sister taxa in the tree. Contrasts are standardized by dividing by their standard deviation calculated as the square root of the sum of branch lengths between the pair of traits.

The phylogenetic tree used in this study was kindly provided by Dr. C. Brochier. This tree was generated from the concatenated sequences of ribosomal proteins from 43 archaeal species for which complete genomes are available [36].

Results

Analysis of the Global GC Content in the Archaea RNase P RNA Component

For the purposes of this study the archaea species were grouped according to their optimal growth temperatures (OGTs). Three groups were defined: the mesophile group contained species with OGTs between 25°C and 45°C (n = 22); the thermophile group included those species with OGTs between 45°C and 75°C (n = 20) and the hyperthermophile group contained the species with OGTs above 75°C (n = 23). We initiated our study by calculating the global GC content in the RNA component of the RNase P from the archaea species in each group (Figure 2).

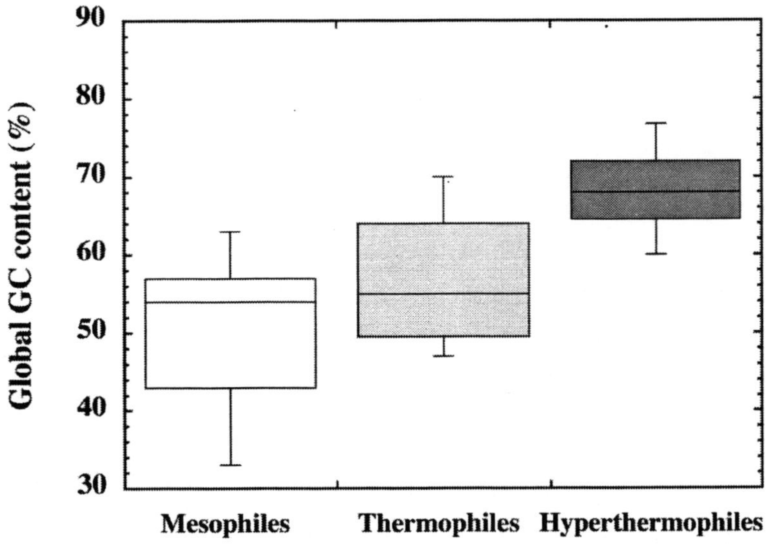

Figure 2. Global GC content in the archaeal RNase P RNA. Boxes extend from the first quartile to the third quartile, with the median values indicated by an horizontal bar within the box. Whiskers extend to the smallest and largest non-outlier values. Temperature ranges are: mesophiles (25-40°C), thermophiles (40-75°C) and hyperthermophiles (>75°C). The number of archaea species included in the analysis are: mesophiles (n = 22), thermophiles (n = 20) and hyperthermophiles (n = 23).

The mean GC content for the mesophile group is 50±5%, for the thermophiles 56±7%, and for the hyperthermophiles 69±6%. We used the Mann-Whitney-Wilcoxon rank sum test (*U*-Test) to compare the average GC content between two temperature groups such as mesophiles versus thermophiles (*U*-Test, $p = 0.05$), thermophiles versus hyperthermophiles (*U*-Test, $p < 0.001$), and mesophiles versus hyperthermophiles (*U*-Test, $p < 0.001$). Therefore, the analysis of the global GC content in the archaeal RNase P RNA component reveals a trend towards higher GC content in the thermophilic and hyperthermophilic species. Thus, the results from this analysis prompted us to look in more detail and to investigate the specificity of these compositional differences.

Compositional Analysis of the Secondary Structures of the Archaeal RNase P as a Function of OGT

The secondary structures of the RNase P RNA component of the 65 archaea species analyzed above were obtained using the RNAalifold and RNA fold software's from the Vienna RNA package [29]. Then, the structures were partitioned into double-stranded stem regions and single-stranded loop regions, and the GC content was calculated for each region. The results of this analysis are summarized in Table 1. The compositional analysis of the RNase P RNA secondary structures indicates that the GC content of the stem regions is highest in the hyperthermophiles, intermediate in the thermophiles and lower in the mesophiles. The Mann-Whitney-Wilcoxon rank sum test demonstrates that the differences among the hypertermophyles and the other two groups are clearly significant: hyperthermophiles versus thermophiles (*U*-Test, $p<0.001$), hypethermophiles versus mesophiles (*U*-Test, $p<0.001$).

Table 1.Compositional analysis of the secondary structure of the archaeal RNase P RNA. Average GC content in the RNase P RNA stems, loops and pseudoknots for mesophiles (25-40°C), thermophiles (40-75°C) and hyperthermophiles (>75°C) archaea. The values correspond to mean ± standard error

Group	Length (bp)	% GC Stems	% GC Loops	% GC Pseudoknots
Hyperthermophiles	311 ± 45	79.8 ± 5.1	39.2 ± 5.2	75.0 ± 2.5
Thermophiles	314 ± 64	57.8 ± 5.0	39.6 ± 5.8	64.2 ± 3.3
Mesophiles	312 ± 40	52.6 ± 8.2	42.8 ± 8.2	65.3 ± 4.6

However, the same analysis shows that the differences in the stems GC content between the thermophiles and mesophiles are not statistically significant (*U*-Test, $p = 0.130$). On the other hand, the GC content in the loop regions does not differ among the three temperature groups, indicating that there is no correlation between the GC content of the loops and the OGT. These results are consistent with previous studies, which have observed that the GC content of stem regions in rRNAs, tRNAs and SRP-RNAs correlates with the OGT, while the loop regions of these molecules show no correlation. In addition, we also analyzed the GC content in the pseudoknots of the RNase P RNA. The comparative analysis shows that the pseudoknots of the RNase P RNAs of the hypertermophilic group have a neatly higher GC content than the other two groups (Table 1). The statistical analysis shows that the differences between the hyperthermophiles and the other two groups are significant (*U*-Test, $p < 0.01$), while the differences between thermophiles and mesophiles are not (*U*-Test, $p = 0.122$). To explore in more detail these patterns, we examined the correlation between the GC content in each individual RNase P RNA structure and the OGT of the species. Thus, we plotted the GC content of the archaeal RNase P RNAs against the OGT of the species (Figure 3). Again, the data were separated in two sets: one set for the stem regions and another for the loop regions. A clearly positive correlation is observed between the GC content of the stem regions and the OGT. However, the increase in the GC content is not linear. The stems GC content seems to increase at a moderate rate for those organisms with OGT between 25-60°C, and then increases dramatically for those organisms with an OGT between 65°C and above.

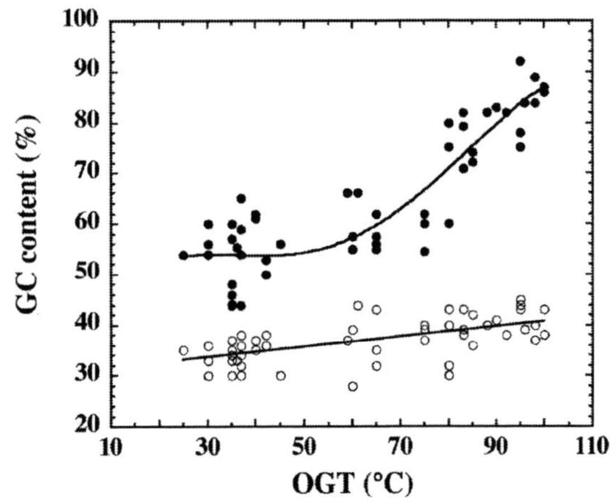

Figure 3. Relationship between the GC content of the archaeal RNase P RNA stems and loops, and the optimal growth temperature. The data for stem regions are shown as black dots, and the data for loop regions are shown as open squares.

Conversely, a linear but much weaker trend is observed for the relationship between GC content of the loop regions and OGT (slope = 0.099, $r = 0.30$, $p < 0.001$). Thus, it appears from this study that the selection for increased GC content at higher OGT operates essentially on the stem regions of the RNase P RNA.

The Increase in the GC Content of the Stems Is More Prominent in the S-Domain of the RNase P RNA

Sequence alignments and consensus RNase P structures have shown that the C-domain of archaeal RNase P RNAs is relatively well conserved while the S-domain displays more variability (20, 37, 38). Thus, to further extend our analysis on the adaptation of this RNA molecule to OGT, we analyzed separately the GC content in the stems and loops of the C and S-domains (Table 2). The results of this study show that the differences in the stems GC content of the three temperature groups are highest in the case of the S-domain. The Kruskal-

Wallis test indicates that when considering the S-domain alone the differences among the three temperature groups are statistically significant ($p < 0.001$). Phylogenetically uncontrolled analysis shows a relatively small correlation between the GC content of the C-domain of the RNase P RNA and OGT (slope = 0.34, $r = 0.60$, $p < 0.001$).

Table 2.Compositional analysis of the secondary structure of the C- and S- domains of the archaeal RNase P RNA. Average GC content in the stems and loops of the C- and S-domains of the archaeal RNase P RNA. Mesophiles (25-40°C), thermophiles (40-75°C) and hyperthermophiles (>75°C). The values correspond to mean ± standard error

Group	C-domain		S-domain	
	% GC Stems	% GC Loops	% GC Stems	% GC Loops
Hyperthermophiles	78.4 ± 4.8	45.1 ± 6.1	82.5 ± 4.9	41.2 ± 5.2
Thermophiles	64.4 ± 3.3	41.1 ± 5.0	60.8 ± 3.8	37.5 ± 7.0
Mesophiles	61.6 ± 3.0	37.8 ± 2.2	39.3 ± 7.5	30.8 ± 5.6

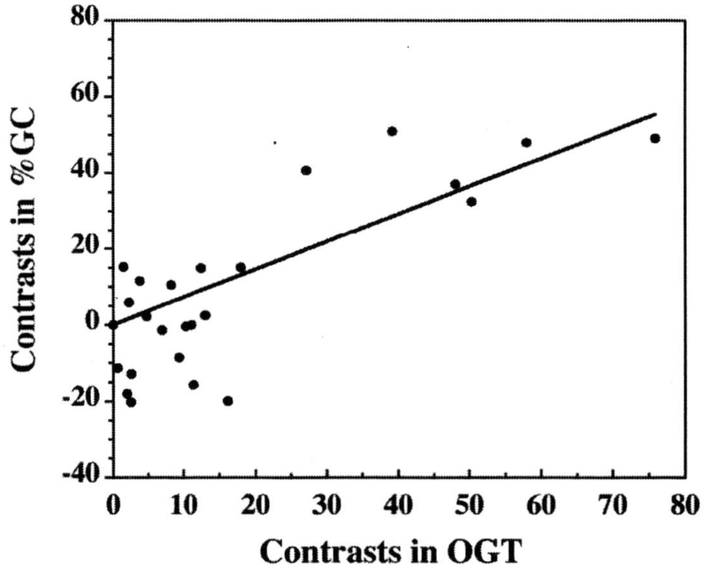

Figure 4. Phylogenetically-independent relationship between the GC content in the stems of the S-domain of the archaeal RNase P RNA and OGT. Correlation between the stems GC content in the S-domain and OGT-independent standardized contrasts (slope = 0.88, r = 0.80, p < 0.001).

Conversely, a good correlation with OGT is found for the S-domain (slope = 0.76, $r = 0.88$, $p < 0.001$). Thus our results indicate an apparent correlation between the stems GC content of the S-domain and the OGT. However, this correlation could be due to shared ancestry between the archaea species considered in our analysis rather than to thermal adaptation. To control for this posibility we generated phylogenetically independent contrasts for the stems GC content in the S-domain and OGT [33]. This analysis shows that a robust correlation (slope = 0.88, r = 0.80, p< 0.001) persists when the phylogenetic relationships are taken into consideration (Figure 4). Therefore, it can be concluded that the variation in the stems GC content of the S-domain among archaea species with different OGT is most probably due to natural selection.

Analysis of Protein Composition Reveals Thermal Adaptation in the RNase P Protein Component

After analyzing the thermal adaptations at the level of the RNase P RNA component we analyzed the protein components of this ribonucleoprotein in order to identify trends, which could be related with the OGT of the archaea species.

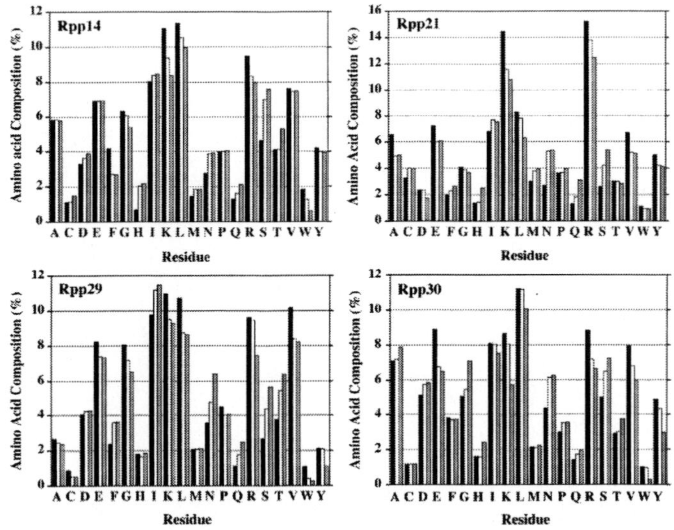

Figure 5. Amino acid composition of the protein components of the archaeal RNase P holoezyme. Bars in black colour are for hyperthermophilic archaea, in white for thermophilic archaea and in grey for mesophilic archaea.

We started this study by analyzing the amino acid composition of the sequences of the four protein components (Rpp30, Rpp29, Rpp21 and Rpp14) of the RNase P holoenzyme (Figure 5).

A number of common trends were identified in the four proteins. Thus, in the hypertermophilic species there is a clear increase in the proportion of the amino acids Lys, Leu, Arg, Val and Trp. In addition in these species we found that the amino acids Glu and Tyr are also clearly increased in the Rpp30, Rpp29 and Rpp21 proteins.

Conversely, we observe that in the four proteins of the hyperthemophiles, the proportion of the amino acids Asn, Ser, Gln and His is decreased.

In terms of overall amino acid properties the analysis shows that in the hypertermophilic species the proportion of charged basic amino acids are increased in the four RNase P protein components.

Also, non-polar amino acids are found to be particularly increased in the Rpp30 and Rpp14 proteins of the hyperthermophilic species.

An interesting feature of this analysis is that in the four protein components of the RNase P of the hyperthermophilic and thermophilic arachaea we observe a very clear decrease in the proportion of tiny and small amino acids.

Finally we calculated the theoretical isoelectric points (iP) of these proteins and found that in the four cases the iP of the hyperthermophilic species is slightly higher than for the thermophilic and mesophilic species (Table 3).

Table 3. Theoretical isoelectric point of the archaeal RNase P protein subunits. Average isoelectric point (iP) of the protein components of the archaeal RNase P holoenzyme. Mesophiles (25-40°C), thermophiles (40-75°C) and hyperthermophiles (>75°C). The values correspond to mean ± standard error

Group	Rpp14	Rpp21	Rpp29	Rpp30
Hyperthermophiles	10.9 ± 0.3	11.8 ± 0.7	11.1 ± 0.5	9.9 ± 0.7
Thermophiles	10.3 ± 0.9	10.9 ± 0.3	10.4 ± 0.8	9.1 ± 0.8
Mesophiles	9.1 ± 0.2	10.0 ± 0.5	9.8 ± 0.2	7.4 ± 0.5

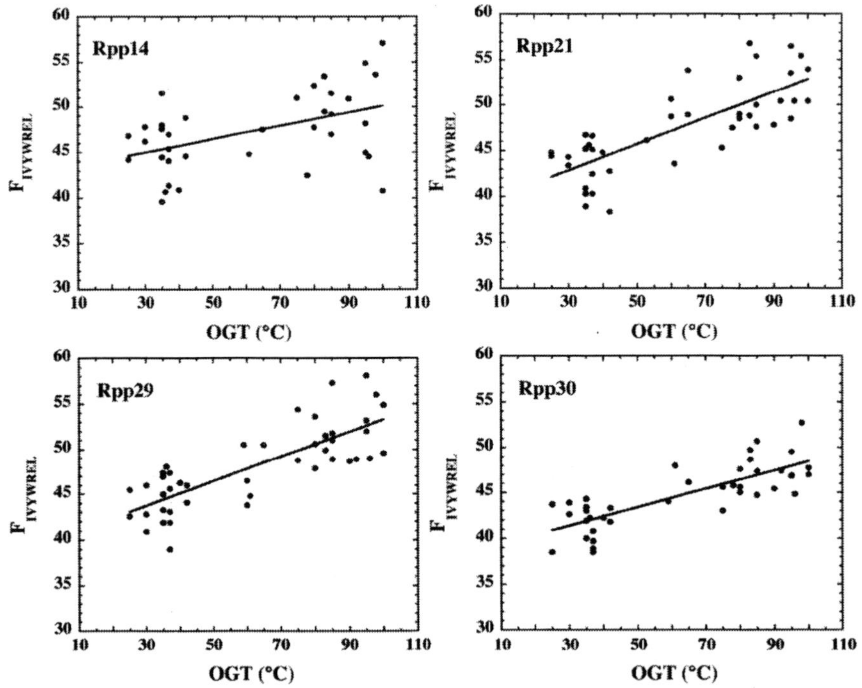

Figure 6. Correlation between the Sum F of fractions of Ile Val, Tyr, Trp, Arg, Glu, and Leu (IVYWREL) amino acids in the four RNase P protein components and the OGT of the archaea species. The slopes and correlation coefficients are: Rpp14 (slope = 0.07, $r = 0.520$, $p < 0.01$), Rpp21 (slope = 0.15, $r = 0.75$, $p < 0.001$), Rpp29 (slope = 0.13, $r = 0.73$, $p < 0.001$) and Rpp 30 (slope = 0.10, $r = 0.77$, $p < 0.001$).

Recently it was suggested that the contents of the seven amino acids Ile, Val, Tyr, Trp, Arg, Glu, Leu (IVYWREL) could be used as an universal predictor for optimal growth temperature (OGT) in prokaryotes [8].

Thus we calculated the IVYWREL fraction for each of the four protein components of the RNase P holoenzyme and plotted it against the OGT of the archaea species (Figure 6).

These analyses indicated that in the RNase P proteins the total occurrences of amino acids I,V,Y,W,R,E,L are indeed clearly correlated with the OGT. Robust correlation coefficients were obtained for the Rpp21 (slope = 0.15, $r = 0.75$, $p< 0.001$), Rpp29 (slope = 0.13, $r = 0.73$, $p< 0.001$) and Rpp 30 (slope = 0.10, $r = 0.77$, $p< 0.001$).

In the case of the Rpp14 the correlation appears less robust (slope = 0.07, $r = 0.520$, $p< 0.01$). Thus, these results are consistent with the findings of Zeldovich and coworkers [8].

Distribution of Amino Acids in the Structure of the RNase P Proteins

In addition to a preference for certain amino acids, there may be differences in their distribution that may provide structural adaptations that affect thermostability. Thus, we analyzed the amino acid composition of hyperthermophilic and mesophilic RNase P proteins in three major secondary structure elements, α-helices, β-sheets, and coiled regions. To this end the sequences of the Rpps were aligned with EXPRESSO. This software uses PDB files to compute structure-based alignments. Then, the alignments were used to predict the secondary structure elements with the Jpred3 software [39]. This procedure provided accurate secondary structure predictions for the four Rpps proteins (Figure 7).

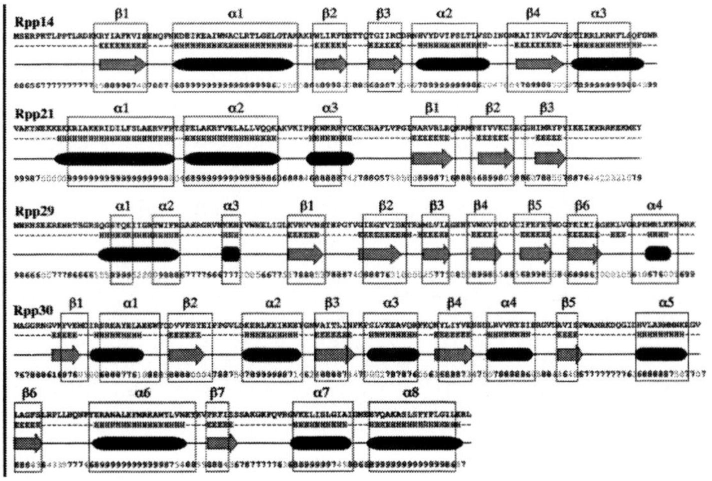

Figure 7. Secondary structure predictions for the RNase P protein components. Sequences for the four RNase P proteins, Rpp14, Rpp21, Rpp29 and Rpp30 of hyperthermophilic and mesophilic archaea were subjected to secondary structure prediction using Jpred3. As an example of the accuracy of these predictions, the figure shows the secondary structure predictions for the RNase P protein components of Pyrococcus furiosus. □-helices and □-strands are represented by cylinders and arrows, respectively. The numbers below the diagram representation indicates the confidence values for the prediction. The boxed regions indicate the position of the secondary structure elements as known from the structures solved by NMR and/or crystallography.

A common trend observed in the RNase P proteins, is that the hyperthermophilic proteins contain significantly less residues in the coiled regions (~10%). Although the majority of amino acids show a similar distribution in the hyperthermophilic or mesophilic proteins, the contents of the amino acid Lys are significantly increased in the three secondary structure elements of the hyperthermophilic proteins.

Also, the amino acids Leu, Arg, Val, Trp and Tyr appear to be particularly increased in the α-helices of the hyperthermophilic proteins. Conversely, the amino acids Asn, Ser, Glu, His and Pro show significantly low frequencies in the coiled regions of the hyperthermophilic proteins. When considering the overall amino acid properties it appears that the RNase P proteins of the hyperthermophiles display a clear increase in the amount of basic charged amino acids in their α-helices. Also, the hyperthermophilic proteins show significantly lower frequencies of tiny and small amino acids in the coiled regions.

Conclusion

To perform their physiological functions the biomolecules must maintain their native conformation. Thus, to occupy new habitats organisms need often to adapt their biomolecules to the prevailing environmental conditions: temperature, pH, salt concentration, etc. When considering the different environmental parameters, temperature appears as particularly crucial because it can penetrate physical barriers and have dramatic destabilizing effects on the structure of all biomolecules. Among the three kingdoms of life, the Archaea stands out as the unique group of organisms who have adapted to a wide range of temperatures: from 0°C to 110°C. The availability of an increasing number of genomes from archaea species thriving at different temperatures provides an invaluable tool to analyze the mechanisms of thermal adaptation at the molecular level. In this study, we have taken advantage of the Archaea genomic datasets to analyze the thermal adaptation of an ancient and universally conserved ribonucleoprotein, the RNase P. In the three kingdoms of life the RNase P is composed of an RNA subunit, the RNase P RNA, and one or more protein subunits (Rpps). The RNase P is an endoribonuclease which is responsible of the 5' endonucleolytic cleavage of transfer RNA precursors (pre-tRNA) to form mature functional tRNAs. Remarkably, the catalytic function can be conducted by the RNA subunit independently of the protein subunits, indicating that the biological activity resides in the RNase P RNA. Thus the RNase P is a true ribozyme [16, 19, 40]. The objective of our study was to analyze the thermal adaptation of the RNase P by conducting a systematic analysis of the compositional variation of its RNA and protein components in mesophilic, thermophilic and hyperthermophilic archaea species. We initiated our study by analyzing the thermal adaptations at the level of the RNase P RNA component.Several studies have shown a positive correlation between the GC content in the stems of structural RNAs and the growth temperature of prokaryotes and eukaryotes [2-5, 41]. Since GC pairs are more stable than AT this led to the proposal of the thermal adaptation hypothesis, which states that GC content in structural nucleic acids should be correlated with the growing temperature of organisms. Our analysis of the global GC content in the archaea RNase P RNA indicates that on average the GC content of the RNase P RNA is higher in the hyperthermophilic (69±6) and thermophilic (56±7) organisms that in the mesophilic ones (50±5). Moreover, the analysis of the GC content of the secondary structures of the RNase P

RNA indicates that the average GC content in the stems is considerably high in the hyperthermophilic species (79.9±5%). However, we do not observe statistically significant differences between termophilic (57.8±5%) and mesophilic (53.6±8%) species. When plotted against the OGT of the species, it appears that the GC content in the stems increases at a moderate rate for those organisms with OGT between 25-60°C, and then increases dramatically for those organisms with an OGT between 65°C and above. The analysis of the RNase P RNA of the few psicrophilic archaea available in the genomic data sets (*Methanococcoides burtonii*, *Methanogenium frigidum*, *Nitrosopumilus maritimus* and *Cenarchaeum symbiosum*) indicates that their stems GC content is similar to that of mesophilic and moderate thermophilic species.This suggests that a minimum stems GC content (~50%) could be required to ensure the proper folding of the RNase P RNA, even at low environmental temperatures. Probably this stems GC content together with the association of the RNase P RNA with its protein cognates provides sufficient stability to the RNase P RNA to ensure its function in a broad range of environmental temperatures (from 4°C to 65°C). However for those organisms thriving at temperatures above 65°C the increase in the stems GC content could represent a necessary advantage to maintain the stability and/or ensure proper folding of the RNase P RNA. Interestingly we have observed that the variation of the GC content in the pseudoknots follows a similar pattern than the GC content in stems. The GC content in the pseudoknots of the RNase P RNA of mesophilic and thermophilic species is relatively important (around 65%), highlighting the importance of these interactions in the folding and/or maintenance of the structure of this macromolecule [20, 21]. However, as for the stems a statistically significant increase in the pseudoknots GC content is observed in hyperthermophilic species (average GC content ~75%). Thus, this observation is consistent with the hypothesis that for those species thriving above 65°C the increase in the GC contents of the double stranded regions of the RNase P RNA has probably provided an adaptative advantage by increasing the stability of the molecule. Interestingly, sequence alignments and consensus RNase P RNA secondary structure predictions have shown that in all three kingdoms the C-domain (containing the catalytic core) is more conserved than the more variable S-domain (involved in pre-tRNA binding). Experimental studies have shown that the two domains can fold independently [42, 43]. Moreover, while the C-domain by itself is not functional or has decreased catalytic activity, the S-domain alone can bind pre-tRNA [44]. Thus, we further extended our study by analyzing separately the relationship between the GC content in the stems and loops of both domains and the OGT of archaea species. Our results show that in hyperthermophilic archaea the stems GC content of the C-domain is clearly higher than in the thermophilic and mesophilic species. However, there are not any statistically significant difference in the GC content of the C-domain between thermophilic and mesophilic archaea. By contrast, we could determine that the GC content in the double stranded regions of the S-domain is higher in the hyperthermophiles, intermediate in the thermophiles and lower in the mesophiles. Moreover, by using phylogenetic non-independence contrasts analysis we could demonstrate that this correlation is not an artefact due to phylogenetic relation but more likely reflects an adaptation to OGT resulting from natural selection. These results are consistent with the work of Baird and coworkers [45]. They showed by using phylogenomic comparative analysis and mutational studies that the S-domain of the RNase P of thermophilic bacteria achieves higher thermal stability by mutating strategic base pairs to GC, decreasing the surface accessibility of the native state, and increasing the amount of structure formation in the native folding transition. An *E.coli* S-domain (mesophile) containing these mutations achieves the same stability and folding cooperatively as the *Thermus Thermophilus* S-domain (thermophile). The observed increase in the GC content of the stems of the archaea RNase P RNA is also consistent with previous studies showing that in prokaryotes, there is a significant correlation between the GC content of structural RNAs (rRNAs, tRNAs and SRP-RNAs) and growth temperature, and that the high GC content is concentrated in the double-stranded regions of the molecule [2, 3, 5]. These studies also revealed that in rRNAs the GC content of the single stranded regions remains relatively low and constant. Similarly, our data indicate that the GC content of the single-stranded regions of the RNase P RNA remains remarkably constant between the three temperature groups (around 40%). It has been proposed that the evolutionary conserved level of guanine (and A) in the unpaired regions of rRNAs could be explained by the fact that these regions could play an essential role in maintaining the secondary and tertiary structure of the molecules [3]. This is probably also the case in RNase P RNA since phylogenetic comparative sequence analysis shows that the majority of the nucleotides conserved in all RNase P RNAs localize in unpaired regions as predicted by secondary structure models [46]. This seems to be particularly true for the nucleotides of the universally conserved regions CRII and CRIII located in the S-domain. This conserved sequences form two interleaving pentaloops, each folding in a T-loop motif. It has been proposed that their high degree of sequence conservation is the reflection of strong structural constraints [47]. Moreover, the loops and bulges, could be not only essential for the appropriate folding of the RNase P RNA itself but could also be involved in the interaction of the RNase P RNA with its pre-tRNA substrate and/or with its cognate proteins. Thus, footprinting experiments with the Rpp29-Rpp14 binary complex decreased the susceptibility to RNase T1 cleavage of a *Pyrococcus furiosus* RNase P RNA derivative containing only the C-domain. Most of the protected sequences correspond to unpaired regions containing the universally conserved nucleotides [48].

After having explored the compositional and structural thermal adaptations of the RNase P RNA component we examined putative adaptations at the level of the protein components of the holoenzyme. The archaeal RNase P contains at least four protein subunits: Rpp14(16 kDa), Rpp21(12 kDa), Rpp29 (12 kDa) and Rpp30 (23 kDa). The structures of the Rpps, as revealed by NMR spectroscopy and/or X-ray crystallography, correspond to common nucleic acid binding protein families: an RRM-like fold (Rpp14), a zinc ribbon (Rpp21), an Sm-like fold (Rpp29), and a TIM barrel (Rpp30). Biochemical experiments of *in vitro* reconstitution, yeast two-hybrid experiments and structural studies have shown that the archaeal Rpps function as two binary complexes: Rpp21-Rpp29 and Rpp14-Rpp30. Kinetic analysis suggest that the two pairs fulfill different roles: Rpp21-Rpp29 increase the affinity of the RNase P RNA for the pre-tRNA substrate, while the Rpp14-Rpp30 enhances the rate of chemical cleavage nearly a 100-fold [22]. Enzymatic foot-printing experiments on the RNase P of *Pyrococcus furiosus* and *Methanocaldococcus jannashii* suggests that the Rpp21-Rpp29 complex bind to the S-domain of the RNase P RNA while the Rpp14-Rpp30 binds to the C-domain [22, 48, 49]. We started our study on the thermal adaptation of the Rpps by comparing the amino acid composition of the four proteins in hyperthermophilic, thermophilic and mesophilic archaea species. The data show that the four Rpps of thermophilic and in particular hyperthermophilic species present a significant increase in the content of the amino acids Lys, Leu, Arg, Val and Trp. Moreover, the amino acids Glu, and Tyr are also significantly increased in the Rpp21, Rpp29 and Rpp30 of hyperthermophiles.

Conversely, the four Rpps of the hyperthermopilic archaea show a marked decrease in the contents of the thermolabile amino acids Thr, Asn, Ser, Gln and His. This results are consistent with previous studies on the thermal adaptation of the proteomes of prokaryotic species (Bacteria and Archaea) which have described that the amino acids Lys, Arg and Glu are more abundant in the thermophilic proteins than in their mesophilic counterparts [6, 7, 50]. These studies also reported, as we observe here with the Rpps, that the higher content of charged amino acids in the thermophilic proteins is compensated by a decrease in the content of polar, non-charged amino acids: Ser, Thr, Asn and Gln [6, 51, 52]. Of the amino acids properties examined, the most remarkable observation was, as mentioned above, the decrease in the thermophilic Rpps of the proportion of uncharged polar amino acids (Ser, Thr, Asn and Gln).

However, we observe also that the four Rpps of the hyperthermophilic archaea show a very clear reduction in the proportion of tiny and small amino acids. This observation has been reported also by other studies analyzing thermal adaptation of proteins [6].

The preference in the hyperthermophilic and thermophilic proteins for those amino acids with larger side chains could be related to their ability to exclude water from the protein interior, to fill cavities, and to reduce the entropic freedom of the protein backbone. Finally, we observe that the isoelectric points of the thermophilic Rpps are slightly higher than those of the mesophilic ones. This is consistent with previous studies showing that the proteomes of thermophilic organisms contain a large fraction of proteins with isoelectric points on the basic range [53], and a general biais in favour of charged rather than polar residues [54, 55].

In a recent study Zeldowich and coworkers [8] have reported that in prokaryotic proteomes the fraction Ile,Val, Tyr, Trp, Arg, Glu ,Leu (IVYWREL) is the most precise predictor of OGT among all possible combinations of amino acids. The I,V,Y,W,R,E,L fraction contains residues of all major types, like, aliphatic and aromatic hydrophobic (Ile, Val, Trp, Leu), polar (Tyr), and charged (Arg, Glu), both basic and acidic. Thus, this predictor measures a very delicate balance between hydrophobic, van der Waals, ionic, and hydrogen-bond interactions in correctly folded proteins. When applied to the Rpps we observed a strong correlation between the IVYWREL and the OGT of the archaea species.

Of the four Rpps, the Rpp14 was the only to show a correlation inferior to $R = 0.7$. Which is consistent with the fact that among the four Rpp proteins, Rpp14 is the one displaying the less marked differences between hyperhermophilic/thermophilic and mesophilic species. To complete our study we generated the secondary structures of the Rpps of hyperthermophilic and mesophilic species and compared the distribution of amino acids in the α–helices, β–sheets and coiled regions.

A common trend appears to be the reduction in the number of amino acids in the coiled regions of the hyperthermophilic Rpp14, Rpp29 and Rpp30. We observe also in the coiled regions a clear decrease in the proportion of small and tiny amino acids, in particular in the C- and N-terminus portions. Taken together this observation suggests that in the hyperthermophilic species these modifications could be necessary to reduce the entropic freedom of the protein backbone. The amino acid Lys, which is significantly increased in the hyperthermophilic Rpps appears to be distributed homogeneously among the three secondary structure elements.

However, when considering the ensemble of basic charged amino acids it appears that their proportion is clearly increased in the α–helices of the hyperthermophilic Rpps.

Overall, our results reveal that the four Rpps of the hyperthermophilic and thermophilic archaea show when compared to the mesophilic species, compositional and structural variations which are consistent with those reported by other studies analyzing thermal adaptation at the protein level.

In conclusion our study demonstrates that in the Archaea both, the RNase P RNA and its four cognate proteins, show compositional and structural variations, which are correlated with the OGT of these organisms. Therefore, to achieve thermal stability of the Archaea RNase P holoenzyme, natural selection has favoured those mutations that increased the stability of each of its individual components. Further studies will be needed to determine the quantitative contribution of each of these individual adaptations to the thermal stability of the ensemble of the holoenzyme.

Acknowledgments

We thank Dr. F. Langa, for the critical reading of the manuscript and his suggestions. Dr. C. Brochier is gratefully acknowledged for providing the phylogenetic tree of the archaea based on concatenated ribosomal proteins.

References

[1] Hickey DA, Singer GA. Genomic and proteomic adaptations to growth at high temperature. *Genome Biol.* 2004;5(10):117.

[2] Galtier N, Lobry JR. Relationships between genomic G+C content, RNA secondary structures, and optimal growth temperature in prokaryotes. *J. Mol. Evol.* 1997 Jun;44(6):632-6.

[3] Wang HC, Hickey DA. Evidence for strong selective constraint acting on the nucleotide composition of 16S ribosomal RNA genes. *Nucleic Acids Res.* 2002 Jun 1;30(11):2501-7.

[4] Nakashima H, Fukuchi S, Nishikawa K. Compositional changes in RNA, DNA and proteins for bacterial adaptation to higher and lower temperatures. *J. Biochem.* 2003 Apr;133(4):507-13.

[5] Wang HC, Xia X, Hickey D. Thermal adaptation of the small subunit ribosomal RNA gene: a comparative study. *J. Mol. Evol.* 2006 Jul;63(1):120-6.

[6] Haney PJ, Badger JH, Buldak GL, Reich CI, Woese CR, Olsen GJ. Thermal adaptation analyzed by comparison of protein sequences from mesophilic and extremely thermophilic Methanococcus species. *Proc. Natl. Acad. Sci. U S A.* 1999 Mar 30;96(7):3578-83.

[7] Mizuguchi K, Sele M, Cubellis MV. Environment specific substitution tables for thermophilic proteins. *BMC Bioinformatics*. 2007;8 Suppl 1:S15.

[8] Zeldovich KB, Berezovsky IN, Shakhnovich EI. Protein and DNA sequence determinants of thermophilic adaptation. *PLoS Comput. Biol.* 2007 Jan 12;3(1):e5.

[9] Tekaia F, Yeramian E, Dujon B. Amino acid composition of genomes, lifestyles of organisms, and evolutionary trends: a global picture with correspondence analysis. *Gene*. 2002 Sep 4;297(1-2):51-60.

[10] Mans RM, Guerrier-Takada C, Altman S, Pleij CW. Interaction of RNase P from Escherichia coli with pseudoknotted structures in viral RNAs. *Nucleic Acids Res*. 1990 Jun 25;18(12):3479-87.

[11] Hartmann RK, Heinrich J, Schlegl J, Schuster H. Precursor of C4 antisense RNA of bacteriophages P1 and P7 is a substrate for RNase P of Escherichia coli. *Proc. Natl. Acad. Sci. U S A*. 1995 Jun 20;92(13):5822-6.

[12] Peck-Miller KA, Altman S. Kinetics of the processing of the precursor to 4.5 S RNA, a naturally occurring substrate for RNase P from Escherichia coli. *J. Mol. Biol*. 1991 Sep 5;221(1):1-5.

[13] Komine Y, Kitabatake M, Yokogawa T, Nishikawa K, Inokuchi H. A tRNA-like structure is present in 10Sa RNA, a small stable RNA from Escherichia coli. *Proc. Natl. Acad. Sci. U S A*. 1994 Sep 27;91(20):9223-7.

[14] Alifano P, Rivellini F, Piscitelli C, Arraiano CM, Bruni CB, Carlomagno MS. Ribonuclease E provides substrates for ribonuclease P-dependent processing of a polycistronic mRNA. *Genes Dev*. 1994 Dec 15;8(24):3021-31.

[15] Altman S, Wesolowski D, Guerrier-Takada C, Li Y. RNase P cleaves transient structures in some riboswitches. *Proc. Natl. Acad. Sci. U S A*. 2005 Aug 9;102(32):11284-9.

[16] Guerrier-Takada C, Gardiner K, Marsh T, Pace N, Altman S. The RNA moiety of ribonuclease P is the catalytic subunit of the enzyme. *Cell*. 1983 Dec;35(3 Pt 2):849-57.

[17] Chen JL, Pace NR. Identification of the universally conserved core of ribonuclease P RNA. *Rna*. 1997 Jun;3(6):557-60.

[18] Lai LB, Vioque A, Kirsebom LA, Gopalan V. Unexpected diversity of RNase P, an ancient tRNA processing enzyme: challenges and prospects. *FEBS Lett*. 2010 Jan 21;584(2):287-96.

[19] Pannucci JA, Haas ES, Hall TA, Harris JK, Brown JW. RNase P RNAs from some Archaea are catalytically active. *Proc. Natl. Acad. Sci. U S A*. 1999 Jul 6;96(14):7803-8.

[20] Haas ES, Brown JW, Pitulle C, Pace NR. Further perspective on the catalytic core and secondary structure of ribonuclease P RNA. *Proc. Natl. Acad. Sci. U S A*. 1994 Mar 29;91(7):2527-31.

[21] Harris JK, Haas ES, Williams D, Frank DN, Brown JW. New insight into RNase P RNA structure from comparative analysis of the archaeal RNA. *RNA*. 2001 Feb;7(2):220-32.

[22] Pulukkunat DK, Gopalan V. Studies on Methanocaldococcus jannaschii RNase P reveal insights into the roles of RNA and protein cofactors in RNase P catalysis. *Nucleic Acids Res*. 2008 Jul;36(12):4172-80.

[23] Fukuhara H, Kifusa M, Watanabe M, Terada A, Honda T, Numata T, et al. A fifth protein subunit Ph1496p elevates the optimum temperature for the ribonuclease P activity from Pyrococcus horikoshii OT3. *Biochem. Biophys. Res. Commun*. 2006 May 12;343(3):956-64.

[24] Huang SL, Wu LC, Liang HK, Pan KT, Horng JT, Ko MT. PGTdb: a database providing growth temperatures of prokaryotes. *Bioinformatics*. 2004 Jan 22;20(2):276-8.

[25] Brown JW. The Ribonuclease P Database. *Nucleic Acids Res*. 1999 Jan 1;27(1):314.

[26] Rosenblad MA, Gorodkin J, Knudsen B, Zwieb C, Samuelsson T. SRPDB: Signal Recognition Particle Database. *Nucleic Acids Res*. 2003 Jan 1;31(1):363-4.

[27] Marszalkowski M, Willkomm DK, Hartmann RK. Structural basis of a ribozyme's thermostability: P1-L9 interdomain interaction in RNase P RNA. *RNA*. 2008 Jan;14(1):127-33.

[28] Zuker M. Mfold web server for nucleic acid folding and hybridization prediction. *Nucleic Acids Res*. 2003 Jul 1;31(13):3406-15.

[29] Gruber AR, Lorenz R, Bernhart SH, Neubock R, Hofacker IL. The Vienna RNA websuite. *Nucleic Acids Res*. 2008 Jul 1;36(Web Server issue):W70-4.

[30] Sperschneider J, Datta A. DotKnot: pseudoknot prediction using the probability dot plot under a refined energy model. *Nucleic Acids Res*. Apr 1;38(7):e103.

[31] Markowitz VM, Szeto E, Palaniappan K, Grechkin Y, Chu K, Chen IM, et al. The integrated microbial genomes (IMG) system in 2007: data content and analysis tool extensions. *Nucleic Acids Res*. 2008 Jan;36(Database issue):D528-33.

[32] Milton JS. *Statistical Methods in Biological and Health Sciences*. New York: McGraw-Hill; 1992.

[33] Felsenstein J. Phylogenies and the comparative method. *American Naturalist*. 1985;125:1-15.

[34] Maddison WP, Maddison DR. Mesquite: A modular system for evolutionary analysis. Version 2.6. http://mesquiteproject.org. 2009.

[35] Midford PE, Garland T, Maddison WP. PDAP Package of Mesquite. Version 1.07. http://mesquiteproject.org. 2005.

[36] Brochier-Armanet C, Boussau B, Gribaldo S, Forterre P. Mesophilic Crenarchaeota: proposal for a third archaeal phylum, the Thaumarchaeota. *Nat. Rev. Microbiol*. 2008 Mar;6(3):245-52.

[37] Pace NR, Brown JW. Evolutionary perspective on the structure and function of ribonuclease P, a ribozyme. *J. Bacteriol*. 1995 Apr;177(8):1919-28.

[38] Siegel RW, Banta AB, Haas ES, Brown JW, Pace NR. Mycoplasma fermentans simplifies our view of the catalytic core of ribonuclease P RNA. *RNA*. 1996 May;2(5):452-62.

[39] Cole C, Barber JD, Barton GJ. The Jpred 3 secondary structure prediction server. *Nucleic Acids Res*. 2008 Jul 1;36(Web Server issue):W197-201.

[40] Kikovska E, Svard SG, Kirsebom LA. Eukaryotic RNase P RNA mediates cleavage in the absence of protein. *Proc. Natl. Acad. Sci. U S A*. 2007 Feb 13;104(7):2062-7.

[41] Varriale A, Torelli G, Bernardi G. Compositional properties and thermal adaptation of 18S rRNA in vertebrates. *Rna*. 2008 Aug;14(8):1492-500.

[42] Pan T. Higher order folding and domain analysis of the ribozyme from Bacillus subtilis ribonuclease P. *Biochemistry*. 1995 Jan 24;34(3):902-9.

[43] Loria A, Pan T. The cleavage step of ribonuclease P catalysis is determined by ribozyme-substrate interactions both distal and proximal to the cleavage site. *Biochemistry*. 1999 Jul 6;38(27):8612-20.

[44] Qin H, Sosnick TR, Pan T. Modular construction of a tertiary RNA structure: the specificity domain of the Bacillus subtilis RNase P RNA. *Biochemistry*. 2001 Sep 18;40(37):11202-10.

[45] Baird NJ, Srividya N, Krasilnikov AS, Mondragon A, Sosnick TR, Pan T. Structural basis for altering the stability of homologous RNAs from a mesophilic and a thermophilic bacterium. *RNA*. 2006 Apr;12(4):598-606.

[46] Schmitz U, Behrens S, Freymann DM, Keenan RJ, Lukavsky P, Walter P, et al. Structure of the phylogenetically most conserved domain of SRP RNA. *Rna*. 1999 Nov;5(11):1419-29.

[47] Krasilnikov AS, Mondragon A. On the occurrence of the T-loop RNA folding motif in large RNA molecules. *RNA*. 2003 Jun;9(6):640-3.

[48] Xu Y, Amero CD, Pulukkunat DK, Gopalan V, Foster MP. Solution structure of an archaeal RNase P binary protein complex: formation of the 30-kDa complex between Pyrococcus furiosus RPP21 and RPP29 is accompanied by coupled protein folding and highlights critical features for protein-protein and protein-RNA interactions. *J. Mol. Biol.* 2009 Nov 13;393(5):1043-55.

[49] Tsai HY, Pulukkunat DK, Woznick WK, Gopalan V. Functional reconstitution and characterization of Pyrococcus furiosus RNase P. *Proc. Natl. Acad. Sci. U S A*. 2006 Oct 31;103(44):16147-52.

[50] Singer GA, Hickey DA. Thermophilic prokaryotes have characteristic patterns of codon usage, amino acid composition and nucleotide content. *Gene*. 2003 Oct 23;317(1-2):39-47.

[51] Chakravarty S, Varadarajan R. Elucidation of factors responsible for enhanced thermal stability of proteins: a structural genomics based study. *Biochemistry*. 2002 Jun 25;41(25):8152-61.

[52] Das R, Gerstein M. The stability of thermophilic proteins: a study based on comprehensive genome comparison. *Funct. Integr. Genomics*. 2000 May;1(1):76-88.

[53] Kawashima T, Amano N, Koike H, Makino S, Higuchi S, Kawashima-Ohya Y, et al. Archaeal adaptation to higher temperatures revealed by genomic sequence of Thermoplasma volcanium. *Proc. Natl. Acad. Sci. U S A.* 2000 Dec 19;97(26):14257-62.

[54] Kumar S, Nussinov R. How do thermophilic proteins deal with heat? *Cell Mol. Life Sci.* 2001 Aug;58(9):1216-33.

[55] Suhre K, Claverie JM. Genomic correlates of hyperthermostability, an update. *J. Biol. Chem*. 2003 May 9;278(19):17198-202.

In: Archaea: Structure, Habitats and Ecological...
Editor: Sakura Y. Katō, pp. 143-161

ISBN 978-1-61761-932-8

Chapter 6

Mechanosensitive Channels in Archaea: Roles, Lipid-Protein Interactions and Future Questions

Daniel Balleza
Unidad de Biofísica (Centro Mixto CSIC-UPV/EHU).
Universidad del País Vasco, Bilbao, Spain

Abstract

Archaea, the third domain of life, include micro-organisms adapted to a variety of environmental conditions, including some of the most hostile locations on Earth. Osmotic regulation is an important issue for a microbe, and mechanosensitive (MS) ion channels are important proteins capable to respond as transducers of mechanical stress by forming non-selective pores in order to deal with drastic osmotic changes in the milieu. As molecular sensors, MS channels respond to increases in membrane tension. When the membrane tension is small, these proteins are in the closed configuration; as the tension grows, the open state becomes ever more frequent, until it dominates at mild or high tensions. It has been proposed that these changes correspond to different channel conformations in a defined protein–lipid interface. This commentary discusses some of the most important aspects in order to understand the function of MS channels in archaebacteria. Special attention is focused on how archaebacterial lipids are attractive targets to study the function of these proteins.

Keywords: Archaea, lipids, mechanosensitive channels.

Abbreviations

BTLs, Bipolar tetraether lipids;
CrenA, Crenarchaeota;
EuryA, Euryarchaeota;
MS, Mechanosensitive.

Introduction

Archaea thrive in a variety of conditions, including hostile environments characterized by anaeobiosis, saturated salt, and both high and low extremes of temperature and pH (Rothschild and Mancinelli 2001). Since their first description (Woese and Fox, 1977), archaea had been considered as extremophiles, however, they have been described in nonextreme habitats, such as ocean waters, freshwater sediments, soils, the rhizosphere, and as endosymbionts (Chaban et al. 2006). According to rRNA sequence comparisons, all known archaea belong to the phyla Crenarchaeota or Euryarchaeota, although two other phyla have been proposed: 'Korarchaeota' and 'Nanoarchaeota' (Huber et al. 2003). Cultivated archaea can be, on the other hand, divided into at least six physiological or metabolic groups: halophiles, thermophiles, acidophiles, methanogens, nitrifiers, and anaerobic methane oxidizers (Valentine 2007). The Euryarchaeota consists of methanogens, extreme halophiles, thermophiles, and extremely acidophilic thermophiles (Liu and Whitman 2008) whereas Crenarchaeota comprise the most thermophilic organism known to date, *Pyrolobus fumarii* and the intensively studied extreme thermoacidophile *Sulfolobusacidocaldarius* (Stetter 1999). The biochemical and structural adaptations that enable archaea to cope with some of these habitats include the chemical structure of their lipids. In contrast to well-studied bacterial lipids, which typically consist of fatty acids esterified to a glycerol moiety, archaeal lipids usually consist of isoprenoidal alcohols that are ether-linked to glycerol (Sprott 1992), and their physicochemical properties have seen only recently revealed (Jacquemet et al. 2009; Chong 2010). They are unique in nature and make possible that some of these micro-organisms deal with extreme habitats (Valentine 2007). There are two major classes of archaeal lipids, the archaeol lipids (diphytanyl glycerol diethers) and the caldarchaeol lipids (dibiphytanyl diglycerol tetraethers or bipolar tetraether lipids, BTLs). These phytanyl chains contain methyl groups at every fourth carbon atom in the backbone.

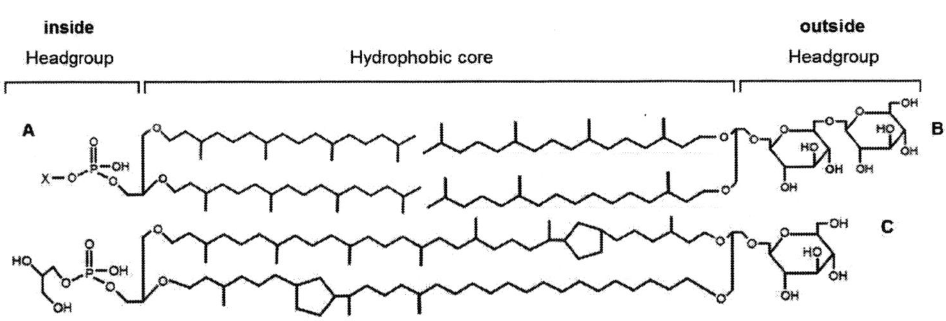

Figure 1. Typical structures of archaeal phospholipids. (A, B) Archaeol lipids, archaetidyl-X, where –X stands for polar head group like ethanolamine (–$CH_2CH_2NH_2$) or serine (–$CH_2CH(COOH)NH_2$). The variety of polar heads may include also phosphor-, glycol-, phosphoglyco-, polyol- or –OH groups, although phosphatidylcholineis rare. (B) Gentiobiosyl archaeol. (C) Caldarchaeol lipids. Main bipolar lipid from *Thermoplasma acidophilum* (Swain et al. 1997). The phytanyl chains contain isoprenoid-like branches (20–40 carbon in length) that can contain cyclopentane rings as in crenarchaeol. These chains are attached via ether bonds to the *sn*-2,3 carbons of the glycerol backbone. Caldarchaeol and isocaldarchaeol have no rings.

The reason for the higher stability of the phytanyl chain could be the reduced segmentary motion of tertiary carbon atoms (*i.e.,* rotation of carbon atoms that are bound to three other C-atoms, resulting in kinks in the acyl chain (Gabriel and Chong 2000). The restriction in hydrocarbon chain mobility also reduces the permeability of the archaeal membrane. Importantly, the ether linkage prevalent in archaeal lipids is chemically more resistant to oxidation and high temperatures than the ester bond, while the distribution of the membrane lipids is asymmetrical, with most of the phosphate groups pointing at the cytoplasm and the sugar moiety exposed at the outer side of the cell (Morii and Koga 1994) (Figure 1). Ether links are also more resistant to high temperatures than ester links, however, in the case of archaeal organisms, the role of neither the head group distribution asymmetry nor the hydrophobic core asymmetry is well understood yet.

Archaeal lipids form bilayers in a similar way as the ester-lipids. However, they assemble in diverse arrangements and combinations forming domains. They can form typical lipid bilayers mixed with domains in which all the lipids have bridging chains that cross the membrane (i.e., monolayers), conferring a high degree of rigidity to the membrane, in a combination of bilayer/monolayer arrangements. Liposomes composed from archaeal lipids (archaeosomes) (Figure 2) are less permeable to ions than bacterial or eukaryotic membranes and show a considerable thermo-stability (Elferink et al. 1994; Mathai et al. 2001; Ulrih et al. 2009). The low permeability of archaeosomes is related with the low expansibility and high rigidity of such membranes, which mainly depend on the structures of BTLs. However, molecular dynamics simulations of membranes formed by archaeal polar lipids and BTLs suggest that water molecules move across these membranes (Shinoda et al. 2005), although archaeosomes made of BTLs lipids show a marked reduction in water permeability, compared with the other lipids (Mathai et al. 2001). A striking feature of these lipids is the occurrence of unusual carbohydrates, in the form of β-D-galactofuranosyl units as cyclopentane rings. The addition of these cyclic structures in the transmembrane portions of archaeal lipids results in an enhanced membrane packing and a reduced membrane fluidity (Ulrih et al. 2009; Chong 2010). This is because by the increase of the cyclization of the C_{40} isoprenoid chains, the lipids can be packed more tightly, which results in a more restricted motion of the lipids and prevents that the membrane becomes too fluid (Albers et al. 2000).

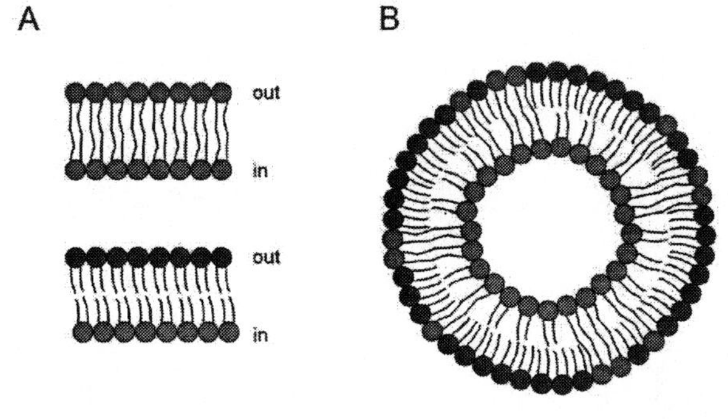

Figure 2. Structure of unilamellar archaeosomes. Conventional diether lipids and bipolar tetraether lipids interact by forming zones of mono- and bilayer into a closed vesicle in an aqueous solution. Since archaeal lipids have important mechanical properties (see text), archaeosomes have considerable stability against extremes of pH, oxidation, and against the action of lipases.

The increasing proportion of cyclopentane rings can be seen as an adaptation to hyperthermophilic and acidic life (Itoh et al. 2001). Thus, the cyclization number of C_{40} chains changed significantly depending on the growth temperature of thermoacidophilic archaea, such as *T. acidophilum* and *S. acidocaldarius* (De Rosa et al. 1980; Uda et al. 2001). In the euryarchaeote *Methanocaldococcus janaschii* increasing temperatures induce the change from diether lipids to the more thermostable BTLs (Sprott et al. 1991). The cyclization of the chains tends to decrease the motion of the lipids but it contributes to acceptable membrane fluidity at elevated growth temperatures. In consequence, the bilayer spanning BTLs decreases membrane fluidity which results in lower ion permeability but those properties increases with the temperature. Consequently, the permeability of the cytoplasmic membrane is a major factor in determining the maximum growth temperature of these microorganisms. In addition, covalent. I

Cells control the fluidity and permeability of their cytoplasmic membrane by modification of its composition. As consequence, the membrane's liquid crystalline state allows optimal functioning of the embedded proteins (Kinnunen 1991). Therefore, biologic membranes are not simply inert physical barriers, but complex and dynamic environments that affect membrane protein structure and function, particularly ion channels (Tillman and Cascio 2003). Importantly, membrane fluidity is an important issue considering the function of MS channels in archaea. MS channels are ubiquitous in all three domains of life (Kloda and Martinac 2001b; 2002) and MS channel activities have been documented in *Haloferax volcanii* (Le Dain et al., 1998), *Thermoplasmaacidophilum* and *T. volcanium* (Kloda and Martinac, 2001a) and *M.jannashii* (Kloda and Martinac, 2001c; Kloda and Martinac, 2001d). Since MS channels play a role in the regulation of turgor pressure, which is essential for division and growth of bacterial cells, it could be expected for archaeal MS channels to perform similar cellular functions and also serve as emergency valves for osmoregulation, as in bacteria (Booth et al. 2007). Environment always changes, and organisms must deal with these changes. For a microbe, dilution of cells into a low osmolarity environment (a hypoosmotic shock) causes a rapid influx of water into the cells down the osmotic gradient. This influx is accompanied by cell swelling, which leads to the linear increase in lateral tension of the membrane, according to its grade of elasticity (Haines et al. 1987). Simultaneously, membrane elasticity depends on the lateral diffusion of phospholipids. The rate varies with the structure of lipids and temperature, and it is within an order of magnitude of $\sim 10^{-8}$ cm^2/s, expanding its area up to a 30-40% (Sackmann and Träuble 1972; Haines 1994). Thus, influx of water generates high turgor pressure that increases the membrane tension, leading to the immediate activation of MS channels, which results in the loss of important osmolytes, such as K^+, glutamate and glycine-betain (Booth et al. 2007).

The Physical Coupling between the Lipids and the Protein in MS Channels

Biological membranes contribute to fundamental functions such as the modulation of embedded proteins and particularly, in prokaryotes, MS channel activity (Perozo and Rees 2003). Consequently, their study at the molecular level is of great interest to fields directly related to many living processes. Most of the membranes studied today are from bacterial or eukaryotic origin (Nagle and Tristram-Nagle 2000; Denich et al. 2003; Lindblom and Orädd 2009). There is now enough experimental data available in the literature to attempt to build membrane models representative of less common supramolecular assemblies such as those found in archaea (Jacquemet et al. 2009; Chong 2010). More unusual is to investigate their unique membrane properties by molecular dynamics simulations (Nicolas 2005) in order to explain archaeal MS channels function (Kloda and Martinac 2001b). Given its apparent simplicity, MscL, the channel of large conductance from bacteria, can be considered the archetype of a mechano-gated channel (Fig. 3). This protein has been studied extensively using a variety of methods at functional and structural levels and now we know that it is gated by increases in membrane tension (Sukharev et al. 1999; Perozo et al. 2006). Importantly, sequences with an important similarity to this protein have been found in genomes from several archaea (Table 1). Nevertheless, this protein is apparently absent in halophiles, which is interesting and consistent with the extremely low probability to down-shock a cell in a hyper-saline milieu. On the other hand, many archaea possess multiple putative homologues of MscS and MscK channels. MscL, MscS, MscK and MscM are all channels gated directly by increases in lateral tension, showing conductances sorted in three categories: 100–150 pS (MscM), 300 and 500 pS (MscS/K), and 1000 pS or higher (MscL) (Perozo et al. 2006). Thus, MscL opens at lytic tensions in emergency circumstances, whereas MscS/K homologs acts at mild tensions and appears to regulate cell turgor during normal life cycle.

Similar to bacterial MscL and MscS, archaeal MS channels are also gated by tension in the membrane and they show similarities in their low selectivity, activation by amphipaths and slight voltage dependence. Together, these proteins constitute a well-coordinated system to regulate cellular turgor and probably prokaryotic growth (Martinac 2004; Martinac et al. 2008).

How MS channels senses tension is a pivotal question. Several studies, including their reconstitution in liposomes with a defined composition of lipids, varying different acyl chains lengths, headgroups and charges, as well as the use of amphipaths, gadolinium and molecular simulation, have allowed to establish important relationships between the chemical structure of membrane lipids and the function of these proteins (Perozo et al. 2002; Elmore and Dougherty 2003; Moe and Blount 2005; Petrov and Martinac 2007; Debret et al. 2008; Vásquez et al. 2008). Overall, these studies have begun to give important molecular insights into how lipid composition affects MS channel gating and how interactions with lipids are determinant for modulate their activity.

Moreover, they indicate how biophysical properties of the membraneare important issues to understand the bacterial adaptability to the environment. Considering the conservation of these proteins between the three domains of life (Kloda and Martinac 2001a; Pivetti et al. 2003; Balleza and Gómez-Lagunas 2009), such findings have led to the suggestion that mechanical force transmitted via the lipid bilayer *per se* is a universal property of MS channels, shared between members of Archaea, Bacteria and Eukarya. Hence, the bilayer mechanism associated to mechanosensitivity mightbe primordial in cellular life and confirm its preservation duringevolution (Pohorille et al. 2005).

Table 1. Complete archaeal genomes sequenced and number of putative MS channels homologs

Species	Accession	MscS-like	Size (aa)	MscL	Size	Division	Habitat of isolation
Aeropyrum pernix K1	NC_000854	2	280	0	–	CrenA	Marine sediments
			286				
Archaeoglobus fulgidus DSM 4304	NC_000917	0	–	0	–	EuryA	Hydrothermal vents
A. profundus DSM 5631	NC_013741	1	272	0	–	EuryA	Hydrothermal vents
Caldivirga maquilingensis IC-167	NC_009954	1	316	0	–	CrenA	Acidic hot springs
Desulfurococcus kamchatkensis 1221n	NC_011766	1	276	0	–	CrenA	Hot springs
Haloarcula marismortui ATCC 43049	NC_006396	2	293	0	–	EuryA	Dead Sea
			371				
Halobacterium salinarum R1	NC_010364	3	300	0	–	EuryA	Hypersaline lakes and salterns
			376				
			259				
Halobacterium sp. NRC-1	NC_002607	3	300	0	–	EuryA	Extreme high salinity conditions
			274				
			259				
Halomicrobium mukohataei DSM 12286	NC_013202	4	429	0	–	EuryA	Soils of salt flats, Argentina
			302				
			316				
			334				
Haloquadratum walsbyi DSM 16790	NC_008212	3	306	0	–	EuryA	Salt lakes
			264				
			381				
Halorhabdus utahensis DSM 12940	NC_013158	5	421	0	–	EuryA	Great Salt Lake, Utah
			297				
			394				
			366				
			314				
Halorubrum lacusprofundi ATCC 49239 1	NC_012029	10	238	0	–	EuryA	Hypersaline lakes, Antarctica
			305				
			262				
			311				
			301				
			372				
			286				
			414				
			400				
			383				
Haloterrigena turkmenica DSM 5511	NC_013743	5	296	0	–	EuryA	Sulfate saline soils, Turkmenia
			310				
			385				
			345				

Species	Accession	MscS-like	Size (aa)	MscL	Size	Division	Habitat of isolation
			408				
Hyperthermus butylicus DSM 5456	NC_008818	2	297	0	–	CrenA	Solfataric sea floor
			300				
Ignicoccus hospitalis KIN4/I	NC_009776	1	257	0	–	CrenA	Hydrothermal systems
Korarchaeum cryptofilum OPF8	NC_010482	1	197	0	–	CrenA	Sediments and hot springs
Metallosphaera sedula DSM 5348	NC_009440	4	291	0	–	CrenA	Termal ponds
			232				
			307				
			256				
Methanobrevibacter smithii ATCC 35061	NC_009515	1	247	0	–	EuryA	Human gut
Methanocaldococcus fervens AG86	NC_013156	3	344	0	–	EuryA	Hydrothermal vents
			349				
			324				
M. jannaschii DSM 2661	NC_000909	3	361	0	–	EuryA	Sea floor surface, chimneys
			350				
			324				
M. vulcanius M7	NC_013407	5	348	0	–	EuryA	Hydrothermal chimneys
			362				
			326				
			270				
			348				
Methanocella paludicola SANAE	NC_013665	4	339	1	101	EuryA	Rice paddy soils
			202				
			155				
			266				
Methanococcoides burtonii DSM 6242	NC_007955	7	320	0	–	EuryA	Anoxic lakes, Antarctica
			279				
			306				
			395				
			361				
			258				
			307				
Methanococcus aeolicus Nankai-3	NC_009635	2	354	0	–	EuryA	Deep marine sediments
			333				
M. maripaludis C6	NC_009975	3	360	0	–	EuryA	Salt marshes, estuaries
			351				

Table 1. (continued)

Species	Accession	MscS-like	Size (aa)	MscL	Size	Division	Habitat of isolation
Methanopyrus kandleri AV19	NC_003551	1	244	0	–	EuryA	Hydrothermal vents
Methanoregula boonei 6A8	NC_009712	1	343	2	125	CrenA	Acidic peat bogs
Methanosaeta thermophila PT	NC_008553	2	268	1	146	EuryA	Anaerobic biofilms
			308				
Methanosarcina acetivorans C2A	NC_003552	6	291	1	101	EuryA	Deep sea hydrothermal vents
			308				
			379				
			385				
			316				
			312				
M. barkeri str. Fusaro 1	NC_007355	6	301	1	101	EuryA	Freshwater lake muds
			298				
			351				
			366				
			313				
			301				
Methanosarcina mazei Go1	NC_003901	7	314	1	77	EuryA	Bovine rumen, human intestine
			295				
			370				
			330				
			318				
			390				
			308				
Methanosphaera stadtmanae DSM 3091	NC_007681	1	253	0	–	EuryA	Human intestine
Methanosphaerula palustris E1-9c	NC_011832	2	288	1	101	EuryA	Organically rich wetland, NY
			344				
Methanospirillum hungatei JF-1	NC_007796	1	342	1	108	EuryA	Sewage sludges
Methanothermobacter thermautotrophicus	NC_000916	1	248	0	–	EuryA	Anaerobic sewage sludges
Nanoarchaeum equitans Kin4-M	NC_005213	2	337	0	–	NanoA	Hydrothermal vents
			323				
Natronomonas pharaonis DSM 2160	NC_007426	5	302	0	–	EuryA	Alcaline salt-saturated lakes
			402				
			426				
			344				
			284				
Nitrosopumilus maritimus SCM1	NC_010085	2	272	0	–	ThaumA	Salt-water aquarium, sea
			275				
Picrophilus torridus DSM 9790	NC_005877	2	299	0	–	EuryA	Dry solfataric fields, Japan

Species	Accession	MscS-like	Size (aa)	MscL	Size	Division	Habitat of isolation
Sulfolobus acidocaldarius DSM 639	NC_007181	3	312 215 296	0	–	CrenA	Terrestrial solfataric springs
Thermococcus gammatolerans EJ3	NC_012804	2	269 376	0	–	EuryA	Hydrothermal chimneys
Thermofilum pendens Hrk 5	NC_008698	2	247 215	0	–	CrenA	Archaeal commensal
Thermoplasma acidophilum DSM 1728	NC_002578	2	286 297	0	–	EuryA	Solfatara fields
Thermoproteus neutrophilus V24Sta	NC_010525	1	189	0	–	CrenA	Hot springs, Iceland

As has been mentioned before, archaeal membranes are different. Indeed, lipids from these organisms are considered to be a specific taxonomic marker (Woese et al. 1978; Woese et al. 1990).

At present, the experimental evidence show that archaeal MS channels can be successfully reconstituted in liposomes of common polar lipids as phosphatidylcholine (PC), phosphatidylethanolamine (PE), or azolectin, a mix of lecithin, cephalin and phosphatidylinositol (Kloda and Martinac 2001a,b,c; Martinac et al. 2010).

However, they can also be studied in native membranes with an interesting higher activation threshold in comparison with azolectin liposomes (Le Dain et al. 1998). Thus, protein-lipid interactions between MS channels and archaeal lipids have not been studied until now and consequently the role of these lipids in channel gating has not been explored.

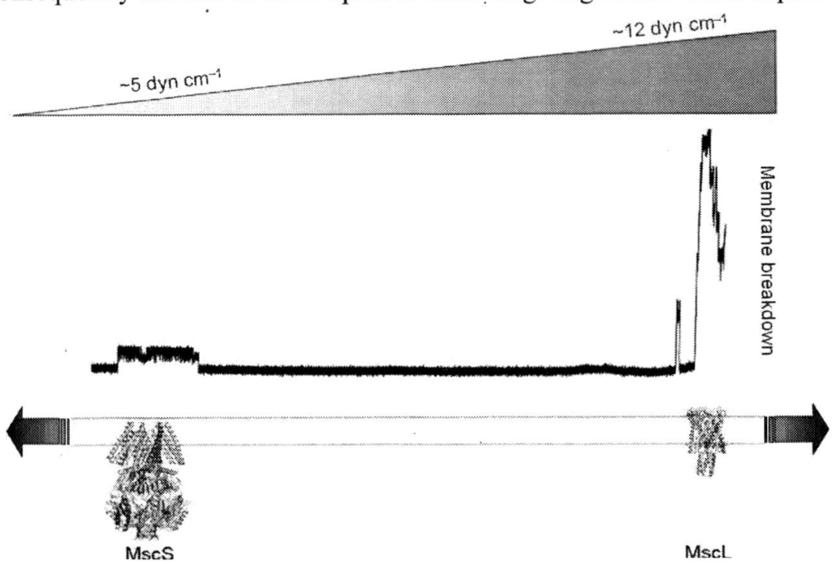

Figure 3. Bacterial MS channels. Comparative structures of the homoheptameric MscS (left) and homopentameric MscL channel (right). A typical current trace from a patch derived from an E. coli spheroplast preparation is depicted. MscS and MscL activities and the tension applied are shown. The negative pressure is generated by suction in the patch electrode.

In addition to BTLs, there is another major class of archaeal lipids, diphytanyl glycerol diethers. Both types of lipids can coexist in native membranes forming domains. They can form bilayers and monolayers, as well as a combination of bilayer/monolayer arrangements (Benvegnu et al. 2008) (see Fig. 2). It has been proposed that phase separation generates intrinsic lateral tension (Kumar et al. 1999). In giant unilamellar vesicles (GUVs) formed by dioleoyl-phosphatidylcholine(DOPC), dipalmitoyl-phosphatidylcholine (DPPC), sphingo-myelin (SM) and cholesterol, different fluid phases are formed and they segregate as domains with characteristic local membrane curvatures (Baumgart et al. 2003). Importantly, fluid/fluid phase coexistence induce line tension at phase boundaries and cholesterol induces lateral and interleaflet lipid redistribution (Baumgart et al. 2005; Honerkamp-Smith et al. 2008), which locally modify the mechanical membrane properties. Furthermore, increases in temperature lead to the formation of crystal liquid, more fluid and homogeneous membranes, disrupting such domains (Baumgart et al. 2005). These phenomena have not been studied in archaeosomes although two-photon fluorescence microscopy studies have revealed important effects of temperature and pH on GUVs composed of archaeal lipids from S. acidocaldarius (Bagatolli et al. 2000).

Since an important adaptation of thermophilic archaea in order to cope with high temperatures is the presence of BTLs in their membranes, the mechanical implications for archaeal MS channels are evident. In thermo-acidophilic archaea such as *Thermoplasma* and *Sulfolobus,* the presence of cyclopentane rings compresses the membrane by significantly shorten the length of hydrocarbon chains (Kloda and Martinac 2001a). This feature provide rigidity and controls width and density of the lipids resulting in a decrease in the fluidity through the reduction of lateral lipid mobility with a high stability to the external mechanical forces (Shinoda et al. 2005). Indeed, it has been showed that membranes of archaeosomes containing BTLs are laterally immobile, as compared to other lipid membranes (Kao et al. 1992) whereas molecular dynamics simulations suggest that cyclohexane rings decreases membrane density, which could be an adaptation to cope with cold temperatures encountered by some planktonic crenarchaea (Sinninghe-Damsté et al., 2002) (Fig. 4). Why cyclopentane and cyclohexane rings, despite their structure similarities, exert an apparent opposite effect on membrane packing has not yet been explained. However, the recent finding of the existence of crenarchaeol in hot spring archaea (Pearson et al. 2004) seems to argue against the hypothesis that cyclohexane rings in BTLs serves as an adaptation of archaea to cold marine environments. Unsaturated etherlipids have also been observed as an adaption to lower temperature (Tolner 1997). Once more, these findings provide a better understanding of the physiological role of archaebacterial lipids. Temperature has a direct influence in membrane fluidity (the inverse of viscosity), which partly comprises the effects of lateral and rotational diffusion of lipids, as well as rotation of lipids around single carbon-carbon bonds (Rubenstein et al. 1979). Lateral tension depends on fluidity and it has been measured in model membranes formed of PC (1.0 ± 0.3 mNm^{-1}) and PE (2.4 ± 0.2 mNm^{-1}) but using BTLs from *S. acidocaldarius* their value increases at least twice (4.3 ± 0.4 mNm^{-1}) (Melikyan et al. 1991).

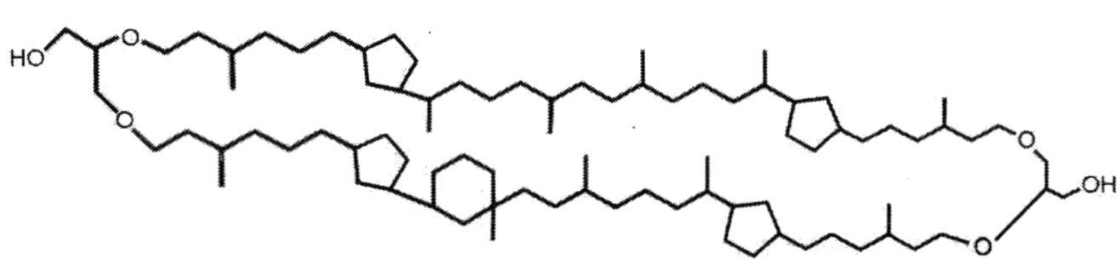

Figure 4. Crenarchaeol, a regioisomer from Cenarchaeum symbiosum.

However, at elevated temperatures (80°C) the flexibility of the chain environment in BTLs membranes approach that of fluid bilayer membranes (Bartucci et al. 2005). Fluidity is a function of temperature and thus, the physical state of membranes can be in a gel, gel-liquid or liquid-crystalline phases (Tolner et al. 1997). Membrane fluidity is also modulated by varying the composition of phospholipids (Cantor 1999a). Bacteria alter their lipid composition such that the membrane fluidity is relatively constant at different growth temperatures (Hazel 1995). On the contrary, in thermophilic archaea, the increase in number of cyclopentane rings groups reduces their membrane fluidity because it decreases with decreasing temperature (Ulrich et al. 2009). Because fluidity is directly related by the rate of lipidic lateral diffusion (Chong 2010) and temperature increases the membrane fluidity, the effect of temperature in archaeal MS channel activation and their role in osmoregulation can be, for now, only speculated and remains to be explored. Moreover, the amount of ciclopentane rings in thermoacidophiles decreases with decreasing pH, indicating that this reduction contribute to increase membrane fluidity at low pHs and to counteract the effect of an increased amount of sugar moieties (Shimada et al. 2008). The relevance of these observations to MS function also is unknown.

Similar to bacteria, a role in osmoregulation as safety valves for archaeal MS channels has been proposed (Kloda and Martinac 2001a). Besides, MS channels may also sense changes in turgor pressure during cell division and cell growth (Csonka and Epstein 1996). The pseudo rigid cell wall is a structural osmoprotectant itself.

In bacteria, it provides a peptidoglycan network that compensates pressure differences (Doyle and Marquis 1994; Koch 2006; Engelhardt 2007b). In archaea, by contrast, protein surface layers (S-layers) provide structural reinforcement to the cell wall and an apparent archaeal periplasmic space with high porosity is observed (Engelhardt 2007b). The mechanical properties of these structures have begun to emerge and show interesting properties in the context of MS channel activation.

Taking as model the structure of the S-layer unit (the *tetrabranchion*) of *Staphylothermus marinus*, it has been possible to model the biophysical properties on vesicles composed with archaeal lipids, including BTLs. In these structures, after a simulated hypo-osmotic shock, the massive water influx increases the vesicle volume, its surface area and diameter, as expected. Interestingly, however, high internal pressure dilates only the uncovered membrane cell, increasing the cell radius, and decreasing the membrane curvature concomitantly, creating rows of bulges (Fig. 5).

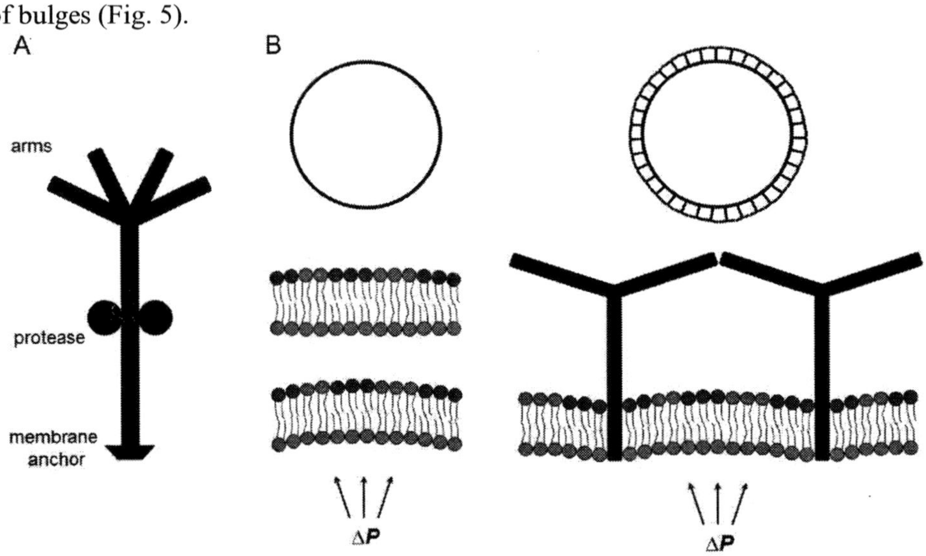

Figure 5. (A) Scheme of the Tetrabrachion structure. (B) Effects of internal pressure on the lipid membrane of model cells, covered by an S-layer of the Crenarchaeota type (right panel) and without any cell wall components (left panel). High internal pressure (ΔP) dilates the cell membrane of the uncovered cell, increases the cell radius and decreases the membrane curvature accordingly. The S-layer anchors hold the interacting membrane lipids in place. Intracellular pressure creates membrane bulges between the anchors and increases the membrane curvature locally. Modified from Engelhardt et al. (2007b).

Since multiple contacts between protein units are usual for S-layers and they are designed to withstand forces of several nN (Engelhardt 2007a), mechanical coupling with MS channels emerge as an important issue to be experimentally tested. Similarly, the protein-lipid interactions and the probable regulation by temperature, pH or salinity remain to be uncovered. On the other hand, S-layers of halobacterianeed divalent cations in order to conserve its integrity (Kessel et al. 1988). Interestingly, cells that are depleted in Mg^{2+} lose the integrity of their S-layer lattice becoming osmosensitive upon dilution of the salt medium and lyse, while cells possessing an intact S-layer are osmoresistant (Engelhardt 2007a). Therefore, a regulatory role by Mg^{2+} in archaeal MS channel activity can also be speculated.

Future Prospects to Be Studied

MS channels have been well documented in archaea. However, in comparison with their counterparts in bacteria, they still remain to be fully understood. Studies using native membranes or functional reconstitution into archaeosomes are scarce. Consequently, considering the peculiar structure of cell membranes and the cell wall on these organisms, some key questions regarding MS channels can now be addressed. First, if their sequence similarity with bacterial homologs reflects functional relatedness, archaeal MS channels play the same role in the osmotic stress response of these organisms? Second, since several copies of MscS-like channels are present in completed archaeal genomes and MscL homologues are practically absent (Table 1), what ecological significance has this genome structure? In other words, are all MscS-like channels functional? Absence of MscL reflects an adaptation to live in high salinity conditions? The tension required to open MscL is about 1.4 times the tension required to activate MscS (Sukharev et al. 1997). EcoMscL gates at a ~12 $dyncm^{-1}$ (Sukharev et al. 1999) while than TbMscL (from *Mycobacterium tuberculosis*) does at ~20 $dyncm^{-1}$ (Moe et al. 1998; 2000). These tension values are near or beyond the lytic limit of many phospholipid bilayer membranes (Evans et al. 2003). In contrast, the midpoint of EcoMscS activation (5.5 $dyncm^{-1}$) is two times lower than that of azolectin-reconstitutedMscL (Sukharev et al 2002). There is a correlation between these parameters and the presence of several copies of MscS-like channels in prokaryotes? Both archaea and bacteria possess multiple homologues of MscS/MscK. This could reflect the need for a variety of MS ion channels with different responses upon challenge with different osmotic cues (Martinac and Kloda 2003). But, they represent an ecological adaptation only to osmotic downshifts? Moreover, absence of MscL reflects that membrane structure in halophiles is incompatible with MscL channel activation?

Clearly, one of the first objectives is to obtain more electrophysiological activity for those putative channels. Third, and more important, how do these channels interact with archaeal lipids (particularly BTLs) and S-layers to sense tension? If they can respond mechanically as their bacterial counterparts, what is the structural significance at level of lipid-protein interactions? To respond these questions, archaeosomes represent a new approach to study MS channel function since they have a high degree of intrinsic rigidity and low permeability (Jacquemet et al. 2009). These model systems can serve as a relatively immobile matrix for studying lipid–protein interactions (De Rosa et al. 1994) and suggests many experiments. These include the study of MS channels in bilayers made of defined archaeal lipid composition. Particularly, the use of BTLs is interesting to determine the tension responsiveness of these proteins, given their tight membrane packing. This approach could provide some light into the BTLs effects on MS channel function. For example, protein-lipid interactions in a monolayer instead a bilayer could detect some unconsidered factor in the hydrophobic mismatch activation of these proteins. Variations in the composition of cell membranes can strongly influence the function of proteins embedded therein (Cantor 1997, 1999b), and since small changes in lateral pressures can induce a large shift in the protein conformational state, the high rigidity of archaeal membranes predicts interesting effects on MS channels. Also, archaeosomes can be used for developing new biotechnological applications (Jacquemet et al. 2009). Finally, ecological relevance can be explored by studying how physical variables as temperature, salinity, pH, and presence of divalent cations could regulate the function of these channels under a controlled hypo-osmotic challenge or during cell growth. As it was mentioned before, the S-layer supported lipid membranes provide an interesting biomechanical material with attractive properties to be explored regarding its flexibility and stability as well as the study of lipid-protein interactions.

To finish, archaeosomes can be considered as interesting protocellular models in order to study mechanosensitivity itself. The properties of some short peptides suggest that their associations, specifically in α-helical conformations are good models for primordial membrane proteins in protocells (Pohorille et al. 2005; Pohorille and Deamer 2009; Balleza 2010, *submitted*). In this regard, valinomycin,a well described natural antibiotic peptide, can be incorporated in archaeal membranes forming channels (Gliozzi et al. 1996).

However, if some channel-forming peptides have mechano-sensory responses in archaeal membranes is still hypothetical although could be anticipated (Polozov et al. 2001). Several studies had allowed the establishment of relationships between the chemical structure andthe physicochemical behaviour of archaeal lipids, particularly BTLs. Although archaea thrive in a variety of environments, particular biotechnological interest arises from extremophiles because adaptations to "hostile" high temperature, salinity and acidity have pronounced effects on the membrane. Lipids and membrane proteins of hyperthermophilic archaea are well adapted to these environmental factors. Their features make them attractive for biotechnical applications.

Nevertheless,further research is still necessary to improve ourknowledge concerning the interactions of proteins with these strange lipids. The proper approaches to solve these questions allow us to understand moreprecisely the function of MS channels in archaea and to enlarge the scope of mechanosensitivity in cells.

Conclusions

The homology founded between archaeal MS channels in comparison with their bacterial counterparts predicts conservation in protein structure and function. Archaeal lipids are, however, significantly different from bacterial lipids and their biophysical properties are just been revealed. Interesting new roles of these molecules interacting with mechano- susceptible membrane proteins can be, now, anticipated. Future research will solve some of these questions but surely will open new ones.

References

Albers SV, van de Vossenberg JL, Driessen AJ, Konings WN. 2000. Adaptations of the archaeal cell membrane to heat stress. *Front. Biosci.* 5: D813-20.

Bagatolli L, Gratton E, Khan TK, Chong PL. 2000. Two-photon fluorescence microscopy studies of bipolar tetraether giant liposomes from thermoacidophilic archaebacteria Sulfolobus acidocaldarius. *Biophys. J.* 79: 416-25.

Balleza D, Gómez-Lagunas F. 2009. Conserved motifs in mechanosensitive channels MscL and MscS. *Eur. Biophys. J.* 38: 1013-1027.

Bartucci R, Gambacorta A, Gliozzi A, Marsh D, Sportelli L. 2005. Bipolar tetraether lipids: chain flexibility and membrane polarity gradients from spin-label electron spin resonance. *Biochemistry* 44: 15017-15023.

Baumgart T, Das S, Webb WW, Jenkins JT. 2005. Membrane elasticity in giant vesicles with fluid phase coexistence. *Biophys. J.* 89: 1067-1080

Baumgart T, Hess ST, Webb WW. 2003. Imaging coexisting fluid domains in biomembrane models coupling curvature and line tension. *Nature* 425: 821-824.

Benvegnu T, Lemiègre L, Cammas-Marion S. 2008. Archaeal Lipids: Innovative Materials for Biotechnological Applications. *Eur. J. Org. Chem.* 28: 4725-4744.

Booth IR, Edwards MD, Black S, Schumann U, Miller S. 2007. Mechanosensitive channels in bacteria: signs of closure? *Nat. Rev. Microbiol.* 5: 431-440 .

Cantor RS. 1997. Lateral Pressures in Cell Membranes: A Mechanism for Modulation of Protein Function. *J. Phys. Chem. B.*101:1723-1725.

Cantor RS. 1999a. Lipid composition and the lateral pressure profile in bilayers. *Biophys. J.* 76: 2625-2639.

Cantor RS. 1999b. The influence of membrane lateral pressures on simple geometric models of protein conformational equilibria. *Chem. Phys. Lipids* 101: 45-56.

Chaban B, Ng SYM, Jarrell KF. 2006. Archaeal habitats – from the extreme to the ordinary. *Can. J. Microbiol.* 52: 73-116.

Chong PL-G. 2010. Archaebacterial bipolar tetraether lipids: Physico-chemical and membrane properties. *Chem. Phys. Lipids* doi:10.1016/j.chemphyslip.2009.12.006.

Csonka LN, Epstein W. 1996. Osmoregulation. In *Escherichia coli* and *Salmonella: Cellular and Molecular Biology,* 2nd ed. (Neidhardt, F. C., Curtiss III, R., Ingraham, J. L.,Lin, E. C. C., Brooks Low, K., Magasanik, B., ed.), ASM Press, Washington DC, pp. 1210–1225.

Debret G, Valadié H, Stadler AM, Etchebest C. 2008. New insights of membrane environment effects on MscL channel mechanics from theoretical approaches. *Proteins* 71: 1183-1196.

Denich TJ, Beaudette LA, Lee H, Trevors JT. 2003. Effect of selected environmental and physico-chemical factors on bacterial cytoplasmic membranes. *J. Microbiol. Methods* 52: 149-82.

De Rosa M, Gambacorta A, Nicolaus B, Sodano S, Bu'Lock JD. 1980. Structural regulations in tetraether lipids of *Caldariella* and their biosynthetic and phyletic implications. *Phytochemistry* 19: 833-836.

De Rosa M, Morana A, Riccio A, Gambacorta A, Trincone A, Incani O 1994. Lipids of the archaea: a new tool for bioelectronics. *Biosens. Bioelectron.* 9: 6699-6675

Doyle RJ, Marquis RE. 1994. Elastic, flexible peptidoglycan and bacterial cell wall properties. *Trends Microbiol.* 2: 57-60.

Elferink MG, de Wit JG, Driessen AJ, Konings WN. 1994. Stability and proton-permeability of liposomes composed of archaeal tetraether lipids. *Biochim. Biophys. Acta* 1193: 247-254.

Elmore DE, Dougherty DA. 2003. Investigating lipid composition effects on the mechanosensitive channel of large conductance (MscL) using molecular dynamics simulations. *Biophys. J.* 85: 1512-1524.

Engelhardt H. 2007a. Are S-layers exoskeletons? The basic function of protein surface layers revisited. *J. Struct. Biol.* 160: 115-124.

Engelhardt H. 2007b. Mechanism of osmoprotection by archaeal S-layers: a theoretical study. *J. Struct. Biol.* 160: 190-199.

Evans E, Heinrich V, Ludwig F, Rawicz W. 2003. Dynamic tension spectroscopy and strength of biomembranes. *Biophys. J.* 85: 23422-2350.

Gabriel JL, Chong PL. 2000. Molecular modeling of archaebacterial bipolar tetraether lipid membranes. *Chem. Phys. Lipids.* 105: 193-200.

Gliozzi A, Robello M, Fittabile L, Relini A, Gambacorta A. 1996. Valinomycin acts as a channel in ultrathin lipid membranes. *Biochim. Biophys. Acta* 1283: 1-3.

Haines TH. 1994. Water transport across biological membranes. *FEBS Lett.* 346: 115-122.

Haines TH, Li W, Green M, Cummins HZ. 1987. The elasticity of uniform, unilamellar vesicles of acidic phospholipids during osmotic swelling is dominated by the ionic strength of the media. *Biochemistry* 26: 5439-5447

Hazel JR. 1995. Thermal adaptation in biological membranes: is homeoviscous adaptation the explanation? *Annu. Rev. Physiol.* 57: 19-42.

Honerkamp-Smith AR, Cicuta P, Collins MD, Veatch SL, den Nijs M, Schick M, Keller SL. 2008. Line tensions, correlation lengths, and critical exponents in lipid membranes near critical points. *Biophys. J.* 95: 236-246.

Huber H, Hohn MJ, Stetter KO, Rachel R. 2003. The phylum Nanoarchaeota: present knowledge and future perspectives of a unique form of life. *Res. Microbiol.* 154: 165-171.

Itoh YH, Sugai A, Uda I, Itoh T. 2001. The evolution of lipids. *Adv. Space Res.* 28: 719-724.

Jacquemet A, Barbeau J, Lemiègre L, Benvegnu T. 2009. Archaeal tetraether bipolar lipids: Structures, functions and applications. *Biochimie* 91: 711-717.

Kao YL, Chang EL, Chong PL. 1992. Unusual pressure dependence of the lateral motion of pyrene-labeled phosphatidylcholine in bipolar lipid vesicles. *Biochem. Biophys. Res. Commun.* 188: 1241-1246.

Kessel M, Buhle EL Jr, Cohen S, Aebi U. 1988. The cell wall structure of a magnesium-dependent halobacterium, *Halobacterium volcanii* CD-2, from the Dead Sea. *J. Ultrastruct. Mol. Struct. Res.* 100: 94-106.

Kinnunen PK. 1991. On the principles of functional ordering in biological membranes. *Chem. Phys. Lipids.*57: 375-399.

Kloda A, Martinac B. 2001a. Mechanosensitive channel of *Thermoplasma*, the cell wall-less archaea: cloning and molecular characterization. *Cell Biochem. Biophys.* 34: 321-347.

Kloda A, Martinac B. 2001b. Mechanosensitive channels in archaea. *Cell Biochem. Biophys.* 34: 349-381.

Kloda A, Martinac B. 2001c. Molecular identification of a mechanosensitive channel in archaea. *Biophys. J.* 80: 229-240.

Kloda A, Martinac B. 2001d. Structural and functional differences between two homologous mechanosensitive channels of *Methanococcus jannaschii*. *EMBO J.* 20: 1888-18896.

Kloda A, Martinac B. 2002. Common evolutionary origins of mechanosensitive ion channels in Archaea, Bacteria and cell-walled Eukarya. *Archaea* 1: 35-44.

Koch AL. 2006. The exocytoskeleton. *J. Mol. Microbiol. Biotechnol.* 11: 115-125.

Kumar PBS, Gompper G, Lipowsky R. 1999.Modulated phases in multicomponent fluid membranes. *Phys. Rev. E* 60: 4610-4618.

Le Dain AC, Saint N, Kloda A, Ghazi A, Martinac B. 1998. Mechanosensitive ion channels of the archaeon *Haloferax volcanii*. *J. Biol. Chem.* 273: 12116-12119.

Lindblom G, Orädd G. 2009. Lipid lateral diffusion and membrane heterogeneity. *Biochim. Biophys. Acta* 1788: 234-244.

Liu Y, Whitman WB. 2008. Metabolic, phylogenetic, and ecological diversity of the methanogenic archaea. *Ann. N. Y. Acad. Sci.* 1125:171-189

Martinac B. 2004. Mechanosensitive ion channels: molecules of mechanotransduction. *J. Cell Sci.*117:2449-2460.

Martinac B, Kloda A. 2003. Evolutionary origins of mechanosensitive ion channels. *Prog. Biophys. Mol. Biol.* 82: 11-24.

Martinac B, Rohde PR, Battle AR, Petrov E, Pal P, Foo AF, Vásquez V, Huynh T, Kloda A. 2010. Studying mechanosensitive ion channels using liposomes. *Methods Mol. Biol.*606: 31-53.

Martinac B, Saimi Y, Kung C. 2008. Ion channels in microbes. *Physiol. Rev.* 88: 1449-1490.

Mathai JC, Sprott GD, Zeidel ML. 2001. Molecular mechanisms of water and solute transport across archaebacterial lipid membranes. *J. Biol. Chem.* 276: 27266-27271.

Melikyan GB, Matinyan NS, Kocharov SL, Arakelian VB, Prangishvili DA, Nadareishvili KG. 1991. Electromechanical stability of planar lipid membranes from bipolar lipids of the thermoacidophilic archebacterium *Sulfolobus acidocaldarius*. *Biochim. Biophys. Acta* 1068: 245-248.

Moe PC, Blount P, Kung C. 1998. Functional and structural conservation in the mechanosensitive channel MscL implicates elements crucial for mechanosensation. *Mol. Microbiol.* 28: 583-592.

Moe PC, Levin G, Blount P. 2000. Correlating a protein structure with function of a bacterial mechanosensitive channel. *J. Biol. Chem.* 275: 311211-31127.

Moe P, Blount P. 2005. Assessment of potential stimuli for mechano-dependent gating of MscL: effects of pressure, tension, and lipid headgroups. *Biochemistry* 44: 12239-12244.

Morii H, Koga Y. 1994. Asymmetrical topology of diether- and tetraether-type polar lipids in membranes of *Methanobacterium thermoautotrophicum* cells. *J. Biol. Chem.* 269: 10492-10497.

Nagle JF, Tristram-Nagle S. 2000. Structure of lipid bilayers. *Biochim. Biophys. Acta* 1469: 159-195.

Nicolas JP. 2005. A molecular dynamics study of an archaeal tetraether lipid membrane: comparison with a dipalmitoylphosphatidylcholine lipid bilayer. *Lipids* 40: 1023-1030.

Pearson A, Huang Z, Ingalls AE, Romanek CS, Wiegel J, Freeman KH, Smittenberg RH, Zhang CL. 2004. Nonmarine crenarchaeol in Nevada hot springs. *Appl. Environ. Microbiol.* 70: 5229–5237.

Perozo E. 2006. Gating prokaryotic mechanosensitive channels.*Nat. Rev. Mol. Cell Biol.* 7: 109-119.

Perozo E, Kloda A, Cortes DM, Martinac B. 2002. Physical principles underlying the transduction of bilayer deformation forces during mechanosensitive channel gating. *Nat. Struct. Biol.* 9: 696-703.

Perozo E, Rees DC. 2003. Structure and mechanism in prokaryotic mechanosensitive channels. *Curr. Opin. Struct. Biol.* 13: 432-442.

Petrov E, Martinac B. 2007. Modulation of channel activity and gadolinium block of MscL by static magnetic fields. *Eur. Biophys. J.* 36: 95-105

Pivetti CD, Yen MR, Miller S, Busch W, Tseng YH, Booth IR, Saier MH Jr. 2003. Two families of mechanosensitive channel proteins. *Microbiol. Mol. Biol. Rev.* 67: 66-85.

Pohorille A, Deamer D. 2009. Self-assembly and function of primitive cell membranes. *Res. Microbiol.* 160: 449-456.

Pohorille A, Schweighofer K, Wilson MA. 2005. The origin and early evolution of membrane channels. *Astrobiology* 5: 1-17.

Polozov IV, Anantharamaiah GM, Segrest JP, Epand RM. 2001. Osmotically induced membrane tension modulates membrane permeabilization by class L amphipathic helical peptides: nucleation model of defect formation. *Biophys. J.* 81: 949-959.

Rothschild LJ, Mancinelli RL. 2001. Life in extreme environments. *Nature* 409: 1092-1101.

Rubenstein JL, Smith BA, McConnell HM. 1979. Lateral diffusion in binary mixtures of cholesterol and phosphatidylcholines. *Proc. Natl. Acad. Sci. USA* 76:15-18.

Sackmann E, Träuble H. 1972. Studies of the crystalline-liquid crystalline phase transition of lipid model membranes. I. Use of spin labels and optical probes as indicators of the phase transition. *J. Am. Chem. Soc.* 28 94: 4482-4491

Shinoda W, Shinoda K, Baba T, Mikami M. 2005. Molecular dynamics study of bipolar tetraether lipid membranes. *Biophys. J.* 89: 3195-3202.

Shimada H, Nemoto N, Shida Y, Oshima T, Yamagishi A. 2008. Effects of pH and temperature on the composition of polar lipids in *Thermoplasma acidophilum* HO-62. *J. Bacteriol.*190: 5404-5411.

Sinninghe-Damste JS, Hopmans, EC, Schouten S, van Duin ACT, Geenevasen JAJ, 2002. Crenarchaeol: the characteristic core glycerol dibiphytanyl glycerol tetraether membrane lipid of cosmopolitan pelagic Crenarchaeota. *J. Lipid. Res.* 43: 1641–1651.

Sprott GD. 1992. Structures of archaebacterial membrane lipids. *J. Bioenerg. Biomembr.* 24: 555-566.

Sprott GD, Meloche M, Richards JC. 1991. Proportions of diether, macrocyclic diether, and tetraether lipids in *Methanococcus jannaschii* grown at different temperatures. *J. Bacteriol.* 173: 3907-3910.

Stetter KO. 1999. Extremophiles and their adaptation to hot environments. *FEBS Lett* 452: 22-25.

Sukharev S. 2002. Purification of the small mechanosensitive channel of *Escherichia coli* (MscS): the subunit structure, conduction, and gating characteristics in liposomes. *Biophys. J.* 83: 290-298.

Sukharev SI, Blount P, Martinac B, Kung C. 1997. Mechanosensitive channels of *Escherichiacoli*: the MscL gene, protein, and activities. *Annu. Rev. Physiol.* 59: 633-657.

Sukharev SI, Sigurdson WJ, Kung C, Sachs F. 1999. Energetic and spatial parameters for gating of the bacterial large conductance mechanosensitive channel, MscL. *J. Gen. Physiol.* 113: 525-540.

Swain M, Brisson JR, Sprott GD, Cooper FP, Patel GB. 1997. Identification of beta-L-gulose as the sugar moiety of the main polar lipid *Thermoplasma acidophilum*. *Biochim. Biophys. Acta* 1345: 56-64.

Tillman TS, Cascio M. 2003. Effects of membrane lipids on ion channel structure and function. *Cell Biochem. Biophys.* 38: 161-190.

Tolner B, Poolman B, Konings WN. 1997. Adaptation of microorganisms and their transport systems to high temperatures. *Comp. Biochem. Physiol. A Physiol.*118: 423-428.

Uda I, Sugai A, Itoh YH, Itoh T. 2001. Variation in molecular species of polar lipids from *Thermoplasma acidophilum* depends on growth temperature. *Lipids* 36: 103-105.

Ulrih NP, Gmajner D, Raspor P. 2009. Structural and physicochemical properties of polar lipids from thermophilic archaea. *Appl. Microbiol. Biotechnol.* 84: 249-260.

Valentine DL. 2007. Adaptations to energy stress dictate the ecology and evolution of the Archaea. *Nat. Rev. Microbiol.* 5: 316-323.

Vásquez V, Sotomayor M, Cordero-Morales J, Schulten K, Perozo E. 2008. A structural mechanism for MscS gating in lipid bilayers. *Science* 321: 1210-1214.

Woese CR, Fox GE. 1977. Phylogenetic structure of the prokaryotic domain: the primary kingdoms. *Proc. Natl. Acad. Sci. USA* 74: 5088-5090.

Woese CR, Kandler O, Wheelis ML. 1990. Towards a natural system of organisms: proposal for the domains Archaea, Bacteria, and Eucarya. *Proc. Natl. Acad. Sci. USA* 87: 4576-4579.

Woese CR, Magrum LJ, Fox GE. 1978. Archaebacteria. *J. Mol. Evol.*11: 245-251.

In: Archaea: Structure, Habitats and Ecological...
Editor: Sakura Y. Katō, pp. 163-174

ISBN 978-1-61761-932-8

Chapter 7

The Nanoarchaeota: Physiology, Genomics and Phylogeny

Ana Casanueva, Marla Tuffin and Don A. Cowan*
Institute for Microbial Biotechnology and Metagenomics,
Department of Biotechnology, University of the Western Cape,
Bellville, 7535, South Africa

Abstract

The Nanoarchaeota, proposed as the fourth sub-division of the Archaea in 2002, are known from a single isolate, "Nanoarchaeum equitans", which exists in a symbiotic association with the hyperthermophilic Crenarchaeote, Ignicoccus hospitalis. The sequencing of both genomes has provided clues on their unusual and very specific association. The discovery of nanoarchaeal sequences in diverse geographical hyperthermophilic locations, and most importantly the identification of nanoarchaeal phylotypes in mesophilic halophilic environments suggest the possibility of either free-living nanoarchaeotes or associations with novel hosts. The "true" phylogenetic position of the Nanoarchaea is still a matter of debate, having been described originally as a separate phylum and later as a member of the phylum Euryarchaeota. The present review will discuss the recent discoveries on this hyperthermophilic organism and its association to Ignicoccus and will analyse the different views on the phylogenetic position of this unique organism.

Keywords: *Nanoarchaeaota*, 16S rRNA, phylogeny, hyperthermophilic.

*Institute for Microbial Biotechnology and Metagenomics, Department of Biotechnology, University of the Western Cape, Private Bag X17, Bellville, 7535, South Africa; Phone: (+2721) 959 2083; Fax: (+2721) 959 3505; E-mail: dcowan@uwc.ac.za

Introduction

Until fairly recently, the domain of the *Archaea* consisted of two distinct branches: the *Crenarchaeota* and the *Euryarchaeota*, determined mostly by sequence comparisons of their 16S rRNA genes (Woese *et al.*, 1990). In the past decade three new sub-divisions have been described within the *Archaea*: the *Korarchaeota*, *Nanoarchaeota* and *Thaumarchaeota* (Barns *et al.*, 1996; Huber *et al.*, 2002; Brochier-Armanet *et al.*, 2008). The sub-division *Nanoarchaeota* is known from a single isolate (Huber *et al.*, 2002), two environmental DNA samples (Hohn *et al.*, 2002), and multiple DNA sequences (McCliment *et al.*, 2006; Casanueva *et al.*, 2008). "*Nanoarchaeum equitans*", the only isolated member of the *Nanoarchaeota*, is a 400 nm diameter coccus typically found associated with the hyperthermophilic chemolithotrophic crenarchaeote *Ignicoccus hospitalis* (Huber *et al.*, 2002; Paper *et al.*, 2007). Here we discuss recent discoveries relating to this organism and its association to *Ignicoccus*, the environmental distribution of putative members of the *Nanoarchaeota*, and analyse the different views on the phylogenetic position of this unique organism.

"Nanoarchaeum Equitans"

The discovery of "*Nanoarchaeum equitans*" has been one of the highlights of this decade for the microbiological community. "*N. equitans*" was discovered following an expedition to the hydrothermal sites at Kolbeinsey Ridge, North of Iceland, at a depth of 106 m (Huber *et al.*, 2002). "*N. equitans*" is a tiny hyperthermophilic coccus with a cell diameter ranging from 350 to 500 nm (Huber *et al.*, 2003). It grows in a symbiotic or parasitic relationship with the crenarchaeote *Ignicoccus hospitalis*, as shown by the fact that nanoarchaeal cells do not grow unless attached to the host's surface (Huber *et al.*, 2002). The "*N. equitans*" 16S rRNA gene failed to amplify using universal and archaeal 16S rRNA gene primers due to several base exchanges in previously "conserved" regions and only nanoarchaeal-specific primers allowed amplification of the gene sequence (Huber *et al.*, 2002; Hohn *et al.*, 2002).

The growth conditions of "*N. equitans*" mimic that of its host, i.e., anaerobic growth with an optimal temperature of 90°C, optimal pH from 5.5-6.0 and salt concentrations ranging from 1 to 4% (Jahn *et al.*, 2008). Microscopic images of "*N. equitans*" show cells with a round, tightly packed cytoplasm surrounded by a cytoplasmic membrane 8 nm wide, a periplasmic space of 20 nm and an outermost S-layer sheath with six fold symmetry and a lattice constant of 15 nm (Huber *et al.*, 2002; Huber *et al.*, 2003; Junglas *et al.*, 2008). The cytoplasm contains bundles of fibres which are either elongated or ring-shaped (Junglas *et al.*, 2008). Cell division involves constriction in the mid-plane of the cocci (Junglas *et al.*, 2008). The "*N. equitans*" genome is one of the smallest non-viral sequenced genomes to date, with a circular chromosome of 490 885 bp (Waters *et al.*, 2003). It contains 552 coding DNA sequences (CDS) and has very high gene density with CDS and RNAs making up 95% of the genome, with little non-coding DNA and very few pseudogenes. One third of all CDS have unknown functions and of these, only 18% have homologs in other genomes. The genome contains single copies of the 5S, 16S and 23S rRNA genes, but these are not arranged in an operon as in most other organisms. One of the most striking features of the "*N. equitans*" genome is the lack of most genes for the biosynthesis of amino acids, nucleotides, cofactors and lipids. There are genes present for incomplete glycolysis, gluconeogenesis, and pentose phosphate pathways and some TCA cycle genes, but other carbon assimilation pathways are completely missing. It is therefore likely that "*N. equitans*" acquires most of its metabolites from its host. However, very few of the annotated genes are identified as transporters. Despite the fact that the "*N. equitans*" genome lacks a large number of metabolic genes, it contains an impressive set of genes involved in DNA repair pathways, as well as a complete set of genes for information processing (replication, transcription and translation). Another unusual characteristic of this genome is the presence of split genes, i.e., two CDS in different regions of the chromosome coding for one gene product. In many cases, each CDS codes for separate functional domains. One example is the reverse gyrase, an enzyme found only in hyperthermophiles, which is encoded by two different genes; NEQ434 encoding the helicase domain and NEQ318 coding for the topoisomerase domain. The original analysis of the "*N. equitans*" genome did not show the presence of the transfer RNA genes (tRNA) for glutamate, histidine, tryptophan and the initiator methionine (Waters *et al.*, 2003). However, a new search identified 9 CDS encoding separate fragments of tRNA genes (Randau *et al.*, 2005a; Randau *et al.*, 2005b). Most of the split tRNA genes were separated into a 5'-half after position 37, which is 1 nucleotide downstream of the anticodon, a common location for tRNA introns. The 3'-half contains a 12-14 bp GC-rich leader sequence which is the reverse complement to the 5'-half downstream sequence. These are thought to anneal and to allow the linkage of the tRNA halves by a splicing endonuclease. Apart from split tRNA genes, "*N. equitans*" also possesses a number of tRNAs which require *cis*-splicing due to the presence of introns. Together, all these tRNA genes are able to read all 61 sense codons (Randau *et al.*, 2005b). The splicing endonuclease performs a dual role in removing tRNA introns and joining tRNA halves (Randau *et al.*, 2005c; Mitchell *et al.*, 2009). This enzyme is a heteromeric endonuclease made up of 2 different subunits with homologs in *Methanopyrus kandleri*, a euryarchaeote, and in all crenarchaeotes.

Another interesting characteristic of The "*N. equitans*" genome is the absence of a ribonuclease P gene (RNase P). The lack of RNase P activity suggests that this function is not provided by one of the CDS with unknown function. RNase P activity is thought to be essential for tRNA 5'-maturation (Randau *et al.*, 2008), and is present in almost all organisms. It has been suggested that the catalytic RNA domain is an ancient functional feature, possibly a remnant of an 'RNA world' and may have been present in the last universal ancestor. "*N. equitans*" is thought to be able to dispense with this activity by having a strict promoter placement upstream of the tRNA genes which allows production of leaderless tRNA transcripts.

A number of "*N. equitans*" genes have been studied in detail, including NEQ288 which codes for an archaeal histone with a unique 4-residue insertion similar to eukaryotic histones (Friedrich-Jahn *et al.*, 2009), and the superoxide reductase gene neelaredoxin, the gene product of which is fully active despite lacking a conserved glutamate found in all superoxide reductases (Rodrigues *et al.*, 2008). One "*N. equitans*" protein which may have wider application is the family B-type DNA polymerase, encoded in 2 separate genes, a large gene (NEQ068) coding for the N-terminus and a smaller gene (NEQ528) for the C-terminus (Choi *et al.*, 2006). The NEQ528 gene product has no activity, while that of NEQ068 has a low 3'-5' exonuclease activity. Together, the protein has polymerase and 3'-5' exonuclease activity.

This enzyme is able to use deaminated bases such as uracil and hypoxanthine and has been successfully used in PCR, producing fewer errors than *Taq* polymerase (Choi *et al.*, 2008).

A comparative analysis of the "*N. equitans*" genome and proteome described features involved in its dual-adaptation to hyperthermophily and obligate parasitism (Das *et al.*, 2006). Hyperthermophilic adaptations included a higher GC content in tRNA and rRNA genes, increased use of aromatic and positively charged amino acid residues in proteins and overrepresentation of purine bases in protein-coding genes. The most striking adaptations to symbiosis (or parasitism) were a large variation in probable membrane associated proteins and a large amount of coding capacity devoted to surface-associated proteins. This clearly suggests the importance of the physical association between the two cell types, and may explain the apparent lack of transporter genes.

Ignicoccus Hospitalis, Host of "Nanoarchaeum Equitans"

All members of the genus *Ignicoccus* are hyperthermophilic, anaerobic, chemolithoautotrophic archaea which gain energy by reduction of elemental sulphur using hydrogen as an electron donor (Huber *et al.*, 2000). They belong to the phylum*Crenarchaeota* and are deeply branched within the *Desulfurococcales*. *Ignicoccus hospitalis* was isolated from a hydrothermal site at Kolbeinsey Ridge, Iceland, and it is the only known host for "*N. equitans*" (Huber *et al.*, 2002; Paper *et al.*, 2007). Its unusual carbon assimilation pathway has recently been resolved and identified as the dicarboxylate/4-hydroxybutyrate autotrophic carbon assimilation cycle (Huber *et al.*, 2008). *Ignicoccus* is the only archaeal genus surrounded by 2 membranes with no rigid cell wall such as an S-layer (Näther and Rachel, 2004; Junglas *et al.*, 2008). The *I. hospitalis* unusual cell envelope consists of an ondulating cytoplasmic membrane with attached vesicles, a periplasmic space which varies in size from 20 to 500 nm within the same cell and also contains vesicles, and an outer membrane. The outer membrane is composed of 2 leaflets with archaeal lipids and a large number of the same thermostable pore-forming protein Imp1227, tightly packed on the membrane (Burghardt *et al.*, 2007). The lipid compositions of the cytoplasmic and the outer membrane differ and it is likely that *I. hospitalis* possesses a specialised system for sorting the different lipids and locating them to the correct membrane. The membrane lipids of *I. hospitalis* and "*N. equitans*" are identical, consisting of archaeol and caldarchaeol lipids (Jahn *et al.*, 2004). This suggests that the lipids are synthesised in *I. hospitalis* and would explain the lack of lipid biosynthetic genes in the genome of "*N. equitans*" (Waters *et al.*, 2003), and thereby direct contact with *I. hospitalis* is necessary for growth of "*N. equitans*" (Huber *et al.*, 2002; Junglas *et al.*, 2008). This almost certainly reflects the necessity of transport of metabolites such as lipids and amino acids from the host to the parasite. Ultrastructure studies of the physical interaction between these two organisms showed that in 83% of the cases "*N. equitans*" cells attached to *I. hospitalis* at a site where the host's cytoplasmic membrane was in contact or very close with its outer membrane with no vesicles present (Junglas *et al.*, 2008). In a minority of instances, the contact was at a site where the periplasmic space was larger than 300 nm and a large number of vesicles were present. In both cases the two cells were either in direct contact or with a 50-60 nm gap bridged by thin fibres, thought to span from the "*N.

equitans" cytoplasm through its S-layer and the *I. hospitalis* membranes, into the host's cytoplasm.

It is yet to be determined definitively whether the relationship between these two organisms is parasitic. The presence of "*N. equitans"* cells did not affect the growth rate of *I. hospitalis* cultures, but did inhibit cell proliferation of individual cells (Jahn *et al.*, 2008). *I. hospitalis* cells with more than two "*N. equitans"* cells attached were not able to form a subculture, possibly due to the inhibition of cell division. Once separated from *I. hospitalis*, it was not possible to reinfect the host with "*N. equitans"*. The dynamics of this unique relationship were studied by analysing growth curves of the co-culture. During *I. hospitalis* log phase, 30% of host cells were associated with 1 to 3 "*N. equitans"* cells. Shortly before *I. hospitalis* reached stationary phase, "*N. equitans"* cells proliferated, with a large number of host cells showing up to 10 "*N. equitans"* cells attached. At *I. hospitalis* stationary phase, more than 80% of *I. hospitalis* cells were covered by 10 "*N. equitans"* cells and 40 to 60% of "*N. equitans"* were found as viable free cells in the supernatant. The complete genome of *I. hospitalis* was recently sequenced and found to be the smallest genome of a free-living organism, with a single circular chromosome of 1 297 538 bp (Podar *et al.*, 2008). The combined genomes of *I. hospitalis* and "*N. equitans"* are smaller than the average free-living prokaryote genome, with both having highly reduced genomes (Forterre *et al.*, 2009). Elements of the *I. hospitalis* genome include the reduced frequency of duplicated genes, the loss of conservation of operon structures and the absence of transposable elements (genome streamlining) (Podar *et al.*, 2008; Forterre *et al.*, 2009). *Ignicoccus* is far removed from the root of the thermophilic *Crenarchaeota* phylogenetic tree (Podar *et al.*, 2008) and genome reduction is thought to have occurred after the divergence from *Desulfurococcales* (Forterre *et al.*, 2009). Evidence of lateral gene transfer between *I. hospitalis* and "*N. equitans"* is shown by the presence of 13 genes where the orthologs are present in both genomes (Podar *et al.*, 2008).

To date, the symbiotic relationship between "*N. equitans"* and *I. hospitalis* is unique in the archaeal world. "*N. equitans"* is not able to infect other hosts and is strictly dependent on *I. hospitalis* for growth (Huber *et al.*, 2002). There is a small possibility that *Nanoarchaea* confer a physiological advantage to *Ignicoccus*, given the high numbers of *Ignicoccus* and nanoarchaeotes in the colonization of nascent hydrothermal vent chimneys (McCliment *et al.*, 2006). Preliminary studies on two novel nanoarchaeal isolates, OP9 and CU1, showed that these were attached to alternative hosts, possibly *Pyrobaculum* (Waters *et al.*, 2003). However, the recalcitrance of *Nanoarchaea* to isolation and culturing has hampered all efforts to understand the wider distribution and ecological role of the genus.

Environmental Distribution of Nanoarchaeotes

It was originally thought that Archaea were restricted to extreme environments. However, members of the Domain have now been shown to be present in almost all environments (Robertson *et al.*, 2005). A similar misconception may apply to *Nanoarchaea*. The highly divergent 16S rRNA sequence of *Nanoarchaea* may have prevented the observation of these organisms in different environments using standard 16S rRNA gene PCR protocols. New archaeal-universal primers (Baker *et al.*, 2003) and nanoarchaeal-specific primers (Huber *et al*., 2002; Hohn *et al*., 2002; Casanueva *et al*., 2008) have allowed the identification of a large number of nanoarchaeal 16S rRNA sequences in diverse environments (Hohn *et al*., 2002; McCliment *et al*., 2006; Casanueva *et al*., 2008). Until recently, all nanoarchaeal sequences were identified in hyperthermophilic sites: *N. equitans* from deep sea hydrothermal vents (Huber *et al*., 2002); OP9 and CU1 uncultured nanoarchaeotes from Yellowstone National Park and Kamchatka, respectively (Hohn *et* al., 2002); nanoarchaeal sequences from Pacific deep-sea hydrothermal vents (McCliment *et al*., 2006); and nanoarchaeal phylotypes from Chinese, New Zealand, Tibet and Chilean hot springs (Casanueva *et al*., 2008). All putative nanoarchaeal 16S rRNA gene sequences group together with high bootstrap support and are more closely related to each other than to sequences from any other archaeal phyla.

The first hint of the possibility of non-hyperthermophilic nanoarchaea was found in deep-sea hydrothermal vent chimneys, where phylogenetically distinct nanoarchaeal sequences dominated the lower temperature regions of the system (4 to 10°C and 6 to 67°C) (McCliment *et al*., 2006). The identification of nanoarchaeal signals in mesophilic hypersaline water and sediment samples from Inner Mongolia and South Africa, which are very unlikely to be contaminated by hydrothermal sources, strongly suggests that "nanoarchaea" may be much more physiologically flexible than previously thought (Casanueva *et al*., 2008). There is also evidence of a wide geographical distribution of members of the *Nanoarchaeota*, with new members being found in Asia, Oceania, South America and Africa. At present it is not known whether these nanoarchaeotes exhibit a "*N. equitans*"-like symbiotic lifestyle or whether they are free-living. While the putative host for nanoarchaeotes OP9 and CU1 is thought to be the hyperthermophilic archaeon *Pyrobaculum*, it is self-evident that the host cells from the hypersaline nanoarchaeotes would be mesophilic extreme halophiles.

Despite the fact that a number of 16S rRNA nanoarchaeal gene sequences have been identified, the placing of the branch within the archaeal phylogenetic tree remains problematic. The discovery of non-hyperthermophilic nanoarchaeotes sheds a different light on the "ancestry" of these organisms.

N. Equitans: Ancestral or Highly Evolved?

The 16S rRNA gene from "*N. equitans*" and the other two uncultured nanoarchaeotes OP9 and CU1 has several base exchanges in universally conserved regions (Huber *et al*., 2002; Hohn *et al*., 2002). The secondary structure of the "*N. equitans*" 16S rRNA gene has standard features of the *Archaea* (Huber *et al*., 2002) and there is no doubt that these organisms are members of this Kingdom. The exact placement within the *Archaea* is, however, an object of ongoing discussion and the exact branching point in the phylogenetic trees is dependent on the molecule(s) being used for the analysis. *Nanoarchaea* were proposed as a new archaeal phylum and considered ancient (Huber *et al*., 2002; Waters *et al*., 2003), but it has been more recently suggested that these organisms are highly evolved members of the *Euryarchaeota* (Brochier *et al*., 2005; Makarova and Koonin, 2005) and even members of the *Crenarchaeota* (Ciccarelli *et al*., 2006).

Nanoarchaea as a Novel Ancestral Phylum

A number of phylogenetic studies using a variety of molecular markers have placed *Nanoarchaea* as a deep branch within the *Archaea* and therefore as a representative of a new phylum of *Archaea*. The 16S rRNA genes of "*N. equitans*", OP9 and CU1 were shown to have 67-80% identities to those of other *Archaea* and 83-100% identity to each other, suggesting that they are phylogenetically distinct from other *Archaea* (Huber *et al.*, 2003). Phylogenetic studies of 16S rRNA genes using different algorithms place them as an isolated, deeply branching lineage of the *Archaea*, although the position of the branch differs depending on the method used (Huber *et al.*, 2002, Hohn *et al.*, 2002, Huber *et al.*, 2003, Branciamore *et al.*, 2008). A similar result was achieved using the 23S rRNA gene as a marker (Huber *et al.*, 2003). An analysis of concatenated amino acids of 35 ribosomal proteins placed "*N. equitans*" at the most deeply branching position within the *Archaea*, suggesting a very early evolutionary divergence (Waters *et al.*, 2003). Brochier *et al.* (2005a) arrived to the same result using 50 concatenated ribosomal proteins from 25 archaeal genomes. The authors suggested that the placement of the branch could be biased by an above average evolutionary rate of "*N. equitans*" genome, causing long branch attraction (LBA, where fast evolving taxa cluster together forming long branches). Using translational ribosomal proteins and transcriptional proteins such as RNA polymerase and transcriptional factors, "*N. equitans*" was grouped as a separate phylum. This was supported by strong bootstrap values, but the long branch length retains some risk of an LBA artefact (Brochier *et al.*, 2005b).

The complete lack of operon structure in the "*N. equitans*" genome (Waters *et al.*, 2003) can be interpreted as an ancestral trait (DiGiulio, 2007). The ribosomal superoperon is conserved in all *Archaea* and *Bacteria*, but in "*N. equitans*" is nearly completely absent with ribosomal genes scattered throughout the genome. Another unique, possibly ancestral feature of its genome is the presence of split genes. The Exon gene theory proposes that the precursors of current genes were mini-genes coding for separate protein modules (Di Giulio 2006a). It is then possible that the split genes in "*N. equitans*" represent ancestral evolutionary genes. This would suggest that split genes gave rise to genes with introns and exons, which in turn gave rise to intron-less genes (Di Giulio 2008). If this was the case, the 3' and 5' termini of "*N. equitans*" split genes, respectively, would represent sites where introns would occur in eukaryotic orthologs. This was found to be the case for a number of "*N. equitans*" split genes: DNA polymerase delta, an RNA binding protein, alanyl tRNA synthetase, RNA polymerase subunit B, topoisomerase and N-acetyltransferase 10, suggesting that these represent the ancestral version of such genes. Interestingly, "*N. equitans*" reverse gyrase, a gene only present in thermophiles, is also split, supporting the theory that the last universal ancestor was a thermophile.

Some "*N. equitans*" tRNA genes are also split, supporting the model that tRNAs originated by direct duplication of RNA hairpin structures (Di Giulio, 2006b). The 'cleavage points' of the "*N. equitans*" tRNA genes also coincides with positions where introns occur in other tRNAs. A phylogenetic analysis of tRNAs in 45 Archaea concluded that tRNA emerged through a combination of separate genes (Fujishima *et al.*, 2008). This would also suggest that these are examples of ancestral rather than evolved genes (Di Giulio, 2006b; DiGiulio, 2007; DiGiulio, 2009).

It has also been noted that the "*N. equitans*" genome shows no evidence of genome degradation, such as the presence of pseudogenes and long intergenic distances, phenomena often seen in highly evolved parasites (Makarova and Koonin, 2005).

Taken together, these characteristics suggest that "*N. equitans*" is an ancestral, primitive organism with a very early evolutionary divergence. However, the genome of "*N. equitans*" is anything but primitive, having modern archaeal type information machinery (Waters *et al.*, 2003). It also lacks genes for central metabolism, biosynthesis and bioenergy which would be expected to be present in the archaeal ancestor.

Nanoarchaea as a Highly Evolved Member of the *Euryarchaea*

Various studies have placed the *Nanoarchaea* as highly evolved members of the *Euryarchaea.* Using 23 concatenated proteins from the small ribosomal subunit from 25 archaeal genomes for construction of phylogenetic trees, "*N. equitans*" was placed within the *Euryarchaeota* with 98% bootstrap value (Brochier *et al.*, 2005a).

The same analysis was performed using each "*N. equitans*" ribosomal protein separately. In each tree, "*N. equitans*" had a very long branch length, suggesting a fast evolutionary rate: 33 of the proteins clustered with the *Euryarchaeota*, while 9 grouped with the *Crenarchaeota*, but this could be due to Lateral Gene Transfer (LGT) from its host*Ignicoccus*. A new analysis removing these 9 proteins firmly placed *N. equitans* within the *Euryarchaea*, as a sister branch to the *Thermococcales*. A BlastP analysis of 563 open reading frames (ORFs) showed no homology for a third of these, 12% with homology to crenarchaeal proteins, 56% with euryarchaeal proteins including 25% to *Thermococcales*. Phylogenetic analyses using elongation factors EF-1α and EF-2, topoisomerase and reverse gyrase as molecular markers all placed "*N. equitans*" within the *Euryarchaeota*, as a late branch within the *Thermococcales*. "*N. equitans*" also shares a large number of signature genes with *Euryarchaeota*, especially *Pyrococcus,* a member of the order *Thermococcales* (Dutihl *et al.*, 2008). Nanoarchaeal histone dimers contain a proline tetrad motif which is also present in euryarchaeal histone dimers but absent from the *Crenarchaea* (Robertson *et al.*, 2005). "*N. equitans*" also encodes 5 of the 9 proteins which are conserved in all *Euryarchaeota* and not found in the *Crenarchaeota*, but it does not encode any of the 16 crenarchaeal genes found only in all *Crenarchaeota* (Makarova and Koonin, 2005). Taken together, these studies suggest that *Nanoarchaea* are highly derived members of the *Euryarchaeota*, but due to their fast evolutionary rate appear to have a basal rooting in the phylogenetic tree.

The genome of *I. hospitalis* is highly reduced and streamlined which suggest it is a highly evolved organism which has specialised for its anaerobic, hyperthermophilic, chemolithoautotrophic lifestyle (Forterre *et al.*, 2009). The close and dependent relationship of "*N. equitans*" with its host would suggest that its genome is also highly evolved and specialised to a parasitic lifestyle. The lack of RNase P in "*N. equitans*", a universally conserved protein in all organisms except in a few highly evolved parasites, and the lack of a large number of metabolism and biosynthesis genes suggest genome reduction in "*N. equitans*" (Waters *et al.*, 2003; Randau *et al.*, 2008). Certain highly evolved parasitic organisms have undergone genome rearrangement which could explain the extent of the operon disruption found in the "*N. equitans*" genome (Makarova and Koonin, 2005).

It has also been shown that parasitic genomes often evolve much faster than their free-living counterparts, their high evolutionary rates obscuring their true positions in phylogenetic trees. Interestingly, if "*N. equitans*" is a highly derived member of the *Euryarchaeota* due to its parasitic lifestyle, it has followed a unique path to parasitism by conserving nearly a complete information processing system while losing most of its metabolic capacity.

Conclusion

The jury is still out on whether "*N. equitans*", and other members of this group, represent early divergent ancestral organisms, or whether these are highly evolved, highly specialised microbes. The identification of 16S rRNA gene sequences from other members of this group in the past decade has not been able to shed any light on the debate. Sequencing of its genome and that of its host, *I. hospitalis*, has demonstrated the highly specialised relationship shared by these organisms but has not been able to elucidate the origin of "*N. equitans*". The isolation of novel nanoarchaeotes, the generation of new genome sequences and the characterisation of their lifestyles, whether free-living or parasitic, would be of crucial importance in the resolution of this issue. Whether an ancestral or a highly evolved parasite, "*N. equitans*" is a unique and interesting organism which has surprised the microbiological world and promises to amaze us even further.

References

Baker GC, Smith JJ, Cowan DA (2003) Review and re-analysis of domain-specific 16S primers. *J. Microbiol. Methods*55:541-555.

Barns SM, Delwiche CF, Palmer JD, Pace NR (1996) Perspectives on archaeal diversity, thermophily and monophyly from environmental rRNA sequences. *Proc. Natl. Acad. Sci. USA*93:9188-9193.

Branciamore S, Gallori E, Di Giulio M (2008) The basal phylogenetic position of Nanoarchaeum equitans (Nanoarchaeota). *Front Biosci.* 13:6886-6892.

Brochier C, Gribaldo S, Zivanovic Y, Confalonieri F, Forterre P (2005a) Nanoarchaea: representatives of a novel archaeal phylum or a fast-evolving euryarchaeal lineage related to Thermococcales? *Genome Biol.* 6:R42.

Brochier C, Forterre P, Gribaldo S (2005b) An emerging phylogenetic core of Archaea: phylogenies of transcription and translation machineries converge following addition of new genome sequences. *BMC Evol. Biol.* 5:36.

Brochier-Armanet C, Boussau B, Gribaldo S, Forterre P (2008) Mesophilic *Crenarchaeota*: proposal for a third archaeal phylum, the *Thaumarchaeaota*. *Nat. Rev. Microbiol.* 6: 245-252.

Burghardt T, Näther DJ, Junglas B, Huber H, Rachel R (2007) The dominating outer membrane protein of the hyperthermophilic Archaeum *Ignicoccus hospitalis*: a novel pore-forming complex. *Mol. Microbiol.* 63:166-176.

Casanueva A, Galada N, Baker GC, Grant WD, Heaphy S, Jones B, Yanhe M, Ventose A, Blamey J and Cowan DA (2008) Nanoarchaeal 16S rRNA gene sequences are widely dispersed in hyperthermophilic and mesophilic halophilic environments. *Extremophiles* 12:651-656.

Choi JJ, Nam KH, Min B, Kim SJ, Söll D and Kwon ST (2006) Protein *Trans*-splicing and Characterization of a Split Family B-type DNA Polymerase from the Hyperthermophilic *Nanoarchaeum equitans*. *J. Mol. Biol.* 356:1093-1106.

Choi JJ, Song JG, Nam KH, Lee JI, Bae H, Kim GA, Sun Y, Kwon ST (2008) Unique substrate spectrum and PCR application of *Nanoarchaeum equitans* family B DNA polymerase. *Appl. Environ. Microbiol.* 74:6563-6569.

Ciccarelli FD, Doerks T, von Mering C, Creevey CJ, Snel B, Bork P (2006) Toward Automatic Reconstruction of a highly Resolved Tree of Life. *Science* 311:1283-1287.

Das S, Paul S, Bag SK and Dutta C (2006) Analysis of *Nanoarchaeum equitans* genome and proteome composition: indications for hyperthermophilic and parasitic adaptation. *BMC Genomics* 7:186.

Di Giulio M (2006a) Nanoarchaeum equitans is a living fossil. *J. Theor. Biol.* 242:257-260.

Di Giulio M (2006b) The non-monophyletic origin of the tRNA molecule and the origin of genes only after the evolutionary stage of the last universal common ancestor (LUCA). *J. Theor. Biol.* 240:343-352.

Di Giulio M (2007) The tree of life might be rooted in the branch leading to Nanoarchaeota. *Gene* 401:108-113.

Di Giulio M (2008) The split genes of Nanoarchaeum equitans are an ancestral character. *Gene* 421:20-26.

Di Giulio M (2009) Formal proof that the split genes of tRNAs of Nanoarchaeum equitans are an ancestral character. *J. Mol. Evol.* 69:505-11.

Dutihl BE, Snel B, Ettema TJG and Huynen MA (2008) Signature genes as a phylogenomic tool. *Mol. Biol. Evol.* 25:1659-1667.

Forterre P, Gribaldo S, Brochier-Armanet C (2009) Happy together: genomic insights into the unique *Nanoarchaeum/Ignicoccus* association. *J. Biol.* 8:7.

Friedrich-Jahn U, Aigner J, Längst, Reeve JN and Huber H (2009) Nanoarchaeal origin of Histone H3? *J. Bacteriol.* 191:1092-1096.

Fujishima K, Sugahara J, Tomita M, Kanai A (2008) Sequence evidence in the archaeal genomes that tRNAs emerged through the combination of ancestral genes as 5' and 3' tRNA halves. *PLoS One* 3:e1622.

Hohn MJ, Hedlund BP, Huber H (2002) Detection of 16S rDNA sequences representing the novel phylum "*Nanoarchaeota*": Indication for a wide distribution in high temperature biotopes. *Syst. Appl. Microbiol.* 25:551-554.

Huber H, Burggraf S, Mayer T, Wyschkony I, Rachel R, Stetter KO (2000) *Ignicoccus* gen. nov., a novel genus of hyperthermophilic, chemolithoautotrophic Archaea, represented by two new species, *Ignicoccusislandicus* sp nov and *Ignicoccus pacificus* sp nov. *Int. J. Syst. Evol. Microbiol.* 50:2093-2100.

Huber H, Hohn MJ, Rachel R, Fuchs T, Wimmer VC, Stetter KO (2002) A new phylum of *Archaea* represented by a nano-sized hyperthermophilic symbiont. *Nature* 417:63-67.

Huber H, Hohn MJ, Stetter KO and Racherl R (2003) The phylum*Nanoarchaeota*: Present knowledge and future perspectives of a unique form of life. *Res. Microbiol.* 154:165-171.

Huber H, Gallenberger M, Jahn U, Eylert E, Berg IA, Kockelkorn D, Eisenreich W, Fuchs G (2008) A dicarboxylate/4-hydroxybutyrate autotrophic carbon assimilation cycle in the hyperthermophilic Archaeum *Ignicoccus hospitalis*. *Proc. Natl. Acad. Sci. U S A* 105:7851-7856.

Jahn U, Summons R, Sturt H, Grosjean E, Huber H (2004) *Composition of the lipids of Nanoarchaeum equitans* and their origin from its host*Ignicoccus* sp. strain KIN4/I. *Arch. Microbiol.* 182:404-413.

Jahn U, Gallenberger M, Paper W, Junglas B, Eisenreich W, Stteter KO, Rachel R, Huber H (2008) *Nanoarchaeum equitans* and *Ignicoccus hospitalis*: New Insights into a Unique Intimate Association of Two Archaea. *J. Bact* 190:1743-1750.

Jahn U, Gallenberger M, Paper W, Junglas B, Eisenreich W, Stetter KO, Rachel R, Huber H (2008) Nanoarchaeum equitans and Ignicoccus hospitalis: new insights into a unique, intimate association of two archaea. *J. Bacteriol.* 190: 1743-1750.

Junglas B, Briegel A, Burghardt T, Walther P, Wirth R, Huber H and Rachel R (2008) *Ignicoccus hospitalis* and *Nanoarchaeum equitans*: ultrastructure, cell-cell interaction, and 3D reconstruction from serial sections of freeze-substituted cells and by electron cryotomography. *Arch. Microbiol.* 190:395-408.

Makarova KS and Koonin EV (2005) Evolutionary and functional genomics of the Archaea. *Curr. Opin. Microbiol.* 8:586:594.

McCliment EA, Voglesonger KM, O'Day PA, Dunn EE, Holloway JR, Cary SC (2006) Colonization of nascent, deep-sea hydrothermal vents by a novel Archaeal and Nanoarchaeal assemblage. *Environ. Microbiol.* 8:114-125.

Mitchell M, Xue S, Erdman R, Randau L, Söll D and Li H (2009) Crystal structure and assembly of the functional *Nanoarchaeum equitans* tRNA splicing endonuclease. *Nucleic. Acids Res.* 37:5793-5802.

Näther DJ and Rachel R (2004) The outer membrane of the hyperthermophilic archaeon *Ignicoccus*: dynamics, ultrastructure and composition. *Biochem. Soc.Trans.* 32:199-203.

Paper W, Jahn u, Hohn MJ, Kronner M, Näther DJ, Burghardt T, Rachel R, Stetter KO, Huber H (2007) *Ignicoccus hospitalis* sp. nov., the host of "*Nanoarchaeum equitans*". *Int. J. Syst. Evol. Microbiol.* 57:803-808.

Podar M, Anderson I, Makarova KS, Elkins JG, Ivanova N, Wall MA, Lykidis A, Mavromatis K, Sun H, Hudson ME, Chen W, Deciu C, Hutchison D, Eads JR, Anderson A, Fernandes F, Szeto E, Lapidus A, Kyrpides NC, Saier MH Jr, Richardson PM, Rachel R, Huber H, Eisen JA, Koonin EV, Keller M, Stetter KO (2008) A genomic analysis of the archaeal system *Ignicoccus hospitalis-Nanoarchaeum equitans*. *Genome Biol.* 9:R158.

Randau L, Münch R, Hohn MJ, Jahn D and Söll D (2005a) *Nanoarchaeum equitans* creates functional tRNAs from separate genes for their 5'- and 3'-halves. *Nature* 433:537-541.

Randau L, Pearson M and Söll D (2005b) The complete set of tRNA species in *Nanoarchaeum equitans*. *FEBS Lett.* 579:2945-2947.

Randau L. Calvin K, Hall M, Yuan J, Podar M, Li H and Söll D (2005c) The heteromeric *Nanoarchaeum equitans* splicing endonuclease cleaves noncanonical bulge-helix-bulge motifs of joined tRNA halves. *Proc. Natl. Acad. Sci. USA* 102:17934-17939.

Randau L, Schröder I and Söll D (2008) Life without RNase P. *Nature* 453:120-123.

Robertson CE, Harris JK, Spear JR, Pace NR (2005) Phylogenetic diversity and ecology of environmental Archaea. *Curr. Opin. Microbiol.* 8:638-642.

Rodrigues JV, Victor BL, Huber H, Saraiva LM, Soares CM, Cabelli DE and Teixeira M (2008) Superoxide reduction by *Nanoarchaeum equitans* neelaredoxin, an enzyme lacking the highly conserved glutamate iron ligand. *J. Biol. Inorg. Chem.* 13:219-228.

Waters E, Hohn MJ, Ahel I, Graham DF, Adams MD, Barnstead M, *et al.* (2003) The genome of *Nanoarchaeum equitans*: Insights into early archaeal evolution and derived parasitism. *Proc. Natl. Acad. Sci. USA* 100:12984-12988.

Woese CR, Kandler O and Wheelis ML (1990) Towards a natural system of organisms: Proposal for the domains archaeal, bacteria and eucarya. *Proc.Natl. Acad.Sci. USA* 87:4576-4579.

In: Archaea: Structure, Habitats and Ecological... ISBN 978-1-61761-932-8
Editor: Sakura Y. Katō, pp. 175-198

Chapter 8

Chaperonin and Prefoldin - Two Molecular Chaperones that Work Cooperatively in Archaea and Eukaryotes

Tamotsu Zako[1,2,*], *Ryo Iizuka*[2,3], *Taro Kanzaki*[2], *Mizuo Maeda*[1] *and Masafumi Yohda*[2]

[1]Bioengineering laboratory, RIKEN Institute, 2-1 Hirosawa, Wako, Saitama, 351-0115 Japan

[2]Department of Biotechnology and Life Science, Tokyo University of Agriculture and Technology, 2-24-16 Naka-cho, Koganei, Tokyo 184-8588, Japan

[3]Laboratory of Bio-Analytical Chemistry, Graduate School of Pharmaceutical Sciences, The University of Tokyo, 7-3-1 Hongo, Bunkyo, Tokyo 113-0033, Japan

Abstract

Prefoldin and group II chaperonins are molecular chaperones found in Archaea and eukaryotic cytosol.

Chaperonins are large double ring complexes of ~1000 kDa enclosing a central cavity in which non-native proteins can refold into their native form. Group II chaperonins act independently of a cofactor that corresponds to GroES of the group I chaperonins.

Instead, the helical protrusion at the tip of the apical domain forms a built-in-lid for the central cavity.

Chaperonins use ATPase cycling to promote the conformational change necessary to open and close the lid. Prefoldin can capture nascent polypeptides or non-native proteins with its coiled coil motifs and, subsequently, transfers it to a group II chaperonin for

*Bioengineering laboratory, RIKEN Institute, 2-1 Hirosawa, Wako, Saitama, 351-0115 Japan. Tex: +81-48-467-9312; Fax: +81-48-462-4658; E-mail: zako@riken.jp.

correct folding. Prefoldin also cooperates with eukaryotic chaperonins in the folding of actin and tubulin.

Because of its biological importance, the manner in which prefoldin and group II chaperonin cooperate to fold a non-native protein has been extensively studied.

Here, we review a series of our studies that describe the structural and biological characteristics of archaeal prefoldin and chaperonin. This chapter also reviews recent reports on archaeal and eukaryotic prefoldin and chaperonin.

Detailed interaction studies between prefoldin and chaperonin allow us to discuss their possible cooperative mechanisms.

1. Group II Chaperonins

Two Distinct Groups of Chaperonins

A number of newly-translated or denaturedproteins, which are referred as non-native proteins, require assistance to convert to a functional three-dimensional conformation.

This process is called protein folding. Proteins known as molecular chaperones are shown to assist the folding process (Figure 1) [1,2].

Chaperonins (CPNs) are ubiquitous molecular chaperones that form double-ring assemblies of about 60 kDa subunits.

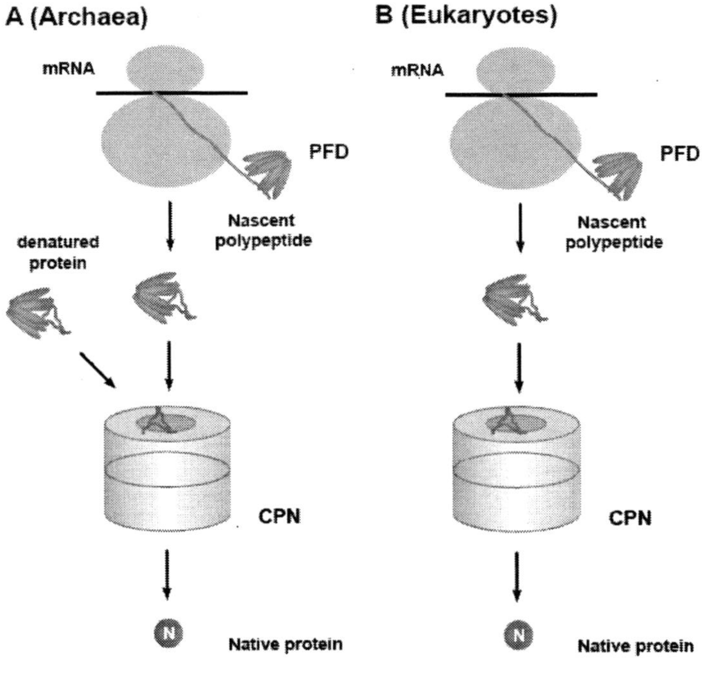

Figure 1. Models for protein folding assisted by prefoldin (PFD) and group II chaperonin (CPN) in Archaea and eukaryotic cytosol. In this model, two molecular chaperones, PFD and CPN in (A) Archaea, (B) Eukaryotic cytosol are shown. PFD can bind to nascent polypeptides and prevent their aggregation. Non-native polypeptides are transferred from PFD to group II CPN for productive folding. In Archaea, denatured protein can be captured by PFD and transferred to CPN.

The resulting structure has a large central cavity in which non-native proteins can undergo productive folding in an ATP-dependent manner (Figure 1) [1,3].

Based on their amino acid sequences and oligomeric structures, they are divided into two groups [4,5].

Group I chaperonins are found in bacteria (GroEL), mitochondria (mitochondrial 60-kDa heat-shock protein, mtHsp60), chloroplasts (Rubisco subunit binding protein, RBP), and some Archaea. They are represented by the *Escherichia coli* chaperonin, GroEL.

Group II CPNs exist in archaea (known as thermosome) and eukaryotic cytosol (known as TCP-1 Ring Complex (TRiC) orChaperonin-Containing TCP-1 (CCT)) [4-7].

Although the basic structure of both group II CPNs is quite similar, there are a number of significant differences between them. We will hereafter discuss the structural and biochemical properties of Group II CPNs in detail, comparing themto Group I CPNs.

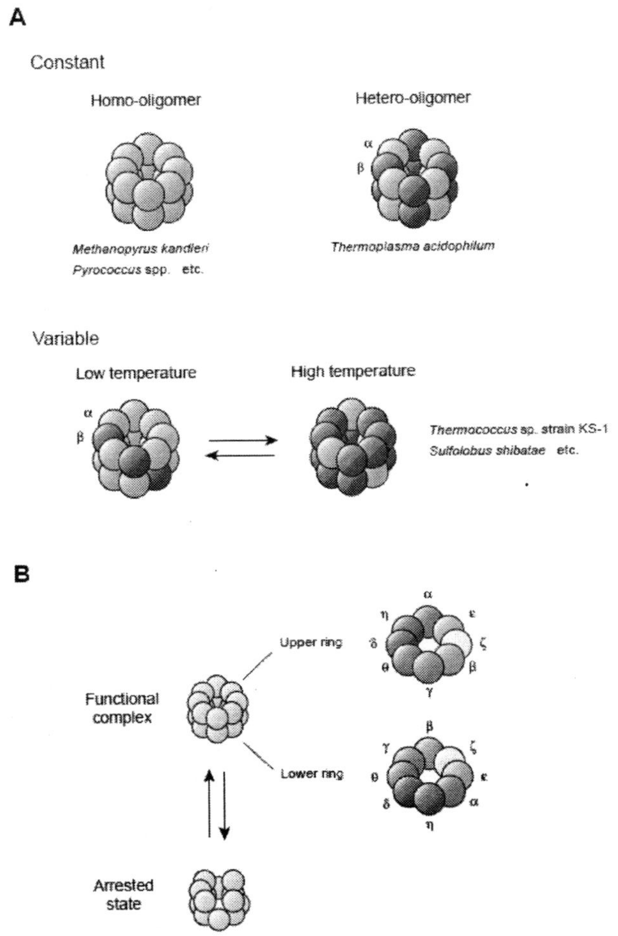

Figure 2. Assembly of group II chaperonin (CPN) complexes. (A) Assembly of archaeal CPN complexes. The subunit composition of CPN complexes is kept constant in some archaea (*upper panel*), whereas the composition changes with growth temperature in several archaea (*lower panel*). (B) Assembly of eukaryotic cytoplasmic CPN CCT. In a functional complex, each ring contains eight different subunits, each with a fixed position in the ring. The orientation of the subunits is as proposed by Martin-Benito *et al.* [18].

Subunits and their Composition

GroEL is composed of 14 identical subunits arranged in two heptameric rings stacked back-to-back [1,3]. In contrast, group II CPNs form toroidal double rings consisting of eight or nine homologous subunits [4] (Figure 2 and Table 1). The precise functional properties and physiological significance of the heterologous subunit composition of group II CPN subunits is still the subject of active investigation. The archaeal CPNs are composed of up to five sequence-related subunits (Table 1). Most archaeal CPNs are composed of one or two sequence-related subunits. *Sulfolobus* species [8,9], *Haloferax volcanii* [10], *Methanosarcina mazei* [7] and *Methanococcoides burtonii* [11] contain three CPNsubunit genes. It was found that there are five CPN subunits (Hsp60-1, -2, -3, -4 and -5) in *Methanosarcina acetivorans* [12]. The subunit composition of CPN complexes remains constant in some Archaea (Figure 2A, *upper*), whereas the composition changes with growth temperature in several species [9,13,14] (Figure 2A, *lower*).

For example, the CPN from the hyperthermophilic archaeon, *Thermococcus* sp. strain KS-1 (*T.* KS-1) is composed of two highly sequence-related subunits, α and β [15]. The *T.* KS-1 CPN exists as a hetero-oligomer with variable subunit composition *in vivo* [14]. The expression of α and β subunits is regulated differently, and only the β subunit is thermally inducible.

Table 1. Properties of archaeal and eukaryotic group II CPNs

	Archaeal CPN	Eukaryotic CPN (CCT)
Rotational symmetry	8 (Most archaea) or 9 (*Sulfolobus* spp.)	8
Subunit species	1 (*Pyrococcus* spp. etc.) 2 (Most archaea) 3 (*Sulfolobus* spp. etc.) 5 (*Methanosarcina acetivorans*)	8 or 9 (Mammalian testis)
Subunit composition	Constant(*Thermoplasma acidophilum* etc.) Variable (*Thermococcus* sp. strain KS-1 etc.)	Variable
Substrate	unkown	Cytoskeletal proteins (Actin, Tublin etc.) β-rich protein ($G_β$-transducin, polyQ etc.)

Thus, the proportion of the β subunit in *T.* KS-1 CPN increases with temperature, and β subunit-rich CPNsare more thermostable than α subunit rich-CPNs [16] (Figure 2A). The hyperthermoacidophilic archaeon, *Sulfolobus shibatae* contains group II CPN composed of three different subunits (α, β and γ). The expression of the α and β genes is increased by heat shock and decreased by cold shock. In contrast, γ subunit expression is undetectable at heat shock temperatures and is low under normal growth conditions, but is induced by cold shock [9].

The halophilic archaeon *Haloferax volcanii* has three group II CPN genes named *cct1*, *cct2* and *cct3*. Large *et al.* [10] showed that the three CPN subunits are all expressed but not to the same level. Interestingly, all three subunits are individually dispensable, but at least one must be present for growth [17].

The eukaryotic cytosolicCPN CCT is the most complex CPN and exists as a hexadecameric complex composed of eight different subunits (α, β, γ, δ, ε, ζ, η and θ, corresponding to CCT1-CCT8 in yeast) (Figure 2B). Recently, the subunit arrangement was determined [18], as shown in Figure 2B. Although the eight subunits share ~30% amino acid sequence identity with each other [19], the sequence of the apical domain(see below), which contains the substrate-binding sites, is less conserved [5,20]. Electron microscopic and cross-linking studies pointed out that the specific subunits of CCT contribute to the binding of substrate proteins [21-23].

For example, β, δ and ε subunits are shown to bind β-actin, whereas all CCT subunits can contribute to the binding of tubulin. These observations support the idea that the specific subunits are involved in binding to substrate proteins. CCT is ubiquitously expressed in all eukaryotic cells, but is much less abundant than other CPNs. Whereas GroEL and archaeal CPN subunits are induced by heat shock, CCThas not been found to be induced by heat shock irrespective of the presence of heat shock element sequences present in all of CCT subunit genes, suggesting that CCT plays a general role in maintaining cell function even in the absence of various stresses [24]. Interestingly, the expression levels of the CCT subunits are strongly correlated with growth rate and are markedly up-regulated at the G_1/S transition through early S phase in cultured cells [25-27].

In addition, Yokota *et al.* also reported that the contents of particular CCT subunit species change concomitantly with substrate binding and folding activities during the cell cycle [27]. There is also a report that cold shock can induce CCT transcription in *Saccharomyces cerevisiae* [28].

Structure and Lid of the Cavity

Crystal structures of GroEL [29] and the group II CPN from an acidothermophilic archaeum, *Thermoplasma acidophilum* [30] indicate that group I and II CPNs share a similar domain arrangement. Each subunit is comprised of three domains: the apical, intermediate and equatorial domains (Figure 3A).

The equatorial domain contains the ATP binding site and is involved in intra- and inter-ring contacts. The apical domain is involved in binding to substrate proteins. The intermediate domain connects the equatorial and apical domains of each subunit and transfers the ATP-induced conformational changes from the equatorial to the apical domain.

The most striking structural difference between them is the lid of the central cavity of the CPN complex.

The co-CPN GroES serves as a lid for group I CPNs and has a heptameric dome-like structure. GroES interacts with one or both GroEL rings in an ATP-regulated fashion, thereby sealing the cavity from the outside. In contrast, group II CPNs work without the cofactor corresponding to GroES.

The crystal structure of the CPN from *T. acidophilum* suggested that the function of GroES is replaced by long helical protrusions from the apical domain, called the helical protrusion (Figure 3A). This region is strictly conserved among group II CPNs, but is absent in group I CPNs [4].

In the crystal structure, the helical protrusions overhang the central cavity (Figure 3B and C). The closed cavities have the hydrophilic surface implicated in protein folding [30].

Taken together with the structural and biochemical analyses [31-34], the protrusion is revealed to act as a built-in lid for the central cavity, substituting for a detachable lid.

Thus far, all crystal structures of group II CPNs are in a closed built-in lid the conformation [30,35] (Figures 3B and C). Irrespective of many efforts to determine the structure of the group II CPNs in the open conformation, experiments have not been successful until now.

It is quite likely that the CPNs are locked in lid-closed conformation under the crystallization conditions [32,36]. There is no crystal structure available for the eukaryotic CCT complex.

The overall structure has been determined at 25-30 Å resolution by electron microscopy [21,22,33] and is similar to those of archaeal CPNs [30,37-39].

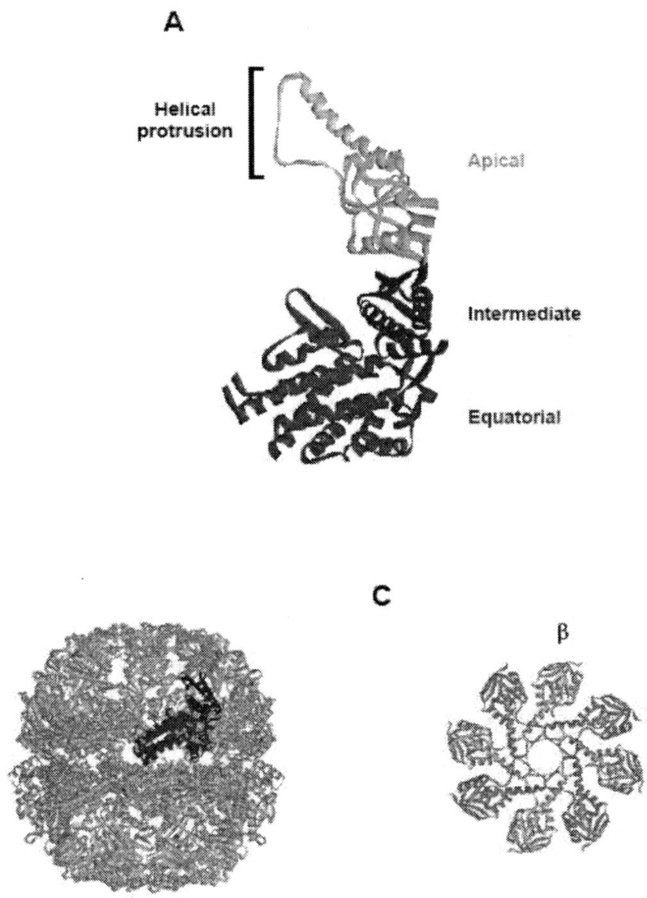

Figure 3. Structure of archaeal group II chaperonin (CPN) from *Thermoplasma acidophilum*. (A) Domain arrangement of α subunit. The subunit has three distinct domains: equatorial (red), intermediate (blue), and apical (green). The helical protrusion, which is unique to group II CPNs, is highlighted (yellow). (B) Side view of the crystal structure of *T. acidophilum*CPN. One subunit has been colored to reveal the domain structure, as shown in (A). (C) Top view of α (green) and β (yellow) subunits apical domains in the crystal structure. The coordinates are from the Protein Data Bank code 1A6D, and figures were drawn with the Viewer Light 5.0 software (Accelrys).

Substrates and their Folding

The molecular chaperone system of Archaea is much simpler than those of other organisms. So far, only six kinds of chaperoneshave been identified in Archaea: group II CPN, Hsp70, prefoldin (PFD) (see below), small heat shock protein, peptidyl-prolyl *cis-trans* isomerase and AAA (ATPase associated with various cellular activities) proteins [40,41]. Intriguingly, Hsp70, an essential molecular chaperone in bacteria and eukaryotes, is absent in all of hyperthermophilic Archaea[40].

The group II CPN is considered to be the only chaperone able to mediate the folding of proteins in hyperthermophilic Archaea. The archaeal CPN subunits are strongly induced by heat shock [9,14,42]. In some cases, this induction can be significant to the extent that CPNs are the major or only proteins synthesized under heat shock conditions. Thus, they are often referred to as "thermosomes". Nevertheless, it is important to realize that they are also expressed at high levels under basal conditions (1-2% of total proteins). Indeed, codon usage analyses predict that group II CPNs are the mostly highly expressed in Archaea [43]. In both basal and stressed conditions, archaeal CPNs should have a broad range of substrates, resembling the behavior of GroEL, although their *in vivo* substrates are not known. Consistent with this observation, *in vitro* studies have shown that many archaeal CPNs capture denatured proteins to prevent aggregation and assist their refolding in an ATP-dependent manner[4,31,44].

In contrast, CCT has a relatively low abundance, less than 0.1% of the soluble proteins [45], and CCT subunits are not induced by heat stress. Unlike GroEL, CCT can bind co-translationally to nascent polypeptides as they emerge from ribosomes [46-48]. Approximately 15% of nascent chains are estimated to be folded with the assistance of CCT by proteomic analysis [49]. While CCT was originally thought to be a chaperone that is highly specialized for the folding of actin and tubulins [50-52], a number of non-cytoskeletal proteins were found to be substrates for CCT [53]. CCT tends to bind to aggregate-prone and β-rich proteins, including proteins containing tryptophan-aspartic acid (WD) repeats, which form a β-propeller domain [54-58], and polyglutamine-expansion proteins [59-61]. CCT binding to β-strands could stabilize a β-sheet-containing domain until it is fully synthesized, at which point concerted release into the protected environment of the CPN cavity would ensure productive folding [53]. Meanwhile, model substrates of GroEL, such as malate dehydrogenase and rhodanese, are not recognized by CCT. This indicates that the mechanism of substrate recognition by CCT is different from that of other CPNs. It is also noteworthy that GroEL binds to its targets much more rapidly than CCT forms complexes with actin or tubulin *in vitro*. GroEL has been shown to recognize non-native substrates in relatively unstructured to partially folded states with exposed hydrophobic surfaces. On the other hand, CCT preferentially binds partially folded or quasi-native intermediates of substrate proteins [21,62,63].In GroEL, the substrate binding sites involve a number of hydrophobic residues that can generally bind the exposed hydrophobic surfaces of non-native proteins [1,3].

Consequently, the helical protrusion was assumed to be involved in substrate binding because of the presence of a large hydrophobic patch in the region [30,64].

However, we have shown that the helical protrusion region is not required for recognition and binding of the substrate protein [31].

Recently, Spiess *et al.* reported that mutations introduced in Helix 11 of the CCT1 subunit apical domain abolish substrate binding [23]. It is speculated that a "true" substrate binding site is in and/or near Helix 11, which is structurally equivalent to the GroEL substrate binding site.

Conformational Change and ProteinFoldingMechanism

A key feature of group II CPNs is their ability to close their cavity and encapsulate the bound substrate, providing a comfortable environment for protein folding. The conformational cycle for opening and closing of the folding chamber is driven by ATP binding and hydrolysis.Based on structural information for group II CPNs available from electron microscopy, X-ray crystallography and small-angle X-ray scattering [21,22,30,31,33,34,37-39,65], three conformational states have been identified in both archaeal and eukaryotic group II CPNs: an open, a spherical closed, and an asymmetric bullet-shaped conformation. These forms are considered to exist in the functional cycle of group II CPNs. Figure4 shows schematic models for the functional cycle of archaeal and eukaryotic group II CPNs. In the absence of nucleotides, both CPNs are maintained in the open conformation, and able to capture non-native polypeptides (Figures 4A, *state i* and 4B, *state I*). Although the exact location of the substrate binding site has not been experimentally determined, it is likely that non-native substrates interact *via* the exposed hydrophobic surface of the apical domain. Subsequently, the addition of ATP leads to a conformational change to the asymmetric structure [34,65,66] (Figures 4A, *state ii* and 4B, *state II*). A slight difference in the conformational change cycle between archaeal and eukaryotic CPNs has been suggested from experiments using phosphate analogues: *i.e.*, ATP binding is sufficient to close the lid of archaeal CPNs, while lid closure of CCT is triggered by the transition state of ATP hydrolysis [34,65].

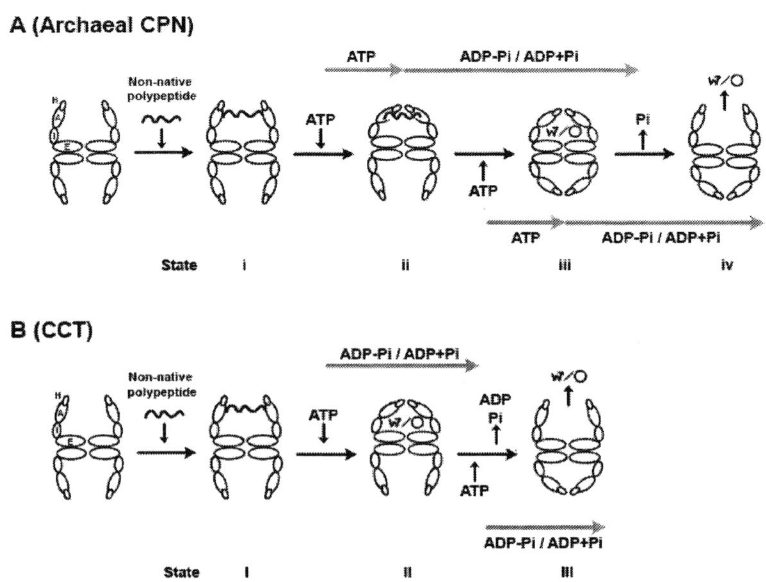

Figure 4. Schematic model for the reaction mechanism of group II chaperonin (CPN). (A) Archaeal CPN. (B) CCT. See text for details. *A*, *I* and *E* refer to the apical, intermediate, and equatorial domains, respectively. *H* represents the helical protrusion.

In addition, asymmetric conformational changes to the symmetric closed conformation have been shown to occur when ATP binds to the other ring in archaeal CPNs [31,32,65]. The substrate is released from the cavity wall into the hydrophilic central cavity where productive folding occurs [65] (Figure 4A, *state iii*). Consequently, the release of γ-phosphate generated by ATP hydrolysis in the folding-active ring triggers opening of the lid and release of the substrate (Figure 4A, *state iv* and 4B, *state III*). The two rings of group II CPNs may alternate as a folding chamber and group II CPNs may have a similar molecular mechanism to GroEL and its partner GroES[1,3].

2. Prefoldin

PFD Was Discovered as a Microtubule Biogenesis-Related Protein

Prefoldin (PFD) is a ~90 kDa molecular chaperone also known as GimC (Genes Involved in Microtubule biogenesisComplex). This chaperone was first identified in a screening study as a novel protein complex encoded by six different *GIM* genes that promotes formation of functionalα- and γ- tubulin [67]. Disruptions of *GIM* genes in yeast cause microtubule defects and a reduced level of α–tubulin. The *gim* null mutants are synthetically lethal with mutated γ-tubulin and are supersensitive towards a microtubule depolymerizing drug. The Gim proteins were shown to interact with each other and to be present in common complexes. In separate studies, PFD was discovered as a denatured-actin-binding protein in rabbit reticulocytes and in bovine testis, and was recognized as a homolog of GimC [68]. Recently, Lundin *et al.* reported that PFD-mediated tubulin biogenesis is important for cell division and cell migration in *Caenorhabditis elegans* [69]. In addition to eukaryotic cells, prefoldin has been identified in Archaea, regardless of the presence of actin or tubulin [67,68,70,71]. Sequence database searches revealed that PFD exists ubiquitously in nearly all eukaryotic and archaeal genomes that have been sequenced thus far.

Subunit Composition

The eukaryotic yeast GimC and bovine PFD complexes consist of six distinct protein subunits termed Gim1-Gim6 and PFD1-PFD6, respectively (Table 2). Closely related homologues of these six proteins are also found inthe eukaryotic genome sequenced thus far, such as animals, plants and fungi. Sequence alignment analysis of PFD subunits showed that these six subunits can be grouped into two separate classes, represented by PFD2/5 (denoted as theαclass) and PFD1/3/4/6 (denoted as the β class) (Table 2) [70]. The difference between these groupings is that the α class subunits contain an insertion of two β-strand (~25 amino acid) insertions not present in the β class subunits. Most archaeal PFD is a hexamer consisting of twoαand four β subunits (Table 2). Interestingly, *Methanocaldococcus jannaschii*, a hyperthermophilic methanogen, contains a second α class subunit, which was up-regulated over 20-fold in response to heat shock, whereas most archaeal PFD exists ubiquitously and not stress-inducible [70,72]. Phylogenetic analysis showed that the most closely related PFD is from thermophilic bacterium*Aquifex aeolicus*, which suggests a horizontal gene transfer[70]. It was also reported that the hyperthermophilic archaeon *Thermococcus kodakaraensis KOD1* contains a second class of α and β subunit genes in its genome [73]. Recently, it was found that *Thermococcus* KS-1 also contains two pairs of α and β subunits (Iizuka *et al.* MS *in prep*).

Table 2. Properties of archaeal and eukaryotic PFDs

	Archaeal PFD	Eukaryotic PFD
Subunit	6 (2×α subunit and 4×β subunit)	6 (2×α-class subunit and 4×β-class subunit)
Subunit species	2 (Most Archaea (α and β) 4 (*Thermococcus* sp. strain KS-1 and KOD1 (α1,α2 and β1,β2))	6
Substrate	unkown	Cytoskeletal proteins (Actin, Tublin etc.)

It seems that the *Thermococcus* strains have gained an improved chaperone system that may provide an advantage for growthin a wider range of temperatures. Curiously enough, *A. aeolicus* contains only the PFD αsubunit and does not have a gene corresponding to the β subunit in its genome [70]. However, it is unclear whether this PFD functions without the corresponding β subunit or whether the β subunit gene was not found in the homology search.

Structure of PFD

The crystal structure of PFD from *M. thermoautophicus* (previously known as *Methanothermobacterium thermoautophicum*) (MtPFD) has been resolved to 2.3 Å resolution(Figure 5A) [74]. The quaternary structure of PFDis composed of a hexamer of two α and four β PFD subunits. PFD has the striking appearance of a jellyfish: its body consists of a double barrel assembly with six long tentacle-like coiled coils protruding from it. The distal regions of the coiled coils expose hydrophobic patches. Each α and β subunit contains two and one central β-hairpin, respectively, that are flanked N- and C-terminally by coiled-coil helices. The coiled-coil helices within each subunit assemble antiparallel to each other, and these two-stranded coiled coils form the cavity. This structure is considered to be conserved for all PFDs since the structures of archaeal (*Pyrococcus horikoshii*) (PhPFD) and eukaryotic (bovine and human) PFDs obtained by cryoelectron microscopy (cryo-EM) are essentially identical [75,76]. The recently determined crystal structure of PhPFD was almost identical with that of MtPFD (Figure 5B) [77].

Coiled coils are versatile structural modules that engage in a variety of cellular activities. The coiled coilalso represents a novel functional motif in chaperones [78]. Besides PFD, the coiled coil motif was found in other molecular chaperones, caseinolyte protease component B (ClpB) and heat shock locus gene product U and V (HslUV). ClpB and HslUV are both AAA+ ATPases and can bind to non-native proteins, suggesting that the coiled coil is more widely used in handling non-native proteins.

Much interestingly, a bacterial chaperone called seventeen-kilodalton protein (Skp) was shown to have a structure intriguingly similar to PFD although they are unrelated in sequence

[79,80]. Skp is located in the periplasm space of Gram-negative bacteria and stabilizes outer membrane protein. Skp was shown to interact with lipopolysaccharides in the outer membrane and help insertion of membrane protein, which also supports the idea that the coiled coil is used in handling non-native proteins.

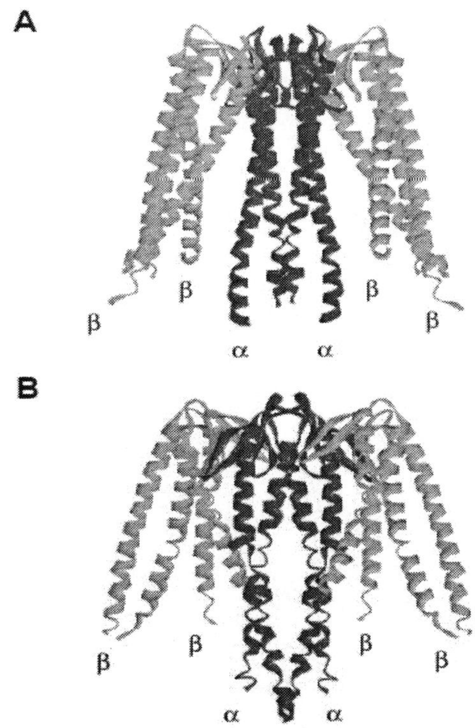

Figure 5. Structure of archaeal prefoldin (PFD). Side view of the crystal structures of (A) MtPFD (PDB code 1FXK) and (B) PhPFD (PDB code 2ZDI).The αand βsubunits are colored in green and red, respectively. Figures were drawn with the Viewer Light 5.0 software (Accelrys).

Eukaryotic PFDs Participate in MicrotubuleBiogenesis

A reduced level of α-tubulin and delayed folding of actin by deletion of *GIM* genes in the yeast strain suggests that PFD/GimC is involved in microtubule biogenesis [67]. Pulse-chase and co-immunoprecipitation experiments in yeast cells have shown that deletion of PFD genes causes a delayed folding of actin, a slowed release of folded actin and an increase in the amount of non-native actin in the cytosol [81]. It was also shown that purified eukaryotic PFD binds to urea-denatured β-actin, but not to native actin. Eukaryotic PFD has also been shown to bind nascent chains of both actin and tubulin at the ribosome co-translationally [82].

The three-dimensional structure of the eukaryotic PFD and denatured actin complex obtained by cryo-EM shows that eukaryotic (human and bovine) PFD encapsulates its substrate inside the cavity via the distal ends of its tentacles [75,76]. Using recombinant human PFD, Simons *et al.* supported this idea by showing that the tips of the tentacles are required to bind target proteins [83]. Using hybrid PFD complexes with truncated coiled-coil tips in the individual subunits, it was shown that truncations in the PFD-3 or -4 subunit abolished actin binding and truncations in the PFD-4 or -5 subunit abolished tubulin binding.

These results suggest that human PFD uses the distal ends of different but overlapping sets of subunits to form stable complexes with different target proteins.

Archaeal PFD Recognizes a Wide Range of Non-Native Proteins

PFD has been identified in Archaea, regardless of the presence of actin or tubulin [67,68,70,71]. Sequence database searches revealed that PFD exists ubiquitously in nearly all Archaea. The molecular mechanism of substrate binding with MtPFD and PhPFD has been well-characterized. Detailed *in vitro* studies have revealed that archaeal PFD is capable of stabilizing a wide range of denatured, non-native proteins such as actin, citrate synthase, conalbumin glucose dehydrogenase, dihydrofolate reductase (DHFR), firefly luciferase, green fluorescent protein (GFP), lysozyme and rhodanese [70,71,84,85], suggesting that archaeal PFD performs a general role in recognizing and assisting the biogenesis of non-native proteins in the cell, whereas substrate proteins of eukaryotic PFD appear to be restricted to only a few proteins, principally actin and tubulin. Archaeal PFD is also efficient in preventing thermally induced aggregation.

Substrate InteractionMechanism of Archaeal PFD

The crystal structures of MtPFD and PhPFD indicate that archaeal PFD exsists as a hexamer with coiled coil tentacles protruding from a double β-barrel (Figure 5A and B) [74]. Each subunit has hydrophobic residues at the very tips of the tentacles. Truncation studies show that the distal ends of the coiled coil tentacles contributed in a concerted manner for binding to non-native proteins [74,84,85]. The tentacles of PFD are approximately 80 Å long and bind to and prevent the aggregation of denatured substrate proteins of varying sizes (14-75 kDa). EM images of GFP-bound PhPFD suggest that PFD functions as a molecular clamp since the tentacles move outward ~12° [84]. Because the tentacles may move independently, it is plausible that PFD can grasp a range of non-native proteins of different sizes and shapes with these independent clamps. Lundin *et al.* also suggested that a hinge region, consisting of the loops connecting the coiled coils to the β-barrel domain, could be responsible for the flexibility of the clamps.

In another recent paper, Martin-Benito *et al.* reported their investigation of molecular strategiesfor archaeal PFD capture of non-native proteins [76]. The cryo-EM study of complexes of PhPFD and substrate proteins of different sizes showed that PhPFD selectively uses an increasing number of subunits to interact with substrate proteins of larger sizes. For example, lysozyme (14 kDa) interacts with a pair of β subunits and GFP (27 kDa) interact with a pair of the β subunits and one of the α subunits, respectively. In the case of a much larger substrate protein, conalbumin (75 kDa), all six PhPFD subunits are used for the interaction. These results also suggest that the β subunits are more important than the α subunits, which is consistent with our results using truncation mutants [85].

Detailed kinetic studies on the interactions between PhPFD and denatured substrate proteins have been carried out [86]. In this study, a new method based on fluorescence microscopy was developed to estimate the equilibrium dissociation constants and the rate constant. In this method, the amount of fluorescently-labeled substrate protein in the PhPFD complex bound to a glass slide was measured. This method was possible because of the following characteristics of PhPFD, namely: 1) tight binding of PhPFD to a glass slide by electrostatic interactions and 2) negligible non-specific binding of the non-native substrate proteins to the glass slide. Using this technique, the equilibrium dissociation constants (K_D) of PhPFD with CS and GFP at 50 °C were determined to be approximately 20 and 5 nM, respectively.

Dissociation rate constants (k_{off}) also can be estimated using the same method by measuring the amount of labeled substrate protein in the complex after adding excess non-labeled substrate [86]. The k_{off} values at 50°C were 5.9×10^{-3} s^{-1} for CS and 3.0×10^{-3} s^{-1} for GFP. This result indicates that the substrate protein-PhPFD complexes are not stable, in other words, the complex is in a dynamic equilibrium of association and dissociation.

The Substrate-InteractionMechanism Differs between Archaeal PFD and Eukaryotic PFD

Cryo-EM studies have also provided important insights into the different substrate binding mechanismsof archaeal and eukaryotic PFDs [75,76]. In the case of eukaryotic PFD, denatured cytoskeletal protein is almost entirely encapsulated inside the chaperone cavity formed by the PFD tentacles. In contrast, for archaeal PFD, denatured substrate proteins, even cytoskeletal proteins, interact with the tips of the tentacles and protrude from them. They are not encapsulated inside the cavity, indicating an obvious difference in the subunit interaction mechanism between archaeal and eukaryotic PFDs.

Whereas archaeal PFD is composed of only two types of subunits, eukaryotic PFD is composed of six different subunits. This suggests that the eukaryotic PFD, with a more complex subunit composition, evolved from a simpler PFD similar to that found in the modern Archaea. Taken together with the fact that archaeal PFD can act on a range of substrate proteins, but eukaryotic PFD seems to be restricted to cytoskeletal proteins (actin and tubulin), this trend toward a higher complexity may correlate with a specialized function of eukaryotic PFD [76]. It was also suggested that whereas nonspecific, hydrophobic interactions are used for substrate recognition by archaeal PFD, more specific interactions based on particular sequences in the PFD and substrate are used by eukaryotic PFD for specific recognition of cytoskeletal proteins.

Other Functions

Interestingly, the amino acid sequence search shows that the human PFD subunits 3,4,5 and 6 were previously individually identified as von Hippel-Lindau (VHL) binding protein, a possible transcription factor C1, a c-myc binding protein and KE2, a hydrophilic protein of unknown function in the major histocompatibility complex, respectively. This suggests that each PFD subunit has its own biological activitybesides molecular chaperone activity. This fact also supports the idea that PFD is a multifunctional protein [68]. Mousnier *et al.* recently reported that the human PFD subunit 3, VHL binding protein, plays a role in the degradation of HIV integrase, the viral enzyme responsible for provirus integration into the host genome, by mediating the polyubiquitylation that leads to subsequent proteasome-dependent degradation, in cooperation with VHL ubiquitin ligase, suggestinga PFD-VHL-proteasome pathway [87].

3. Mechanism of Substrate Transfer from PFD to CPN

PFD Transfers Substrate Protein to CPN

As described above, eukaryotic PFD was identified as a molecular chaperone that functions in the folding of cytoskeletal proteins. Since CCT had already been shown to be required for biogenesis of the major cytoskeletal proteins (actin and α- and β-tubulin) as described above, eukaryotic PFD was thought to act cooperatively in their biogenesis. Siegers *et al.* showed that CCT-mediated productive actin folding is accelerated at least 5-fold in the presence of eukaryotic PFD [81]. Vainberg *et al.* showed that PFD transfers its bound denatured actin to CCT in a nucleotide-independent manner *in vitro* [68]. Detailed *in vitro* studies reveal that eukaryotic PFD can bind to nascent polypeptides at the ribosome and maintain them in a folding-incomplete state until it can deliver them to CCT for folding in a protected compartment (Figure 1) [81,82]. Importantly, archaeal PFD can also transfer denatured protein to eukaryotic CCT [70] or archaeal CPN [71,86,88,89], suggesting that this substrate transfer to CPN is a general role for PFD. The molecular mechanism of substrate transfer elucidated from biochemical and structural studies is described below.

PFD Binds to CPN Like the "Lid of a Pod"

In order to understand the substrate transfer mechanism from PFD to CPN, structural analysis of the PFD-CPN complex is essential. Martin-Benito *et al.* showed by cryo-EM study that eukaryotic PFD binds to CCT like the "lid of a pod", which also confirms the physical interaction between the two chaperones (Figure 6).

In the three-dimensional reconstruction model, PFD seems to interact with the tentacle tips to the apical domain of CCT. Importantly, a similar complex structure between archaeal PFD and CPN was also observed using the EM technique, confirming that the transfer mechanism is basically conserved between the eukaryotic and archaeal systems [76].

It should be noted, however, that there is a significant difference between these complexes. The eukaryotic PFD can bind one or both rings of CCT simultaneously [75], whereas the archaeal PFD (PhPFD) was only observed to form asymmetric complexes with PhCPN [76], possibly due to difference in the conformational change model as described above.

Binding Sites for Interaction between PFD and CPN

The two-dimensional average EM images of the eukaryoticCPN-PFD complexes showed that the interaction between the two chaperones occurs at the level of the outer regions of the PFD tentacles and the inner surface of the apical domains of the CPN [75]. It seems that eukaryotic PFD inserts its tips into the cavity of CCT.

Furthermore, a three-dimensional reconstruction model also revealed that PFD interacts with two specific subunits in each of the CCT rings, placed in a 1-4 arrangement.

In Archaea, the interaction between the two chaperones has been characterized in more detail with biochemical experiments and EM analysis [76,85,86,88-90].

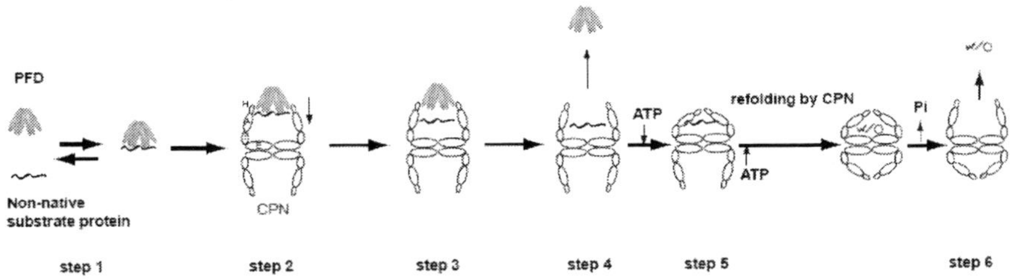

Figure 6. Schematic model for the transfer mechanism of non-native substrate proteins from archaeal prefoldin (PFD) to group II chaperonin (CPN). See text for details. *A*, *I* and *E* refer to the apical, intermediate, and equatorial domains, respectively. *H* represents the helical protrusion.

The two-dimensional average EM images of the archaeal CPN-PFD complex (PhCPN-PhPFD) showed that, similarly to the eukaryotic system, the tentacles of PhPFD use thier coiled-coil tips to interact with the CPN apical domains. However, PhPFD do not penetrate into the archaeal CPN cavity as much as those of the eukaryotic PFD enter the CCT cavity, whereas the eukaryotic PFD seems to insert its tips into the cavity of CCT, implying a difference in the transfer mechanism between archaeal and eukaryotic systems.

To identify the interaction sites and important amino acid residues between archaeal PFD and CPN in more detail, interaction studies using their various mutants have been carried out [85,86,89]. Truncation analyses of PhPFD subunits demonstrated that both N- and C-terminal regions of the outer β subunits are more important for CPN interactions than the two central α subunits, which is consistent with EM observation. The PhPFD β subunit tentacles are also important for substrate binding. Kinetic analyses suggest that PFD-CPN binding is multivalent, rather than a simple one-to-one binding model, consistent with multiple binding sites between the two chaperones [85]. Interestingly, there is likely to be cooperation between α and β subunits during interaction with CPN, suggesting that PhPFD α and β subunits bind to CPN in a concerted manner.

On the other hand, truncation mutants of CPNhave demonstrated that helical protrusion of the apical domain is important for PFD binding, consistent with the EM observation [76,89]. Interestingly, two acidic amino acid residues located in the helical protrusion, namely the 250th and 256th residues, have been shown to be important for strong binding to PFD [89]. Corresponding basic amino acid residues are also present in the distal region of β subunit tentacles, which are conserved in various archaeal PFDs. This observation suggests that electrostatic interactionsare important for the interaction between archaeal PFD and CPN.

Interaction between Archaeal PFD and CPN Is Important for Substrate Transfer

An important observation for elucidating the mechanism of substrate transfer was obtained from kinetic study of the interactionbetween PhPFD and non-native substrate proteins [86]. It was shown that substrate protein and PhPFD are in a dynamic equilibrium of association and dissociation, as described above. Importantly, dissociation of substrate protein from the PhPFD complex was accelerated by CPN, suggesting a possible transfer mechanism

(Figure 6). This facilitated release of substrate protein would be caused by the formation of a ternary complex of prefoldin, substrate protein and CPN, as shown by Martin-Benito *et al.* for the archaeal PFD-CPN complex [76].

One possible mechanism for the facilitated substrate release is a conformational change in PFD. An outward conformational change in archaeal PFD upon substrate protein binding was observed by Lundin *et al.*, as described above. So, it is possible to assume that a similar outward conformational change in the PFD tentacles upon CPN binding results in faster release of the substrate protein. Another possible explanation is a loss in the cooperativity of PFD toward the substrate protein. Since the six tentacles of PFD have been shown to bind to substrate proteins in a concerted manner [74,84,85], it is likely that the cooperative binding ability of PFD to substrate proteins decreases upon CPN binding to one or more of the PFD tentacles, facilitating the release of substrate proteins from PFD.

Another important evidence of substrate transfer was obtained by measurement of the substrate transfer rate constant using GFP as a substrate [89]. In this experiment, a lag time observed during the initial period of GFP refolding, which is thought to reflect the transfer of GFP from PhPFD to CPN, was used for evaluation of the transfer rate constant. This lag time was shortened when CPN mutants, which have faster binding rate constants to PhPFD, were used, indicating that there was a positive correlation between the transfer rate constant of denatured GFP from PFD to CPN and binding rate constant between PFD and CPN.

Elucidated Mechanisms of Substrate Transfer from Archaeal PFD to CPN

Figure 6 shows a schematic reaction mechanismfor the transfer of substrate protein from archaeal PFD to CPNas well as the subsequent refolding by CPN. First, substrate protein is captured by PFD, possibly by hydrophobic interactions (*step 1*). PFD and substrate proteins are in a dynamic equilibrium of association and dissociation. Then PFD-substrate complex binds to CPNlike "the lid of a pot" (*step 2*) [76]. Archaeal PFD may utilize electrostatic interactions to bind to the helical protrusion region of CPN. The dissociation of substrate is accelerated about 5-fold upon binding of CPN. Consequently, the released substrate proteins should enter the cavity of CPN (*step 3*). PFD would then be released from CPN simultaneously with the release of substrate from PFD into CPN (*step* 4), since a ternary complex of substrate protein, PFD and CPN is thought to be unstable and to exist only transiently [75,76]. Subsequently, ATPbinding leads to the closed conformation of CPNin which productive refolding occurs (*step 5*) [32,65]. Consequently, the release of the γ–phosphate generated by ATP hydrolysis in the folding-active ring triggers opening of the lid and release of the substrate (*step 6*).

Comparison of the Mechanism of Substrate Transferfrom PFD to CPN between Eukaryotic and Archaeal Systems

For eukaryotic systems, a model depicting the interaction between denatured actin, PFD and CCT based on the three-dimensional reconstructionof structures of these binary complexes has been presented, although a ternary complex has not been identified as described above [75]. According to this model, the PFD-actin complex interacts with specific

CCT subunits and presents the CCT-binding regions of partially unfolded actin to the apical domains of specific CCT subunits. This is consistent with the identified CCT binding domain present in actin, as assessed using an alanine-scan experiment [91]. These findings suggest a mechanism whereby PFD recognizes and binds actin in an open, quasi-folded conformation via the tips of its tentacles. Subsequently, the PFD-actin complex formed binds to CCT, and PFD is then displaced by CCT binding. CCT then stabilizes this intermediate actin conformation, and folding proceeds to completion following the conformational changes promoted by the cytosolicCPN upon nucleotide binding and hydrolysis. In a recent study by Stirling *et al.*, phosducin-like protein 3 (PhLP3), identified as a novel CCT-binding protein, was shown to inhibit CCT-mediated actin and tubulin folding [92]. It was suggested that the balance of PhLP3 and PFD is important for CCT to modulate the levels of folded actin and tubulin *in vivo*.

Since cryo-EM imagesof the archaeal PFD-CPN complex have not been successfully obtained, it is suggested that the archaeal PFD-CPN complex is less stable than the eukaryotic one [76]. Comparison of the amino acid sequence present in the CPNbinding region of PFD and vice versa showed that this difference comes from electrostatic properties of the binding regions; *i.e.*, the eukaryotic complex is much more electrostatic than the archaeal complex. Hence, a new notion arises that the electrostatic properties of PFD and CPN are important for substrate specificities. Coevolution of the archaeal PFD-CPN system towards a system like that of eukaryotes,whichis capable of interacting two classes of cytoskeletal proteins, actins and tubulins,might also include changes in the electrostatic properties of the responsible region.

References

[1] Hartl, F.U. and Hayer-Hartl, M. (2002). Molecular chaperones in the cytosol: from nascent chain to folded protein. *Science 295*, 1852-8.

[2] Young, J.C., Agashe, V.R., Siegers, K. and Hartl, F.U. (2004). Pathways of chaperone-mediated protein folding in the cytosol. *Nat. Rev. Mol. Cell Biol. 5*, 781-91.

[3] Horwich, A.L., Fenton, W.A., Chapman, E. and Farr, G.W. (2007). Two families of chaperonin: physiology and mechanism. *Annu. Rev. Cell Dev. Biol. 23*, 115-45.

[4] Gutsche, I., Essen, L.O. and Baumeister, W. (1999). Group II chaperonins: new TRiC(k)s and turns of a protein folding machine. *J. Mol. Biol. 293*, 295-312.

[5] Kim, S., Willison, K.R. and Horwich, A.L. (1994). Cystosolic chaperonin subunits have a conserved ATPase domain but diverged polypeptide-binding domains. *Trends Biochem. Sci. 19*, 543-8.

[6] Conway de Macario, E. and Macario, A.J. (2003). Molecular biology of stress genes in methanogens: potential for bioreactor technology. *Adv. Biochem. Eng. Biotechnol. 81*, 95-150.

[7] Klunker, D., Haas, B., Hirtreiter, A., Figueiredo, L., Naylor, D.J., Pfeifer, G., Muller, V., Deppenmeier, U., Gottschalk, G., Hartl, F.U. and Hayer-Hartl, M. (2003). Coexistence of group I and group II chaperonins in the archaeon Methanosarcina mazei. *J. Biol. Chem. 278*, 33256-67.

[8] Archibald, J.M., Logsdon, J.M. and Doolittle, W.F. (1999). Recurrent paralogy in the evolution of archaeal chaperonins. *Curr. Biol. 9*, 1053-6.

[9] Kagawa, H.K., Yaoi, T., Brocchieri, L., McMillan, R.A., Alton, T. and Trent, J.D. (2003). The composition, structure and stability of a group II chaperonin are temperature regulated in a hyperthermophilic archaeon. *Mol. Microbiol. 48*, 143-56.

[10] Large, A.T., Kovacs, E. and Lund, P.A. (2002). Properties of the chaperonin complex from the halophilic archaeon Haloferax volcanii. *FEBS Lett. 532*, 309-12.

[11] Goodchild, A., Raftery, M., Saunders, N.F., Guilhaus, M. and Cavicchioli, R. (2005). Cold adaptation of the Antarctic archaeon, Methanococcoides burtonii assessed by proteomics using ICAT. *J. Proteome. Res. 4*, 473-80.

[12] Maeder, D.L., Macario, A.J. and de Macario, E.C. (2005). Novel chaperonins in a prokaryote. *J. Mol. Evol. 60*, 409-16.

[13] Izumi, M., Fujiwara, S., Takagi, M., Fukui, K. and Imanaka, T. (2001). Two kinds of archaeal chaperonin with different temperature dependency from a hyperthermophile. *Biochem. Biophys. Res. Commun. 280*, 581-7.

[14] Yoshida, T., Ideno, A., Hiyamuta, S., Yohda, M. and Maruyama, T. (2001). Natural chaperonin of the hyperthermophilic archaeum, Thermococcus strain KS-1: a hetero-oligomeric chaperonin with variable subunit composition. *Mol. Microbiol. 39*, 1406-13.

[15] Yoshida, T., Yohda, M., Iida, T., Maruyama, T., Taguchi, H., Yazaki, K., Ohta, T., Odaka, M., Endo, I. and Kagawa, Y. (1997). Structural and functional characterization of homo-oligomeric complexes of alpha and beta chaperonin subunits from the hyperthermophilic archaeum Thermococcus strain KS-1. *J. Mol. Biol. 273*, 635-45.

[16] Yoshida, T., Ideno, A., Suzuki, R., Yohda, M. and Maruyama, T. (2002). Two kinds of archaeal group II chaperonin subunits with different thermostability in Thermococcus strain KS-1. *Mol. Microbiol. 44*, 761-9.

[17] Kapatai, G., Large, A., Benesch, J.L., Robinson, C.V., Carrascosa, J.L., Valpuesta, J.M., Gowrinathan, P. and Lund, P.A. (2006). All three chaperonin genes in the archaeon Haloferax volcanii are individually dispensable. *Mol. Microbiol. 61*, 1583-97.

[18] Martin-Benito, J., Grantham, J., Boskovic, J., Brackley, K.I., Carrascosa, J.L., Willison, K.R. and Valpuesta, J.M. (2007). The inter-ring arrangement of the cytosolic chaperonin CCT. *EMBO. Rep. 8*, 252-7.

[19] Kubota, H., Hynes, G. and Willison, K. (1995). The chaperonin containing t-complex polypeptide 1 (TCP-1). Multisubunit machinery assisting in protein folding and assembly in the eukaryotic cytosol. *Eur. J. Biochem. 230*, 3-16.

[20] Archibald, J.M., Blouin, C. and Doolittle, W.F. (2001). Gene duplication and the evolution of group II chaperonins: implications for structure and function. *J. Struct. Biol. 135*, 157-69.

[21] Llorca, O., Martin-Benito, J., Ritco-Vonsovici, M., Grantham, J., Hynes, G.M., Willison, K.R., Carrascosa, J.L. and Valpuesta, J.M. (2000). Eukaryotic chaperonin CCT stabilizes actin and tubulin folding intermediates in open quasi-native conformations. *Embo. J. 19*, 5971-9.

[22] Llorca, O., McCormack, E.A., Hynes, G., Grantham, J., Cordell, J., Carrascosa, J.L., Willison, K.R., Fernandez, J.J. and Valpuesta, J.M. (1999). Eukaryotic type II chaperonin CCT interacts with actin through specific subunits. *Nature 402*, 693-6.

[23] Spiess, C., Miller, E.J., McClellan, A.J. and Frydman, J. (2006). Identification of the TRiC/CCT substrate binding sites uncovers the function of subunit diversity in eukaryotic chaperonins. *Mol. Cell 24*, 25-37.

[24] Kubota, H. (2002). Function and regulation of cytosolic molecular chaperone CCT. *Vitam. Horm. 65*, 313-31.

[25] Yokota, S., Yamamoto, Y., Shimizu, K., Momoi, H., Kamikawa, T., Yamaoka, Y., Yanagi, H., Yura, T. and Kubota, H. (2001). Increased expression of cytosolic chaperonin CCT in human hepatocellular and colonic carcinoma. *Cell Stress Chaperones 6*, 345-50.

[26] Yokota, S., Yanagi, H., Yura, T. and Kubota, H. (1999). Cytosolic chaperonin is up-regulated during cell growth. Preferential expression and binding to tubulin at G(1)/S transition through early S phase. *J. Biol. Chem. 274*, 37070-8.

[27] Yokota, S., Yanagi, H., Yura, T. and Kubota, H. (2001). Cytosolic chaperonin-containing t-complex polypeptide 1 changes the content of a particular subunit species concomitant with substrate binding and folding activities during the cell cycle. *Eur. J. Biochem. 268*, 4664-73.

[28] Somer, L., Shmulman, O., Dror, T., Hashmueli, S. and Kashi, Y. (2002). The eukaryote chaperonin CCT is a cold shock protein in Saccharomyces cerevisiae. *Cell Stress Chaperones 7*, 47-54.

[29] Braig, K., Otwinowski, Z., Hegde, R., Boisvert, D.C., Joachimiak, A., Horwich, A.L. and Sigler, P.B. (1994). The crystal structure of the bacterial chaperonin GroEL at 2.8 A. *Nature 371*, 578-86.

[30] Ditzel, L., Lowe, J., Stock, D., Stetter, K.O., Huber, H., Huber, R. and Steinbacher, S. (1998). Crystal structure of the thermosome, the archaeal chaperonin and homolog of CCT. *Cell 93*, 125-38.

[31] Iizuka, R., So, S., Inobe, T., Yoshida, T., Zako, T., Kuwajima, K. and Yohda, M. (2004). Role of the helical protrusion in the conformational change and molecular chaperone activity of the archaeal group II chaperonin. *J. Biol. Chem. 279*, 18834-9.

[32] Iizuka, R., Yoshida, T., Shomura, Y., Miki, K., Maruyama, T., Odaka, M. and Yohda, M. (2003). ATP binding is critical for the conformational change from an open to closed state in archaeal group II chaperonin. *J. Biol. Chem. 278*, 44959-65.

[33] Llorca, O., Martin-Benito, J., Grantham, J., Ritco-Vonsovici, M., Willison, K.R., Carrascosa, J.L. and Valpuesta, J.M. (2001). The 'sequential allosteric ring' mechanism in the eukaryotic chaperonin-assisted folding of actin and tubulin. *Embo. J. 20*, 4065-75.

[34] Meyer, A.S., Gillespie, J.R., Walther, D., Millet, I.S., Doniach, S. and Frydman, J. (2003). Closing the folding chamber of the eukaryotic chaperonin requires the transition state of ATP hydrolysis. *Cell 113*, 369-81.

[35] Shomura, Y., Yoshida, T., Iizuka, R., Maruyama, T., Yohda, M. and Miki, K. (2004). Crystal structures of the group II chaperonin from Thermococcus strain KS-1: steric hindrance by the substituted amino acid, and inter-subunit rearrangement between two crystal forms. *J. Mol. Biol. 335*, 1265-78.

[36] Gutsche, I., Holzinger, J., Rossle, M., Heumann, H., Baumeister, W. and May, R.P. (2000). Conformational rearrangements of an archaeal chaperonin upon ATPase cycling. *Curr. Biol. 10*, 405-8.

[37] Nitsch, M., Walz, J., Typke, D., Klumpp, M., Essen, L.O. and Baumeister, W. (1998). Group II chaperonin in an open conformation examined by electron tomography. *Nat. Struct. Biol. 5*, 855-7.

[38] Schoehn, G., Hayes, M., Cliff, M., Clarke, A.R. and Saibil, H.R. (2000). Domain rotations between open, closed and bullet-shaped forms of the thermosome, an archaeal chaperonin. *J. Mol. Biol. 301*, 323-32.

[39] Schoehn, G., Quaite-Randall, E., Jimenez, J.L., Joachimiak, A. and Saibil, H.R. (2000). Three conformations of an archaeal chaperonin, TF55 from Sulfolobus shibatae. *J. Mol. Biol. 296*, 813-9.

[40] Laksanalamai, P., Whitehead, T.A. and Robb, F.T. (2004). Minimal protein-folding systems in hyperthermophilic archaea. *Nat. Rev. Microbiol. 2*, 315-24.

[41] Maruyama, T., Suzuki, R. and Furutani, M. (2004). Archaeal peptidyl prolyl cis-trans isomerases (PPIases) update 2004. *Front Biosci. 9*, 1680-720.

[42] Phipps, B.M., Hoffmann, A., Stetter, K.O. and Baumeister, W. (1991). A novel ATPase complex selectively accumulated upon heat shock is a major cellular component of thermophilic archaebacteria. *Embo. J. 10*, 1711-22.

[43] Karlin, S., Mrazek, J., Ma, J. and Brocchieri, L. (2005). Predicted highly expressed genes in archaeal genomes. *Proc. Natl. Acad. Sci. U S A 102*, 7303-8.

[44] Kusmierczyk, A.R. and Martin, J. (2003). Nucleotide-dependent protein folding in the type II chaperonin from the mesophilic archaeon Methanococcus maripaludis. *Biochem. J. 371*, 669-73.

[45] Lewis, V.A., Hynes, G.M., Zheng, D., Saibil, H. and Willison, K. (1992). T-complex polypeptide-1 is a subunit of a heteromeric particle in the eukaryotic cytosol. *Nature 358*, 249-52.

[46] Etchells, S.A., Meyer, A.S., Yam, A.Y., Roobol, A., Miao, Y., Shao, Y., Carden, M.J., Skach, W.R., Frydman, J. and Johnson, A.E. (2005). The cotranslational contacts between ribosome-bound nascent polypeptides and the subunits of the hetero-oligomeric chaperonin TRiC probed by photocross-linking. *J. Biol. Chem. 280*, 28118-26.

[47] Frydman, J., Nimmesgern, E., Ohtsuka, K. and Hartl, F.U. (1994). Folding of nascent polypeptide chains in a high molecular mass assembly with molecular chaperones. *Nature 370*, 111-7.

[48] McCallum, C.D., Do, H., Johnson, A.E. and Frydman, J. (2000). The interaction of the chaperonin tailless complex polypeptide 1 (TCP1) ring complex (TRiC) with ribosome-bound nascent chains examined using photo-cross-linking. *J. Cell Biol. 149*, 591-602.

[49] Thulasiraman, V., Yang, C.F. and Frydman, J. (1999). In vivo newly translated polypeptides are sequestered in a protected folding environment. *Embo. J. 18*, 85-95.

[50] Frydman, J., Nimmesgern, E., Erdjument-Bromage, H., Wall, J.S., Tempst, P. and Hartl, F.U. (1992). Function in protein folding of TRiC, a cytosolic ring complex containing TCP-1 and structurally related subunits. *Embo. J. 11*, 4767-78.

[51] Gao, Y., Thomas, J.O., Chow, R.L., Lee, G.H. and Cowan, N.J. (1992). A cytoplasmic chaperonin that catalyzes beta-actin folding. *Cell 69*, 1043-50.

[52] Yaffe, M.B., Farr, G.W., Miklos, D., Horwich, A.L., Sternlicht, M.L. and Sternlicht, H. (1992). TCP1 complex is a molecular chaperone in tubulin biogenesis. *Nature 358*, 245-8.

[53] Spiess, C., Meyer, A.S., Reissmann, S. and Frydman, J. (2004). Mechanism of the eukaryotic chaperonin: protein folding in the chamber of secrets. *Trends Cell Biol. 14*, 598-604.

[54] Feldman, D.E., Spiess, C., Howard, D.E. and Frydman, J. (2003). Tumorigenic mutations in VHL disrupt folding in vivo by interfering with chaperonin binding. *Mol. Cell 12*, 1213-24.

[55] Gavin, A.C., Bosche, M., Krause, R., Grandi, P., Marzioch, M., Bauer, A., Schultz, J., Rick, J.M., Michon, A.M., Cruciat, C.M., Remor, M., Hofert, C., Schelder, M., Brajenovic, M., Ruffner, H., Merino, A., Klein, K., Hudak, M., Dickson, D., Rudi, T., Gnau, V., Bauch, A., Bastuck, S., Huhse, B., Leutwein, C., Heurtier, M.A., Copley, R.R., Edelmann, A., Querfurth, E., Rybin, V., Drewes, G., Raida, M., Bouwmeester, T., Bork, P., Seraphin, B., Kuster, B., Neubauer, G. and Superti-Furga, G. (2002). Functional organization of the yeast proteome by systematic analysis of protein complexes. *Nature 415*, 141-7.

[56] Ho, Y., Gruhler, A., Heilbut, A., Bader, G.D., Moore, L., Adams, S.L., Millar, A., Taylor, P., Bennett, K., Boutilier, K., Yang, L., Wolting, C., Donaldson, I., Schandorff, S., Shewnarane, J., Vo, M., Taggart, J., Goudreault, M., Muskat, B., Alfarano, C., Dewar, D., Lin, Z., Michalickova, K., Willems, A.R., Sassi, H., Nielsen, P.A., Rasmussen, K.J., Andersen, J.R., Johansen, L.E., Hansen, L.H., Jespersen, H., Podtelejnikov, A., Nielsen, E., Crawford, J., Poulsen, V., Sorensen, B.D., Matthiesen, J., Hendrickson, R.C., Gleeson, F., Pawson, T., Moran, M.F., Durocher, D., Mann, M., Hogue, C.W., Figeys, D. and Tyers, M. (2002). Systematic identification of protein complexes in Saccharomyces cerevisiae by mass spectrometry. *Nature 415*, 180-3.

[57] Kubota, S., Kubota, H. and Nagata, K. (2006). Cytosolic chaperonin protects folding intermediates of Gbeta from aggregation by recognizing hydrophobic beta-strands. *Proc. Natl. Acad. Sci. U S A 103*, 8360-5.

[58] Siegers, K., Bolter, B., Schwarz, J.P., Bottcher, U.M., Guha, S. and Hartl, F.U. (2003). TRiC/CCT cooperates with different upstream chaperones in the folding of distinct protein classes. *Embo. J. 22*, 5230-40.

[59] Behrends, C., Langer, C.A., Boteva, R., Bottcher, U.M., Stemp, M.J., Schaffar, G., Rao, B.V., Giese, A., Kretzschmar, H., Siegers, K. and Hartl, F.U. (2006). Chaperonin TRiC promotes the assembly of polyQ expansion proteins into nontoxic oligomers. *Mol. Cell 23*, 887-97.

[60] Kitamura, A., Kubota, H., Pack, C.G., Matsumoto, G., Hirayama, S., Takahashi, Y., Kimura, H., Kinjo, M., Morimoto, R.I. and Nagata, K. (2006). Cytosolic chaperonin prevents polyglutamine toxicity with altering the aggregation state. *Nat. Cell Biol. 8*, 1163-70.

[61] Tam, S., Geller, R., Spiess, C. and Frydman, J. (2006). The chaperonin TRiC controls polyglutamine aggregation and toxicity through subunit-specific interactions. *Nat. Cell Biol. 8*, 1155-62.

[62] Gomez-Puertas, P., Martin-Benito, J., Carrascosa, J.L., Willison, K.R. and Valpuesta, J.M. (2004). The substrate recognition mechanisms in chaperonins. *J. Mol. Recognit. 17*, 85-94.

[63] Tian, G., Vainberg, I.E., Tap, W.D., Lewis, S.A. and Cowan, N.J. (1995). Specificity in chaperonin-mediated protein folding. *Nature 375*, 250-3.

[64] Klumpp, M., Baumeister, W. and Essen, L.O. (1997). Structure of the substrate binding domain of the thermosome, an archaeal group II chaperonin. *Cell 91*, 263-70.

[65] Iizuka, R., Yoshida, T., Ishii, N., Zako, T., Takahashi, K., Maki, K., Inobe, T., Kuwajima, K. and Yohda, M. (2005). Characterization of archaeal group II chaperonin-ADP-metal fluoride complexes: implications that group II chaperonins operate as a "two-stroke engine". *J. Biol. Chem. 280*, 40375-83.

[66] Reissmann, S., Parnot, C., Booth, C.R., Chiu, W. and Frydman, J. (2007). Essential function of the built-in lid in the allosteric regulation of eukaryotic and archaeal chaperonins. *Nat. Struct. Mol. Biol. 14*, 432-40.

[67] Geissler, S., Siegers, K. and Schiebel, E. (1998). A novel protein complex promoting formation of functional alpha- and gamma-tubulin. *Embo. J. 17*, 952-66.

[68] Vainberg, I.E., Lewis, S.A., Rommelaere, H., Ampe, C., Vandekerckhove, J., Klein, H.L. and Cowan, N.J. (1998). Prefoldin, a chaperone that delivers unfolded proteins to cytosolic chaperonin. *Cell 93*, 863-73.

[69] Lundin, V.F., Srayko, M., Hyman, A.A. and Leroux, M.R. (2008). Efficient chaperone-mediated tubulin biogenesis is essential for cell division and cell migration in C. elegans. *Dev. Biol. 313*, 320-34.

[70] Leroux, M.R., Fandrich, M., Klunker, D., Siegers, K., Lupas, A.N., Brown, J.R., Schiebel, E., Dobson, C.M. and Hartl, F.U. (1999). MtGimC, a novel archaeal chaperone related to the eukaryotic chaperonin cofactor GimC/prefoldin. *Embo. J. 18*, 6730-43.

[71] Okochi, M., Yoshida, T., Maruyama, T., Kawarabayasi, Y., Kikuchi, H. and Yohda, M. (2002). Pyrococcus prefoldin stabilizes protein-folding intermediates and transfers them to chaperonins for correct folding. *Biochem. Biophys. Res. Commun. 291*, 769-74.

[72] Boonyaratanakornkit, B.B., Simpson, A.J., Whitehead, T.A., Fraser, C.M., El-Sayed, N.M. and Clark, D.S. (2005). Transcriptional profiling of the hyperthermophilic methanarchaeon Methanococcus jannaschii in response to lethal heat and non-lethal cold shock. *Environ. Microbiol. 7*, 789-97.

[73] Fukui, T., Atomi, H., Kanai, T., Matsumi, R., Fujiwara, S. and Imanaka, T. (2005). Complete genome sequence of the hyperthermophilic archaeon Thermococcus kodakaraensis KOD1 and comparison with Pyrococcus genomes. *Genome Res. 15*, 352-63.

[74] Siegert, R., Leroux, M.R., Scheufler, C., Hartl, F.U. and Moarefi, I. (2000). Structure of the molecular chaperone prefoldin: unique interaction of multiple coiled coil tentacles with unfolded proteins. *Cell 103*, 621-32.

[75] Martin-Benito, J., Boskovic, J., Gomez-Puertas, P., Carrascosa, J.L., Simons, C.T., Lewis, S.A., Bartolini, F., Cowan, N.J. and Valpuesta, J.M. (2002). Structure of eukaryotic prefoldin and of its complexes with unfolded actin and the cytosolic chaperonin CCT. *Embo. J. 21*, 6377-86.

[76] Martin-Benito, J., Gomez-Reino, J., Stirling, P.C., Lundin, V.F., Gomez-Puertas, P., Boskovic, J., Chacon, P., Fernandez, J.J., Berenguer, J., Leroux, M.R. and Valpuesta, J.M. (2007). Divergent substrate-binding mechanisms reveal an evolutionary specialization of eukaryotic prefoldin compared to its archaeal counterpart. *Structure 15*, 101-10.

[77] Ohtaki, A., Kida, H., Miyata, Y., Ide, N., Yonezawa, A., Arakawa, T., Iizuka, R., Noguchi, K., Kita, A., Odaka, M., Miki, K. and Yohda, M. (2008) Structure and Molecular Dynamics Simulation of Archaeal Prefoldin: The Molecular Mechanism for Binding and Recognition of Nonnative Substrate Proteins. *J. Mol. Biol. 376, 1130-41*

[78] Martin, J., Gruber, M. and Lupas, A.N. (2004). Coiled coils meet the chaperone world. *Trends Biochem. Sci. 29*, 455-8.

[79] Korndorfer, I.P., Dommel, M.K. and Skerra, A. (2004). Structure of the periplasmic chaperone Skp suggests functional similarity with cytosolic chaperones despite differing architecture. *Nat. Struct. Mol. Biol. 11*, 1015-20.

[80] Walton, T.A. and Sousa, M.C. (2004). Crystal structure of Skp, a prefoldin-like chaperone that protects soluble and membrane proteins from aggregation. *Mol. Cell 15*, 367-74.

[81] Siegers, K., Waldmann, T., Leroux, M.R., Grein, K., Shevchenko, A., Schiebel, E. and Hartl, F.U. (1999). Compartmentation of protein folding in vivo: sequestration of non-native polypeptide by the chaperonin-GimC system. *Embo. J. 18*, 75-84.

[82] Hansen, W.J., Cowan, N.J. and Welch, W.J. (1999). Prefoldin-nascent chain complexes in the folding of cytoskeletal proteins. *J. Cell Biol. 145*, 265-77.

[83] Simons, C.T., Staes, A., Rommelaere, H., Ampe, C., Lewis, S.A. and Cowan, N.J. (2004). Selective contribution of eukaryotic prefoldin subunits to actin and tubulin binding. *J. Biol. Chem. 279*, 4196-203.

[84] Lundin, V.F., Stirling, P.C., Gomez-Reino, J., Mwenifumbo, J.C., Obst, J.M., Valpuesta, J.M. and Leroux, M.R. (2004). Molecular clamp mechanism of substrate binding by hydrophobic coiled-coil residues of the archaeal chaperone prefoldin. *Proc. Natl. Acad. Sci. U S A 101*, 4367-72.

[85] Okochi, M., Nomura, T., Zako, T., Arakawa, T., Iizuka, R., Ueda, H., Funatsu, T., Leroux, M. and Yohda, M. (2004). Kinetics and binding sites for interaction of prefoldin with group II chaperonin: contiguous non-native substrate and chaperonin binding sites in archaeal prefoldin. *J. Biol. Chem. 279*, 31788-31795.

[86] Zako, T., Iizuka, R., Okochi, M., Nomura, T., Ueno, T., Tadakuma, H., Yohda, M. and Funatsu, T. (2005). Facilitated release of substrate protein from prefoldin by chaperonin. *FEBS Lett. 579*, 3718-3724.

[87] Mousnier, A., Kubat, N., Massias-Simon, A., Segeral, E., Rain, J.C., Benarous, R., Emiliani, S. and Dargemont, C. (2007). von Hippel Lindau binding protein 1-mediated degradation of integrase affects HIV-1 gene expression at a postintegration step. *Proc. Natl. Acad. Sci. U S A 104*, 13615-20.

[88] Okochi, M., Matsuzaki, H., Nomura, T., Ishii, N. and Yohda, M. (2005). Molecular characterization of the group II chaperonin from the hyperthermophilic archaeum Pyrococcus horikoshii OT3. *Extremophiles 9*, 127-34.

[89] Zako, T., Murase, Y., Iizuka, R., Yoshida, T., Kanzaki, T., Ide, N., Maeda, M., Funatsu, T. and Yohda, M. (2006). Localization of prefoldin interaction sites in the hyperthermophilic group II chaperonin and correlations between binding rate and protein transfer rate. *J. Mol. Biol. 364*, 110-20.

[90] Zako, T., Funatsu, T. and Yohda, M. (2004). Kinetic analysis of interactions between archaeal prefoldin and chaperonin. *Recent Res. Devel. Biophys. 3*, 475-83.

[91] Neirynck, K., Waterschoot, D., Vandekerckhove, J., Ampe, C. and Rommelaere, H. (2006). Actin interacts with CCT via discrete binding sites: a binding transition-release model for CCT-mediated actin folding. *J. Mol. Biol. 355*, 124-38.

[92] Stirling, P.C., Cuellar, J., Alfaro, G.A., El Khadali, F., Beh, C.T., Valpuesta, J.M., Melki, R. and Leroux, M.R. (2006). PhLP3 modulates CCT-mediated actin and tubulin folding via ternary complexes with substrates. *J. Biol. Chem. 281*, 7012-21.

In: Archaea: Structure, Habitats and Ecological… ISBN 978-1-61761-932-8
Editor: Sakura Y. Katō, pp. 199-219

Chapter 9

Primary Production and Carbon Cycling in the Deep Sea: Archaeal Carbon Fixation and Methane Oxidation

The discovery of nonthermophilic archaea (crenarchaeota and euryarchaeota) in the open ocean and coastal water using culture-independent molecular technologies was reported in two seminal papers in the early 1990's (DeLong, 1992; Fuhrman et al., 1992), and this corrected our previous biased view that crenarcheota only existed in extreme environments such as hydrothermal vents or terrestrial thermal springs. Since then, research on nonthermophilic archaea has been intensely pursued intensely and it is now clear that these organisms occur widely in low temperature marine and terrestrial environments (see reviews by DeLong, 1998; Schleper et al., 2005). Overall, in marine water columns, nonthermophilic archaea are dominated by Group I crenarchaeota and Group II euryarchaeota. In marine sediments, not only Group I crenarchaeota and Group II euryarchaeota are present but also a variety of other archaeal species, including the Marine Benthic Group crenarchaeota, Miscellaneous Crenarchaeotal Group, and an unidentified euryarchaeota group (Vetriani et al., 1999; Inagaki et al., 2006; Biddle et al., 2006). In addition, anaerobic methane-oxidizing archaea (ANME groups) widely occur in association with gas hydrates, cold seeps, and organic rich thermal sediments (e.g., Boetius et al., 2000; Orphan et al., 2001; Mills et al., 2003; Knittel et al., 2005) as well as in anoxic water bodies like the Black Sea and Cariaco Basin (Madrid et al., 2001; Durisch-Kaiser et al., 2005; Schubert et al., 2006). This section reviews recent observations of the occurrence and diversity, abundance, metabolic pathways, and contributions to carbon and nitrogen cycles of nonthermophilic archaea in the deep oceans and marine sediments.

Occurrence and Diversity of Nonthermophilic Archaea in Water Column and Marine Sediments

The first detection of PCR-based archaeal 16S rRNA genes in the marine plankton included a few sequences that are closely related to known hyperthermophilic species (Furman et al., 1992; DeLong, 1992). This observation was initially uncertain as some of these archaea were questioned as being allochthonous thermophiles transported from a putative hydrothermal vent. Soon after, however, DeLong and co-workers discovered high abundance of archaea in cold aerobic Arctic water (DeLong et al., 1994) and also a crenarchaeal symbiont in a cold marine sponge (Preston et al., 1996), crystallizing the fact that the newly discovered archaea are native to the low-temperature marine environment. Since then, a large number of novel archaeal phylotypes have been detected in almost every niche of the marine biotope, ranging from surface to abyssal depths of the Pacific and Atlantic Oceans as well as in deep marine sediments (see review by Schelper et al., 2005). So far, the vast majority of planktonic archaea fall into the group I. 1A crenarchaeota and group II euryarchaeota. It has been observed that group I. 1A crenarcheota generally increase in abundance at greater depths (>200 m); whereas, the group II euryarchaeota are predominant in the photic zone (<200 m) but also are found in significant abundance in deeper waters (Massana et al., 1997; Karner et al., 2001; Herndl et al., 2005; DeLong et al., 2006; Teira et al., 2006). Nonthermophilic archaea in marine sediments are mostly reported from cold-seep and gas hydrate-related environments where methane serves as an abundant energy source fueling the anaerobic oxidation of methane (AOM) coupled to sulfate reduction (Hinrichs et al., 1999; Orphan et al., 2001). In diffusion prone settings of deep sea marine sediments, where methane and sulfate fluxes are low, the AOM process is identified in a sediment horizon, namely the sulfate-methane transition zone, into which both methane and sulfate diffuse (Biddle et al., 2006). In cold seeps and gas hydrates, the archaeal community is dominated by the uncultured ANME groups of euryarchaeota that are known to mediate the AOM (*e.g.,* Inagaki et al., 2003; Biddle et al., 2006; Inagaki et al., 2006; Kendall et al., 2007). In particular, ANME-1 and ANME-2 phylotypes are most abundant although their relative distributions varies from location to location. For example, ANME- 1 is predominant in the Black Sea, whereas ANME-2 is predominant at Hydrate Ridge (Knittel et al., 2005; Nauhaus et al., 2005). In the Gulf of Mexico, both ANME-1 and ANME-2 are conspicuous (Lanoil et al., 2001; Mills et al., 2003; Lloyd et al., 2006; Reed et al., 2006; Pi et al., in rivision). In the diffusion prone underlying the highly productive surface water off the coast of Peru, however, the sedimentary archaeal communities are dominated by the Marine Benthic Group B and the Miscellaneous Crenarchaeotal Group, whereas methanotrophic archaea such as ANME-1 and ANME-2 are not detected (Biddle et al., 2006). Other non-cold seep or non-gas hydrate deep-sea marine sediments are also dominated by nonthermophilic crenarchaeota and nonANME groups of euryarchaeota (e.g., Vetriani et al., 1999; Inagaki et al., 2003; Lanoil et al.,2005; Wang et al., 2005; Inagaki et al., 2006). In particular, marine benthic group B and Thermoplasmatales appear to be cosmopolitan in normal marine sediments (Orphan et al., 2001; Teske, 2006).

Abundance of Nonthermophilic Archaea in the Water Column and Marine Sediments

The observation of very large numbers of nonthermophilic archaea in the open ocean adds another dimension to the importance of these organisms in the marine ecosystem. An early estimate of the abundance of planktonic archaea was made in temperate coastal marine waters and polar seas, which ranged from about 10 to 30% of the total prokaryotic biomass(DeLong, 1992). In 2001, a major study was undertaken in the Pacific Ocean, which demonstrated the dominance of archaea in the mesopelagic zone of this water body (Karner et al., 2001). Monthly sampling within a 12-month period showed that pelagic crenarchaeota comprised a large fraction of the total marine picoplankton below 150 m (the bottom of the euphotic zone) and increased with depth, reaching about 40% of the total DNA-containing picoplankton detected using DAPI staining and fluorescent isothiocyanate multiple labeled probes (Karner et al., 2001). Estimates based on this study indicate that the world's oceans contain approximately 1.3 x 1028 archaeal cells, which is of the same order of magnitude as bacteria (3.1 x 1028 cells) (Karner et al., 2001). Another comprehensive study in the Atlantic Ocean, demonstrated the consistently higher abundance of archaea than bacteria below the 100-m depth, using an improved catalyzed reporter deposition-fluorescence in situ hybridization method and specific oligonucleotide probes (Herndl et al., 2005). Estimates of the archaeal production in the mesopelagic and bathypelagic North Atlantic showed 13-27% contributions to the total prokaryotic production in the oxygen minimum layer, 41-84% in Labrador sea water, and 10-20% in the North Atlantic deep water (Herndl et al., 2005).

While group I crenarchaeota commonly comprise the majority of deep water planktonic archaea, it was observed in one instance that a deeply branching crenarchaeal group related to a hot spring clade (pSL 12) was significantly more abundant than the group I crenarchaeota below the euphotic zone in the North Pacific subtropical gyre (Mincer et al., 2007), suggesting that certain environmental conditions may stimulate the growth of multiple species of crenarchaeota. Another observation involves the unusual abundance of group I crenarchaeota in the suboxic layers of the Black Sea, which makes up to 98% of the total archaeal 16S rRNA gene copies (Coolen et al., 2007), suggesting that these archaeal organisms may be microaerophilic and may be able to tolerant suboxic conditions. Estimates of archaea in marine sediments are complicated by the heterogeneous distribution of these organisms in the porous medium. Geochemical proxies, however, have indicated that 90% of the methane produced in the marine sediments is consumed before it reaches the water column (Valentine and Reeburgh, 2000), suggesting that a large population of methane-oxidizing archaea exists in the sediments and plays a significant role in controlling the flux of methane into the atmosphere. Other studies have estimated that the abundance of archaea in marine sediments varies between 0.01% and 30% of total biomass with high values often occurring near the seafloor (Vetriani et al., 1999; Inakagi et al., 2006).

Metabolic Pathways of Nonthermophilic Archaea in the Water Column and Marine Sediments

The physiology and biochemistry of nonthermophilic archaea have also begun to be elucidated as a result of developments in environmental genomics (Venter et al., 2004; Schleper et al., 2005; Hallam et al., 2006a) and the genomic analysis of a single crenarchaeal phylotype Cenarchaeum symbiosium in a marine sponge (Preston et al., 1996; Hallam et al., 2006b). The recently isolated pure culture of the mesophilic crenarchaeon, Nitrosopumilus maritimus, adds new insight into the physiology of non-thermophilic crenarchaeota (Könneke et al., 2005). Mounting evidence indicates that some of the nonthermophilic crenarchaeota use ammonia as a major energy source for autotrophic growth (Francis et al., 2005; Könneke et al., 2005; Nicol and Schleper, 2006; Wuchter et al., 2006). Other molecular and geochemical studies show that nonthermophilic crenarchaeota can take up amino acids and organic carbon, indicating that some of them are heterotrophs or mixotrophs (Biddle et al., 2006; Herndl et al., 2005; Ingalls et al., 2006; Ouverney et al., 2000; Teira et al., 2006). One recent study showed that some of the benthic crenarchaeota may oxidize methane using it as an energy source and other organic carbon as the carbon source (Biddle et al., 2006).

On the other hand, ANME groups of the euryarchaeota oxidize methane using a reverse methanogenesis pathway (Boetius et al., 2000; Hallam et al., 2003, 2004; Valentine, 2000). While no isolate is available of any of the ANME group members for detailed biochemical studies, genomic analysis of the AOM community in Eel River Basin sediment could distinguish the ANME-1 group from the ANME-2 group based on G+C content and depth of the whole genome shotgun coverage, Importantly, Hallam et al., (2004) demonstrated that nearly all the genes typically associated with methane production are present in the ANME-1 group, except a step encoded by the mer gene. It was proposed that the loss of mcr activity in ANME-1 could promote AOM by increasing the activation barrier for conversion of methylene- H4MPT to methyl-H4MPT. Overall, the association of most of the genes for methanogenesis in the ANME groups strongly supports the hypothesis that AOM proceeds as a reversed pathway of methane production.

The Group II euryarchaeota are known to contain light-harvesting proteins (PRs), suggesting that these novel euryarchaeota may perform light-dependent cellular energy metabolism (DeLong et al., 2006; Frigaard et al., 2006).

Contribution of Nonthermophilic Archaea to Carbon and Nitrogen Cycles in the Deep Ocean

A picture is emerging that the majority of the nonthermophilic archaea in the deep ocean are chemoautotrophs and may use ammonia as a major energy source. Herndl et al., (2005) presented leucine-incorporation evidence that the actively growing archaea in the oxygenated water column, which numerically dominate the prokaryotic community in the North Atlantic below a depth of 100 m to 2435-2790 m in the North Atlantic Deep Water, use bicarbonate or CO2 as a carbon source. Independently, Ingalls et al., (2006) quantified archaeal autotrophy in the subtropical North Pacific gyre using natural radiocarbon and demonstrated, based on an isotopic mass balance model, that 83% of total archaeal metabolism is autotrophy that fixes ^{14}C-depleted dissolved inorganic carbon in the deep water; whereas, the remaining metabolism may contribute to heterotrophic consumption of the ^{14}C-enriched modern organic carbon. Based on the depth-integrated mean C fixation rate (0.014 fmol C archaeon^{-1} day^{-1}) and on the assumption that the global ocean contains approximately 1.3 x 10^{28} cells of archaea, Herndl et al., (2005) derived a global inorganic carbon fixation rate of 6.55 x 10^{13} mol C year^{-1}. This global carbon fixation rate is compared with that derived from archaeal nitrification, which is estimated to be 3.3 x 10^{13} mol C year^{-1} (assuming all of the ammonium would be oxidized by crenarchaeota fixing one carbon atom for every 10 nitrogen molecules oxidized) (Wuchter et al., 2006).

The agreement between these estimated rates suggest that archaeal nitrification coupled to inorganic carbon fixation may contribute profoundly to the biogeochemical cycles of carbon and nitrogen in the global ocean (Figure 1). Methane formation by methanogens in the dark ocean is also a major sink of CO_2; however, its contribution to carbon fixation is two orders of magnitude lower (~8 x 10^{11} mol C year^{-1}) than that from crenarchaeota in the global ocean (Herndl et al., 2005). In marine sediments, methanogens are the sole source of biogenic methane below the sulfate reduction zone; whereas within the sulfate reduction zone, methane-oxidizing archaea in consortia with sulfate-reducing bacteria are mainly responsible for methane consumption (Figure 1). Although heterotrophic crenarchaeota are known to perform oxidation of methane (Biddle et al., 2006), their relative contribution to methane consumption is unknown. In either case, CO_2 from anaerobic methane oxidation is largely sequestrated into carbonate minerals that constitute a major fraction of the marine sediment or form carbonate mounds above the ocean floor. This has been observed frequently in the Gulf of Mexico (Formolo et al., 2004; Sassen et al., 2004) as well as other methane-rich marine sediments (e.g., Aloisi et al., 2002), where gas hydrates and cold seeps are found abundantly. Globally, a huge amount of methane (> 10 trillion tons; Zimmer, 2001) may be sequestrated in the hydrate structure in marine sediment or hydrate mounds, which are known as an attractive alternative energy resource beyond petroleum hydrocarbons and coals.

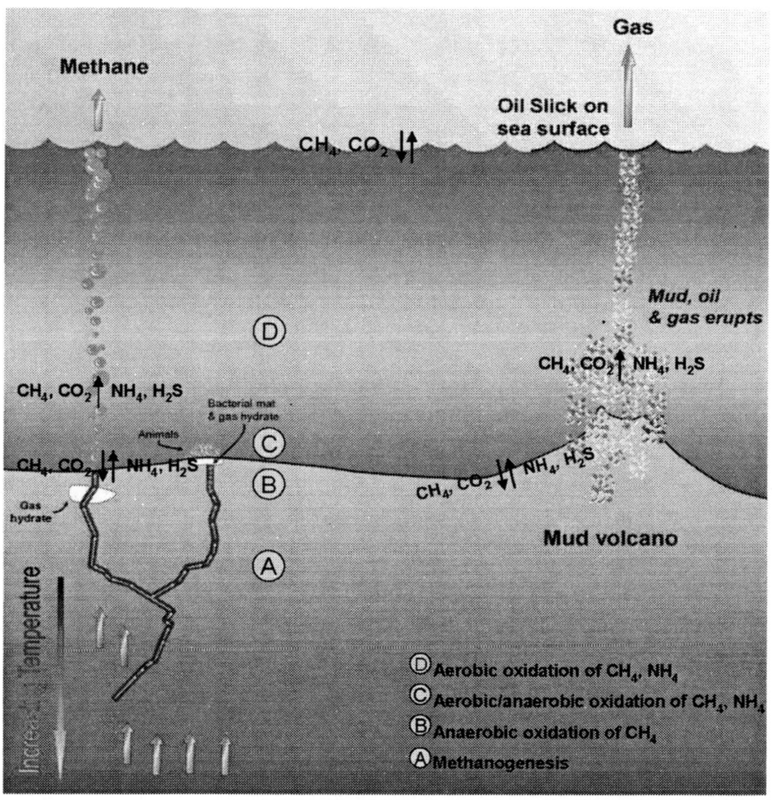

Figure 1. A model of carbon, nitrogen, and sulfur cycling in the deep ocean and deep-sea marinesediments affected by gas hydrates or cold seeps (Modified from Whelan et al., 2005).

References

Aloisi, G., Bouloubassi, I., Heijs, S. K., Pancost, R. D., Pierre, C., Damste, J. S. S., Gottschal, J. C., Forney, L. J., and Rouchy, J. M. (2002). CH_4-consuming microorganisms and the formation of carbonate crusts at cold seeps. *Earth and Planetary Science Letters*, 203, 195-203.

Anbar, A. D. and Knoll, A. H. (2002). Proterozoic ocean chemistry and evolution: A bioinorganic bridge? *Science*, 297, 1137-1142.

Anderson, N. L. and Anderson, N. G. (1998). Proteome and proteomics: New technologies, new concepts, and new words. *Electrophoresis*, 19, 1853-1861.

Angly, F. E., Felts, B., Breitbart, M., Salamon, P., Edwards, R. A., Carlson, C., Chan, A. M., Haynes, M., Kelley, S., Liu, H., Mahaffy, J. M., Mueller, J. E., Nulton, J., Olson, R., Parsons, R., Rayhawk, S., Suttle, C. A., and Rohwer, F. (2006). The marine viromes of four oceanic regions. *Plos Biology*, 4, 2121-2131.

Arce, F. T., Avci, R., Beech, I. B., Cooksey, K. E., and Wigglesworth-Cooksey, B. (2004). live bioprobe for studying diatom-surface interactions. *Biophysical Journal*, 87, 4284-4297.

Auerbach, I.D., Sorensen, C., Hansma, H. G., and Holden, P. A. (2000). Physical morphology and surface properties of unsaturated *Pseudomonas putida* biofilms. *Journal of Bacteriology*, 182, 3809-3815.

Azam, F., Fenchel, T., Field, J. G., Gray, J. S., Meyer-Reil, L. A., and Tingstad, F. (1983). The ecological role of water-column microbes in the sea. *Marine Ecology Progress Series*, 10, 257-263.

Bahatyrova, S., Frese, R. N., Siebert, C. A., Olsen, J. D., van der Werf, K. O., van Grondelle, R., Niederman, R. A., Bullough, P. A., and Hunter, C. N. (2004). The native architecture of a photosynthetic membrane. *Nature*, 430, 1058-1062.

Bauer, J. E., Williams, P. M., and Druffel, E. R. M. (1992). ^{14}C activity of dissolved organic carbonfractions in the northcentral Pacific and Sargasso Sea. *Nature*, 357, 667-670.

Beatty, J. T. (2002). On the natural selection and evolution of the aerobic phototrophic bacteria. *Photosynthesis Research*, 73, 109-114.

Béjà, O., Aravind, L., Koonin, E. V., Suzuki, M. T., Hadd, A., Nguyen, L. P., Jovanovich, S., Gates, C. M., Feldman, R. A., Spudich, J. L., Spudich, E. N., and DeLong, E. F. (2000). Bacterial rhodopsin: Evidence for a new type of phototrophy in the sea. *Science*, 289, 1902-1906.

Béjà, O., Spudich, E. N., Spudich, J. L., Leclerc, M., and DeLong, E. F. (2001). Proteorhodopsin phototrophy in the ocean. *Nature*, 411, 786-789.

Béjà, O., Suzuki, M. T., Heidelberg, J. F., Nelson, W. C., Preston, C. M., Hamada, T., Eisen, J. A., Fraser, C. M., and DeLong, E. F. (2002). Unsuspected diversity among marine aerobic anoxygenic phototrophs. *Nature*, 415, 630-633.

Bergh, O., Borsheim, K. Y., Bratbak, G., and Heldal, M. (1989). High abundance of viruses found in aquatic environments. Nature, 340, 467–468.

Beveridge, T. J. and Graham, L. L. (1991). Surface layers of bacteria. *Microbiology and Molecular Biology Reviews*, 55, 684-705.

Beveridge, T. J., Makin, S. A., Kadurugamuwa, J. L., and Li, Z. (1997). Interactions between biofilms and the environment. *FEMS Microbiology Reviews*, 20, 291-303,.

Beveridge, T. J. (1999). Structures of Gram-Negative Cell Walls and Their Derived Membrane Vesicles. *Journal of Bacteriology*, 181, 4725-4733.

Biddle, J. F., Lipp, J. S., Lever, M. A., Lloyd, K. G., Sorensen, K. B., Anderson, R., Fredricks, H. F., Elvert, M., Kelly, T. J., Schrag, D. P., Sogin, M. L., Brenchley, J. E., Teske, A., House, C. H., and Hinrichs, K. U. (2006). Heterotrophic Archaea dominate sedimentary subsurface ecosystems off Peru. *Proceedings of the NationalAcademy of Sciences of the United States of America*, 103, 3846-3851.

Binder, B. J., Chisholm, S. W., Olson, R. J., Frankel, S. L., and Worden, A. Z. (1996). Dynamics of picophytoplankton, ultraphytoplankton and bacteria in the central equatorial Pacific. *Deep-Sea Research II*, 43, 907-931.

Binnig, G., Quate, C. F., and Gerber, C. (1986). Atomic force microscope. *Physical Review Letters*, 56, 930 - 933.

Blackstock, W. P. and Weir, M. P. (1999). Proteomics: quantitative and physical mapping of cellular proteins. *Trends in Biotechnology*, 17, 121-127.

Boetius, A., Ravenschlag, K., Schubert, C. J., Rickert, D., Widdel, F., Gieseke, A., Amann, R., Jorgensen, B. B., Witte, U., and Pfannkuche, O. (2000). A marine microbial consortium apparently mediating anaerobic oxidation of methane. *Nature*, 407, 623-626.

Boschker, H. T. S., Nold, S. C., Wellsbury, P., Bos, D., de Graaf, W., Pel, R., Parkes, R. J., and Cappenberg, T. E. (1998). Direct linking of microbial populations to specific biogeochemical processes by C-13-labelling of biomarkers. *Nature*, 392, 801-805.

Bralower, T. J., Silva, I. P., and Malone, M. J. (2002). New evidence for abrupt climate change in the Cretaceous and Paleogene: An Ocean Drilling Program expedition to Shatsky rise, northwest Pacific. *GSA Today*, 12, 4-10.

Brocks, J. J. and Pearson, A. (2005) Building the biomarker tree of life. In *Molecular Geomicrobiology*. Chantilly: Mineralogical Soc America, pp. 233-258.

Brussow, H., Canchaya, C., and Hardt, W. D. (2004). Phages and the evolution of bacterial pathogens: From genomic rearrangements to lysogenic conversion. *Microbiology and Molecular Biology Reviews*, 68, 560-602.

Buck, K. R., Chavez, F. P., and Campbell, L. (1996). Basin-wide distributions of living carbon components and the inverted trophic pyramid of the central gyre of the north Atlantic Ocean, summer 1993. *Aquatic Microbial Ecology*, 10, 283-298.

Butt, H. J., Jaschke, M., and Ducker, W. (1995). Measuring surface forces in aqueous electrolyte solution with the atomic force microscope. *Bioelectrochemistry and Bioenergetics*, 38, 191-201.

Camesano, T. A., Natan, M. J., and Logan, B. E. (2000). Observation of changes in bacterial cell morphology: using tapping mode atomic force microscopy. *Langmuir*, 16, 4563-4572.

Campbell, L., Liu, H. B., Nolla, H. A., and Vaulot, D. (1997). Annual variability of phytoplankton and bacteria in the subtropical North Pacific Ocean at Station ALOHA during the 1991-1994 ENSO event. *Deep-Sea Research Part I-Oceanographic Research Papers,* 44, 167-and.

Canchaya, C., Proux, C., Fournous, G., Bruttin, A., and Brussow, H. (2003). Prophage genomics. *Microbiology and Molecular Biology Reviews*, 67, 473-473.

Chada, V. G. R., Sanstad, E. A., Wang, R., and Driks, A. (2003). Morphogenesis of *Bacillus* spore surfaces. *Journal of Bacteriology*, 185, 6255-6261.

Chen, F., Lu, J. R., Binder, B. J., Liu, Y. C., and Hodson, R. E. (2001). Application of digital image analysis and flow cytometry to enumerate marine viruses stained with SYBR gold. *Applied and Environmental Microbiology*, 67, 539-545.

Chen, F., Wang, K., Kan, J. J., Suzuki, M. T., and Wommack, K. E. (2006). Diverse and unique picocyanobacteria in Chesapeake Bay, revealed by 16S-23S rRNA internal transcribed Spacer sequences. *Applied and Environmental Microbiology*, 72, 2239-2243.

Chen, F., Wang, K., Stewart, J., and Belas, R. (2006). Induction of multiple prophages from a marine bacterium: a genomic approach. *Applied and Environmental Microbiology*, 72, 4995-5001.

Cochran, P. K., Kellogg, C. A., and Paul, J. H. (1998). Prophage induction of indigenous marine lysogenic bacteria by environmental pollutants. *Marine Ecology-Progress Series*, 164, 125-133.

Cochran, P. K. and Paul, J. H. (1998). Seasonal abundance of lysogenic bacteria in a subtropical estuary. *Applied and Environmental Microbiology*, 64, 2308-2312.

Comita, P. B. and Gagosian, R. B. (1983). Membrane lipid from deep sea hydrothermal vent methanogen: A new macrocyclic glycerol diether. *Science*, 222, 1329-1331.

Coolen, M. J. L., Abbas, B., van Bleijswijk, J., Hopmans, E. C., Kuypers, M. M. M., Wakeham, S. G., and Damste, J. S. S. (2007). Putative ammonia-oxidizing Crenarchaeota in suboxic waters of the Black Sea: a basin-wide ecological study using 16S ribosomal and functional genes and membrane lipids. *Environmental Microbiology*, 9, 1001-1016.

Curtis, T. P., Sloan, W. T., and Scannell, J. W. (2002). Estimating prokaryotic diversity and its limits. *Proceedings of the NationalAcademy of Sciences of the United States of America*, 99, 10494-10499.

Damste, J. S. S., Rijpstra, W. I. C., Hopmans, E. C., Prahl, F. G., Wakeham, S. G., and Schouten, S. (2002). Distribution of membrane lipids of planktonic Crenarchaeota in the Arabian seat. *Applied and Environmental Microbiology*, 68, 2997-3002.

Damste, J. S. S., Strous, M., Rijpstra, W. I. C., Hopmans, E. C., Geenevasen, J. A. J., van Duin, A. C. T., van Niftrik, L. A., and Jetten, M. S. M. (2002). Linearly concatenated cyclobutane lipids form a dense bacterial membrane. *Nature*, 419, 708-712.

Delong, E. F. (1992). Archaea in coastal marine environments. *Proceedings of the National Academy of Sciences of the United States of America*, 89, 5685-5689.

DeLong, E. F., Wu, K. Y., Prezelin, B. B., and Jovine, R. V. (1994). High abundance of Archaea in Antarctic marine picoplankton. *Nature*, 371, 695-697.

DeLong, E. F., King, L. L., Massana, R., Cittone, H., Murray, A., Schleper, C., and Wakeham, S. G. (1998). Dibiphytanyl ether lipids in nonthermophilic crenarchaeotes. *Applied and Environmental Microbiology*, 64, 1133-1138.

DeLong, E. F., Preston, C. M., Mincer, T., Rich, V., Hallam, S. J., Frigaard, N. U., Martinez, A., Sullivan, M. B., Edwards, R., Brito, B. R., Chisholm, S. W., and Karl, D. M. (2006). Community genomics among stratified microbial assemblages in the ocean's interior. *Science*, 311, 496-503.

Doktycz, M. J., Sullivan, C. J., Hoyt, P. R., Pelletier, D. A., Wu, S., and Allison, D. P. (2003). AFM imaging of bacteria in liquid media immobilized on gelatin coated mica surfaces. *Ultramicroscopy*, 97, 209-216.

Dufrêne, Y. F. (2002). Atomic Force Microscopy, a Powerful Tool in Microbiology. *Journal of Bacteriology*, 184, 5205-5213.

Dufrêne, Y. F. (2004). Using nanotechniques to explore microbial surfaces. *Nature*, 2, 451-460.

Dufrêne, Y. F. (2004). Refining our perception of bacterial surfaces with the atomic force microscope. *Journal of Bacteriology*, 186, 3283-3285.

Dufresne, A., Salanoubat, M., Partensky, F., Artiguenave, F., Axmann, I. M., Barbe, V., Duprat, S., Galperin, M. Y., Koonin, E. V., Le Gall, F., Makarova, K. S., Ostrowski, M., Oztas, S., Robert, C., Rogozin, I. B., Scanlan, D. J., de Marsac, N. T., Weissenbach, J., Wincker, P., Wolf, Y. I., and Hess, W. R. (2003). Genome sequence of the cyanobacterium *Prochlorococcus marinus* SS120, a nearly minimal oxyphototrophic genome. *Proceedings of the NationalAcademy of Sciences of the United States of America*, 100, 10020-10025.

Dumitrescu, M. and Brassell, S.C. (2003) Biomarkers confirm cyanobacteria as a major source of planktonic organic matter during the Early Aptian. Geological Society of America (GSA) Seattle Annual Meeting (November 2-5, 2003, Abstracts with Programs, pp. 391). URL: *http://gsa.confex.com/gsa/2003AM/finalprogram/abstract_67241.htm*.

Durisch-Kaiser, E., Klauser, L., Wehrli, B., and Schubert, C. (2005). Evidence of intense archaeal and bacterial methanotrophic activity in the black sea water column. *Applied and Environmental Microbiology*, 71, 8099-8106.

Emerson, R. J. and Camesano, T. A. (2004). Nanoscale investigation of pathogenic microbial adhesion to a biomaterial. *Applied and Environmental Microbiology*, 70, 6012-6022.

Fang, H. H. P., Chan, K. Y., and Xu, L. C. (2000). Quantification of bacterial adhesion forces using atomic force microscopy (AFM). *Journal of Microbiological Methods*, 40, 89-97.

Feely, R. A., Sabine, C. L., Takahashi, T., and Wanninkhof, R. (2001). Uptake and Storage of carbon dioxide in the Ocean: the Global CO2 Survey. *Oceanography*, 14, 18-32.

Fenchel, T. (2001). Ecology - Marine bugs and carbon flow. Science, 292, 2444-2445.

Fichmann, J. and Westermeier, R. (1999) *2-D proteome analysis protocols*. New Jersey: Humana Press.

Fletcher, M., 1996. Bacterial attachment in aquatic environments: a diversity of surfaces and adhesion strategies. In: Fletcher, M. (Eds.), *Bacterial Adhesion: Molecular and Ecological Diversity* (pp. 1–24), New York: John Wiley and Sons.

Formolo, M. J., Lyons, T. W., Zhang, C. L., Kelley, C., Sassen, R., Horita, J., and Cole, D. R. (2004). Quantifying carbon sources in the formation of authigenic carbonates at gas hydrate sites in the Gulf of Mexico. *Chemical Geology*, 205, 253-264.

Francis, C. A., Roberts, K. J., Beman, J. M., Santoro, A. E., and Oakley, B. B. (2005). Ubiquity and diversity of ammonia-oxidizing archaea in water columns and sediments of the ocean. *Proceedings of the NationalAcademy of Sciences of the United States of America*, 102, 14683-14688.

Freeman, K. H., Hayes, J. M., Trendel, J. M., and Albrecht, P. (1990). Evidence from carbon isotope measurements for diverse origins of sedimentary hydrocarbons. *Nature*, 343, 254-256.

Friedrich, M. W. (2006). Stable-isotope probing of DNA: insights into the function of uncultivated microorganisms from isotopically labeled metagenomes. *Current Opinion in Biotechnology*, 17, 59-66.

Frigaard, N. U., Martinez, A., Mincer, T. J., and DeLong, E. F. (2006). Proteorhodopsin lateral gene transfer between marine planktonic Bacteria and Archaea. *Nature*, 439, 847-850.

Fuhrman, J. A., McCallum, K., and Davis, A. A. (1992). Novel major archaebacterial group from marine plankton. *Nature*, 356, 148-149.

Fuhrman, J. A. and Suttle, C. A. (1993). Viruses in marine planktonic systems. *Oceanography*, 6, 50-62.

Fuhrman, J. A. (1999). Marine viruses and their biogeochemical and ecological effects. *Nature*, 399, 541-548.

Gaboriaud, F., Bailet, S., Dague, E., and Jorand, F. (2005). Surface Structure and Nanomechanical Properties of *Shewanella putrefaciens* Bacteria at Two pH values (4 and 10) Determined by Atomic Force Microscopy. *Journal of Bacteriology*, 187, 3864-3868.

Gad, M. and Ikai, A. (1995). Method for immobilizing microbial cells on gel surface for dynamic AFM studies. *Biophysical Journal*, 69, 2226-2233.

Georicke, R. and Welschmeyer, N. A. (1993). The marine prochlorophyte *Prochlorococcus* contributes significantly to phytoplankton biomass and primary production in the Sargasso Sea. *Deep-Sea Research*, 2283-2294.

Giovannoni, S. J., Bibbs, L., Cho, J. C., Stapels, M. D., Desiderio, R., Vergin, K. L., Rappe, M. S., Laney, S., Wilhelm, L. J., Tripp, H. J., Mathur, E. J., and Barofsky, D. F. (2005). Proteorhodopsin in the ubiquitous marine bacterium SAR11. *Nature*, 438, 82-85.

Giovannoni, S. J., Tripp, H. J., Givan, S., Podar, M., Vergin, K. L., Baptista, D., Bibbs, L., Eads, J., Richardson, T. H., Noordewier, M., Rappe, M. S., Short, J. M., Carrington, J. C., and Mathur, E. J. (2005). Genome streamlining in a cosmopolitan oceanic bacterium. *Science*, 309, 1242-1245.

Glaeser, J. and Overmann, J. (2003). Characterization and in situ carbon metabolism of phototrophic consortia. *Applied and Environmental Microbiology*, 69, 3739-3750.

Gobler, C. J., Hutchins, D. A., Fisher, N. S., Cosper, E. M., and Sanudo-Wilhelmy, S. A. (1997). Release and bioavailability of C, N, P, Se, and Fe following viral lysis of a marine chrysophyte. *Limnology and Oceanography*, 42, 1492-1504.

Hafenbradl, D., Keller, M., Thiericke, R., and Stetter, K. O. (1993). A novel unsaturated archaeal ether core lipid from the hyperthermophile *Methanopyrus kandleri*. *Systematics and Applied Microbiology*, 16, 165-169.

Hallam, S. J., Girguis, P. R., Preston, C. M., Richardson, P. M., and DeLong, E. F. (2003). Identification of methyl coenzyme M reductase A (mcrA) genes associated with methane-oxidizing archaea. *Applied and Environmental Microbiology*, 69, 5483-5491.

Hallam, S. J., Putnam, N., Preston, C. M., Detter, J. C., Rokhsar, D., Richardson, P. M., and DeLong, E. F. (2004). Reverse methanogenesis: Testing the hypothesis with environmental genomics. *Science*, 305, 1457-1462.

Hallam, S. J., Konstantinidis, K. T., Putnam, N., Schleper, C., Watanabe, Y., Sugahara, J., Preston, C., de la Torre, J., Richardson, P. M., and DeLong, E. F. (2006). Genomic analysis of the uncultivated marine crenarchaeote Cenarchaeum symbiosum. *Proceedings of the NationalAcademy of Sciences of the United States of America*, 103, 18296-18301.

Hallam, S. J., Mincer, T. J., Schleper, C., Preston, C. M., Roberts, K., Richardson, P. M., and DeLong, E. F. (2006). Pathways of carbon assimilation and ammonia oxidation suggested by environmental genomic analyses of marine Crenarchaeota. *Plos Biology*, 4, 520-536.

Hayes, J. M., Freeman, K. H., Popp, B. N., and Hoham, C. H. (1990). Compound-specific isotopic analyses: A novel tool for reconstruction of ancient biogeochemical processes. *Organic Geochemistry*, 16, 1115-1128.

Hedges, J. I. (1992). Global biogeochemical cycles: progress and problems. *Marine Chemistry*, 39, 67-93.

Hedrick, D. B., Gucker, J. B., and White, D. C. (1991). Archaebacterial ether lipid diversity analyzed by supercritical fluid chromatography: integration with a bacterial lipid protocol. *Journal of Lipid Research*, 32, 659-666.

Herndl, G. J., Reinthaler, T., Teira, E., van Aken, H., Veth, C., Pernthaler, A., and Pernthaler, J. (2005). Contribution of Archaea to total prokaryotic production in the deep Atlantic Ocean. *Applied and Environmental Microbiology*, 71, 2303-2309.

Higgins, M. J., Sader, J. E., Mulvaney, P., and Wetherbee, R. (2003). Probing the surface of living diatoms with atomic force microscopy: the nanostructure and nanomechanical properties of the mucilage layer. *Journal of Phycology*, 39, 722-734.

Hinrichs, K. U., Hayes, J. M., Sylva, S. P., Brewer, P. G., and DeLong, E. F. (1999). Methane-consuming archaebacteria in marine sediments. *Nature*, 398, 802-805.

Hinterdorfer, P., Baumgartner, W., Gruber, H. J., Schilcher, K., and Schindler, H. (1996). Detection and localization of individual antibody-antigen recognition events by atomic force microscopy. *Proceedings of the NationalAcademy of Sciences of the United States of America*, 93, 3477-3481.

Hopmans, E. C., Schouten, S., Pancost, R. D., van der Meer, M. T. J., and Damste, J. S. S. (2000). Analysis of intact tetraether lipids in archaeal cell material and sediments by high performance liquid chromatography/atmospheric pressure chemical ionization mass spectrometry. *Rapid Communications in Mass Spectrometry*, 14, 585-589.

House, C. H., Schopf, J. W., and Stetter, K. O. (2003). Carbon isotopic fractionation by Archaeans and other thermophilic prokaryotes. *Organic Geochemistry*, 34, 345-356.

Huber, R., Wilharm , T., Huber, D., Trincone, A., Burgraf, S., Konig, H., Rachel, R., Rochinger, I., Fricke, H., and Stetter, K. O. (1992). *Aquifex pyrophilus* gen. nov. sp. nov., represents a novel group of marine hyperthermophilic hydrogen-oxidizing bacteria. *Systematics and Applied Microbiology*, 15, 340-351.

Inagaki, F., Suzuki, M., Takai, K., Oida, H., Sakamoto, T., Aoki, K., Nealson, K. H., and Horikoshi, K. (2003). Microbial communities associated with geological horizons in coastal subseafloor sediments from the Sea of Okhotsk. *Applied and Environmental Microbiology*, 69, 7224-7235.

Inagaki, F., Nunoura, T., Nakagawa, S., Teske, A., Lever, M., Lauer, A., Suzuki, M., Takai, K., Delwiche, M., Colwell, F. S., Nealson, K. H., Horikoshi, K., D'Hondt, S., and Jorgensen, B. B. (2006). Biogeographical distribution and diversity of microbes in methane hydrate-bearing deep marine sediments, on the Pacific Ocean Margin. *Proceedings of the NationalAcademy of Sciences of the United States of America*, 103, 2815-2820.

Ingalls, A. E., Shah, S. R., Hansman, R. L., Aluwihare, L. I., Santos, G. M., Druffel, E. R. M., and Pearson, A. (2006). Quantifying archaeal community autotrophy in the mesopelagic ocean using natural radiocarbon. *Proceedings of the NationalAcademy of Sciences of the United States of America*, 103, 6442-6447.

Ivanova, E. P., Alexeeva, Y. V., Pham, D. K., Wright, J. P., and Nicolaul, D. V. (2006). ATP level variations in heterotrophic bacteria during attachment on hydrophilic and hydrophobic surfaces. *International Microbiology*, 9, 37-46.

Jahnke, L. L., Summons, R. E., Dowling, L. M., and Zahiralis, K. D. (1995). Identification of methanotrophic lipid biomarkers in cold-seep mussel gills-Chemical and isotopic analysis. *Applied and Environmental Microbiology*, 61, 576-582.

Jahnke, L. L., Eder, W., Huber, R., Hope, J. M., Hinrichs, K. U., Hayes, J. M., Des Marais, D. J., Cady, S. L., and Summons, R. E. (2001). Signature lipids and stable carbon isotope analyses of octopus spring hyperthermophilic communities compared with those of Aquificales representatives. *Applied and Environmental Microbiology*, 67, 5179-5189.

Jiang, S. C. and Paul, J. H. (1996). Occurrence of lysogenic bacteria in marine microbial communities as determined by prophage induction. *Marine Ecology Progress Series*, 142, 27-28.

Jean, W., Lorraine, E., Lawrence, C., Steven, L., Harry, R. (2005). Surface and subsurface manifestations of gas movement through a N-S stransect of the Gulf of Mexico, *Marine and Petroleum Geology*, 22, 479-497.

Jiao, N. Z., Yang, Y. H., Koshikawa, H., and Watanabe, M. (2002). Influence of hydrographic conditions on picoplankton distribution in the East China Sea. *Aquatic Microbial Ecology*, 30, 37-48.

Jiao, N. Z., Sieracki, M. E., Zhang, Y., and Du, H. L. (2003). Aerobic anoxygenic phototrophic bacteria and their roles in marine ecosystems. *Chinese Science Bulletin*, 48, 1064-1068.

Jiao, N. Z., Yang, Y. J., and Luo, T. W. (2004). Membrane potential based characterization by flow cytometry of physiological states in an aerobic anoxygenic phototrophic bacterium. *Aquatic Microbial Ecology*, 37, 149-158.

Jiao, N. Z., Yang, Y. H., Hong, N., Ma, Y., Harada, S., Koshikawa, H., and Watanabe, M. (2005). Dynamics of autotrophic picoplankton and heterotrophic bacteria in the East China Sea. *Continental Shelf Research*, 25, 1265-1279.

Jiao, N. Z., Feng, F. Y., and Wei, B. (2006). Proteorhodopsin - A new path for biological utilization of light energy in the sea. *Chinese Science Bulletin*, 51, 889-896.

Jiao, N. Z., Zhang, Y., and Chen, Y. (2006). Time series observation based InfraRed Epifluorescence Microscopic (TIREM) approach for accurate enumeration of bacteriochlorophyll-containing microbes in marine environments. *Journal of Microbiological Methods*, 65, 442-452.

Jiao N. Z. (2006). *Marine Microbial Ecology*. Beijing, China: Science Press.

Jiao, N. Z., Yang, Y. H., Zeng, Y. H., Hong, N., Liu, R. L. Chen, F., and Wang, P. X. (2007). Distinct Distribution Pattern of Abundance and Diversity of Aerobic Anoxygenic Phototrophic Bacteria in the Global Ocean. *Environmental Microbiology*, 9(12):3091-3099.

Kan, J., Hanson, T. E., Ginter, J. M., Wang, K., and Chen, F. (2005). Metaproteomic analysis of Chesapeake Bay microbial communities. *Saline Systems*, 1, 7.

Karner, M. B., DeLong, E. F., and Karl, D. M. (2001). Archaeal dominance in the mesopelagic zone of the Pacific Ocean. *Nature*, 409, 507-510.

Kasas, S., Gotzos, V., and Celio, M. R. (1993). Obeservation of living cells using the atomic force microscope. *Biophysical Journal*, 64, 539-544.

Kendall, M. M., Wardlaw, G. D., Tang, C. F., Bonin, A. S., Liu, Y. T., and Valentine, D. L. (2007). Diversity of Archaea in marine sediments from Skan Bay, Alaska, including cultivated methanogens, and description of *Methanogenium boonei* sp nov. *Applied and Environmental Microbiology*, 73, 407-414.

Kerr, R. A. (2002). Evolution - Could poor nutrition have held life back? *Science*, 297, 1104-1105.

Knittel, K., Losekann, T., Boetius, A., Kort, R., and Amann, R. (2005). Diversity and distribution of methanotrophic archaea at cold seeps. *Applied and Environmental Microbiology*, 71, 467-479.

Koblizek, M., Béjà, O., Bidigare, R. R., Christensen, S., Benitez-Nelson, B., Vetriani, C., Kolber, M. K., Falkowski, P. G., and Kolber, Z. S. (2003). Isolation and characterization of *Erythrobacter* sp strains from the upper ocean. *Archives of Microbiology*, 180, 327-338.

Kolber, Z. S., Van Dover, C. L., Niederman, R. A., and Falkowski, P. G. (2000). Bacterial photosynthesis in surface waters of the open ocean. *Nature*, 407, 177-179.

Kolber, Z. S., Plumley, F. G., Lang, A. S., Beatty, J. T., Blankenship, R. E., VanDover, C. L., Vetriani, C., Koblizek, M., Rathgeber, C., and Falkowski, P. G. (2001). Contribution of aerobic photoheterotrophic bacteria to the carbon cycle in the ocean. *Science*, 292, 2492-2495.

Könneke, M., Bernhard, A. E., de la Torre, J. R., Walker, C. B., Waterbury, J. B., and Stahl, D. A. (2005). Isolation of an autotrophic ammonia-oxidizing marine archaeon. *Nature*, 437, 543-546.

Kuznetsov, Y. G., Malkin, A. J., Lucas, R. W., Plomp, M., and McPherson, A. (2001). Imaging of viruses by atomic force microscopy. *Journal of General Virology*, 82, 2025-2034.

Kuznetsov, Y. G., Gurnon, J. R., Etten, J. L., and McPherson, A. (2005). Atomic force microscopy investigation of a chlorella virus, PBCV-1. *Journal of Structural Biology*, 149, 256-263.

Lang, A. S. and Beatty, J. T. (2007). Importance of widespread gene transfer agent genes in alpha-proteobacteria. *Trends in Microbiology*, 15, 54-62.

Langworthy, T. A., Holzer, G., Zeikus, J. G., and Tornabene, T. G. (1983). Iso-and anteiso-branched glycerol diethers of the thermophilic anaerobes *Thermo- desulfotobacterium commune*. *Systematics and Applied Microbiology*, 4, 1-17.

Lanoil, B. D., Sassen, R., La Duc, M. T., Sweet, S. T., and Nealson, K. H. (2001). Bacteria and Archaea physically associated with Gulf of Mexico gas hydrates. *Applied and Environmental Microbiology*, 67, 5143-5153.

Lanoil, B. D., La Duc, M. T., Wright, M., Kastner, M., Nealson, K. H., and Bartlett, D. (2005). Archaeal diversity in ODP legacy borehole 892b and associated seawater and sediments of the Cascadia Margin. *FEMS Microbiology Ecology*, 54, 167-177.

Lead, J. R., Muirhead, D., and Gibson, C. T. (2005). Characterization of freshwater natural aquatic colloids by atomic force microscopy (AFM). *Environmental Science and Technology*, 39, 6930 -6936.

Li, W. K. W. and Wood, A. M. (1988). Vertical distribution of North Atlantic ultraphytoplankton: analysis by flow cytometry and epifluorescence microscopy. *Deep-Sea Research*, 35, 1615-1638.

Lloyd, K. G., Lapham, L., and Teske, A. (2006). Anaerobic methane-oxidizing community of ANME-1b archaea in hypersalineGulf of Mexico sediments. *Applied and Environmental Microbiology*, 72, 7218-7230.

Lower, S. K., Tadanier, C. J., and Hochella, M. F. (2000). Measuring interfacial and adhesion forces between bacteria and mineral surfaces with biological force microscopy. *Geochimica. Et. Cosmochimica. Acta.*, 64, 3133-3139.

Lower, S. K., Hochella, M. F., and Beveridge, T. J. (2001). Bacterial Recognition of Mineral Surfaces: Nanoscale Interactions Between *Shewanella* and α-FeOOH. *Science*, 292, 1360 - 1363.

Lu, J., Chen, F., and Hodson, R. E. (2001). Distribution, isolation, host specificity, and diversity of cyanophages infecting marine *Synechococcus* spp. in river estuaries. *Applied and Environmental Microbiology*, 67, 3285-3290.

Macko, S. A., Engel, M. H., and Qian, Y. (1994). Early diagenesis and organic matter preservation—A molecular stable carbon isotope perspective. *Chemical Geology*, 114, 365-379.

Madigan M. T., and Martinko J. M. (2006). *Brock Biology of Microorganisms* (11th Ed.). New Jersey: Prentice Hall.

Madrid, V. M., Taylor, G. T., Scranton, M. I., and Chistoserdov, A. Y. (2001). Phylogenetic diversity of bacterial and archaeal communities in the anoxic zone of the CariacoBasin. *Applied and Environmental Microbiology*, 67, 1663-1674.

Man, D. L., Wang, W. W., Sabehi, G., Aravind, L., Post, A. F., Massana, R., Spudich, E. N., Spudich, J. L., and Béjà, O. (2003). Diversification and spectral tuning in marine proteorhodopsins. *Embo Journal*, 22, 1725-1731.

Man-Aharonovich, D., Sabehi, G., Sineshchekov, O. A., Spudich, E. N., Spudich, J. L., and Béjà, O. (2004). Characterization of RS29, a blue-green proteorhodopsin variant from the Red Sea. *Photochemical and Photobiological Sciences*, 3, 459-462.

Marie, D., Brussaard, C. P. D., Thyrhaug, R., Bratbak, G., and Vaulot, D. (1999). Enumeration of marine viruses in culture and natural samples by flow cytometry. *Applied and Environmental Microbiology*, 65, 45-52.

Massana, R., Murray, A. E., Preston, C. M., and DeLong, E. F. (1997). Vertical distribution and phylogenetic characterization of marine planktonic Archaea in the Santa Barbara Channel. *Applied and Environmental Microbiology*, 63, 50-56.

McElfresh, M., Baesu, E., Balhorn, R., Belak, J., and Allen, M. J. (2002). Combining constitutive materials modeling with atomic force microscopy to understand the mechanical properties of living cells. *Proceedings of the NationalAcademy of Sciences of the United States of America*, 99, 6493-6497.

McFeters, G. A., Yu, F. P., Pyle, B. H., and Stewart, P. S. (1995). Physiological assessment of bacteria using fluorochromes. *Journal of Microbiological Methods*, 21, 1-13.

Michel, H. and Oesterhelt, D. (1980). Electrochemical proton gradient across the cell membrane of *Halobacterium halobium*: effect of N,N9-dicyclohexylcarbodiimide, relation to intracellular adenosine triphosphate, adenosine diphosphate, and phosphate concentration, and influenceof the potassium gradient. *Biochemistry*, 19, 4607~4614.

Miller, W. and Moran, M. A. (1997). Interaction of photochemical and microbial processes in the degradation of refractory dissolved organic matter from a coastal marine environment. *Limnology and Oceanography*, 42, 1317-1324.

Mills, H. J., Hodges, C., Wilson, K., MacDonald, I. R., and Sobecky, P. A. (2003). Microbial diversity in sediments associated with surface-breaching gas hydrate mounds in the Gulf of Mexico. *FEMS Microbiology Ecology*, 46, 39-52.

Mincer, T. J., Church, M. J., Taylor, L. T., Preston, C., Kar, D. M., and DeLong, E. F. (2007). Quantitative distribution of presumptive archaeal and bacterial nitrifiers in MontereyBay and the North Pacific Subtropical Gyre. *Environmental Microbiology*, 9, 1162-1175.

Mopper, K., Zhou, X., Kieber, R. J., Kieber, D. J., Sikorski, R. J., and Jones, R. D. (1991). Photochemical degradation of dissolved organic carbonand its impact on the oceanic carbon cycle. *Nature*, 353, 60-62.

Mozes, N., Handley, P. S., Busscher, H. J., and Rouxhet, P. G. (1991). *Microbial cell surface analysis: structural and physicochemical methods*. New York: VCH Publishers.

Müller, D. J., Baumeister, W., and Engel, A. (1999). Controlled unzipping of a bacterial surface layer with atomic force microscopy. *Proceedings of the NationalAcademy of Sciences of the United States of America*, 96, 13170-13174.

Müller, S., Ullrich, S., Lösche, A., Loffhagen, N., and Babel, W. (2000). Flow cytometric techniques to characterise physiological states of *Acinetobacter calcoaceticus*. *Journal of Microbiological Methods*, 25, 79-86.

Nauhaus, K., Treude, T., Boetius, A., and Kruger, M. (2005). Environmental regulation of the anaerobic oxidation of methane: a comparison of ANME-I and ANME-II communities. *Environmental Microbiology*, 7, 98-106.

Nicol, G. W. and Schleper, C. (2006). Ammonia-oxidising Crenarchaeota: important players in the nitrogen cycle? *Trends in Microbiology*, 14, 207-212.

Noble, R. T. and Fuhrman, J. A. (1998). Use of SYBR Green I for rapid epifluorescence counts of marine viruses and bacteria. *Aquatic Microbial Ecology*, 14, 113-118.

Nogawa, H. and Tanoue, E. (2003). Dissolved organic matter in oceanic waters. *Journal of Oceanography*, 59, 129-147.

Novo, D., Perlmutter, N. G., Hunt, R. H., and Shapiro, H. M. (1999). Accurate flow cytometric membrane potential measurement in bacteria using diethyloxacarbocyanine and a ratiometric technique. *Cytometry*, 35, 55-63.

Novo, D. J., Perlmutter, N. G., Hunt, R. H., and Shapiro, H. M. (2000). Multiparameter flow cytometric analysis of antibiotic effects on membrane potential, membrane permeability, and bacterial counts of Staphylococcus aureus and Micrococcus luteus. *Antimicrobial Agents and Chemotherapy*, 44, 827-834.

O'Farrell, P. H. (1975). High resolution two-dimensional electrophoresis of proteins. *Journal of Biological Chemistry*, 250, 4007-4021.

Orphan, V. J., Hinrichs, K. U., Ussler, W., Paull, C. K., Taylor, L. T., Sylva, S. P., Hayes, J. M., and Delong, E. F. (2001). Comparative analysis of methane-oxidizing archaea and sulfate-reducing bacteria in anoxic marine sediments. *Applied and Environmental Microbiology*, 67, 1922-1934.

Orphan, V. J., House, C. H., Hinrichs, K. U., McKeegan, K. D., and DeLong, E. F. (2001). Methane-consuming archaea revealed by directly coupled isotopic and phylogenetic analysis. *Science*, 293, 484-487.

Orphan, V. J., House, C. H., Hinrichs, K. U., McKeegan, K. D., and DeLong, E. F. (2002). Multiple archaeal groups mediate methane oxidation in anoxic cold seep sediments. *Proceedings of the NationalAcademy of Sciences of the United States of America*, 99, 7663-7668.

Ouverney, C. C. and Fuhrman, J. A. (2000). Marine planktonic Archaea take up amino acids. *Applied and Environmental Microbiology*, 66, 4829-4833.

Palenik, B., Brahamsha, B., Larimer, F. W., Land, M., Hauser, L., Chain, P., Lamerdin, J., Regala, W., Allen, E. E., McCarren, J., Paulsen, I., Dufresne, A., Partensky, F., Webb, E. A., and Waterbury, J. (2003). The genome of a motile marine *Synechococcus*. *Nature*, 424, 1037-1042.

Palenik, B., Ren, Q. H., Dupont, C. L., Myers, G. S., Heidelberg, J. F., Badger, J. H., Madupu, R., Nelson, W. C., Brinkac, L. M., Dodson, R. J., Durkin, A. S., Daugherty, S. C., Sullivan, S. A., Khouri, H., Mohamoud, Y., Halpin, R., and Paulsen, I. T. (2006). Genome sequence of *Synechococcus* CC9311: Insights into adaptation to a coastal environment. *Proceedings of the NationalAcademy of Sciences of the United States of America*, 103, 13555-13559.

Pancost, R. D., Freeman, K. H., Patzkowsky, M. E., Wavrek, D. A., and Collister, J. W. (1998). Molecular indicators of redox and marine photoautotroph composition in the late Middle Ordovician of Iowa, USA. *Organic Geochemistry*, 29, 1649-1662.

Pancost, R. D., Freeman, K. H., and Wakeham, S. G. (1999). Controls on the carbon-isotope compositions of compounds in Peru surface waters. *Organic Geochemistry*, 30, 319-340.

Pancost, R. D., Bouloubassi, I., Aloisi, G., and Damste, J. S. S. (2001). Three series of non-isoprenoidal dialkyl glycerol diethers in cold-seep carbonate crusts. *Organic Geochemistry*, 32, 695-707.

Pancost, R. D., Hopmans, E. C., and Damste, J. S. S. (2001). Archaeal lipids in Mediterranean cold seeps: Molecular proxies for anaerobic methane oxidation. *Geochimica Et Cosmochimica Acta*, 65, 1611-1627.

Pancost, R. D. and Damste, J. S. S. (2003). Carbon isotopic compositions of prokaryotic lipids as tracers of carbon cyclingin diverse settings. *Chemical Geology*, 195, 29-58.

Pearson, A., McNichol, A. P., Benitez-Nelson, B. C., Hayes, J. M., and Eglinton, T. I. (2001). Origins of lipid biomarkers in Santa MonicaBasin surface sediment: A case study using compound-specific Delta C-14 analysis. *Geochimica. Et. Cosmochimica. Acta.*, 65, 3123-3137.

Pearson, A., Huang, Z., Ingalls, A. E., Romanek, C. S., Wiegel, J., Freeman, K. H., Smittenberg, R. H., and Zhang, C. L. (2004). Nonmarine crenarchaeol in Nevada hot springs. *Applied and Environmental Microbiology*, 70, 5229-5237.

Pelling, A. E., Li, Y., Shi, W., and Gimzewski, J. K. (2005). Nanoscale visualization and characterization of Myxococcus xanthus cells with atomic force microscopy. *Proceedings of the NationalAcademy of Sciences of the United States of America*, 102, 6484-6489.

Pi Y., Ye Q., Jiang H., Pearson A., Li S., Noakes J., Culp R., Dong H., and Zhang C. L. (2007). Lipids and phylogenetic characterization of Archaea associated with gas hydrates in the Gulf of Mexico. *FEMS Microbiology Ecology*, In revision.

Pradet-Balade, B., Boulme, F., Beug, H., Mullner, E. W., and Garcia-Sanz, J. A. (2001). Translation control: bridging the gap between genomics and proteomics? *Trends in Biochemical Sciences*, 26, 225-229.

Preston, C. M., Wu, K. Y., Molinsky, T. F., and DeLong, E. F. (1996). A psychrophilic crenarchaeote inhabits a marine sponge: *Cenarchaeum symbiosum* gen. nov. *Proceedings of the NationalAcademy of Sciences of the United States of America*, 93, 6241-6246.

Proctor, L. M. and Fuhrman, J. A. (1990). Viral mortality of marine bacteria and cyanobacteria. *Nature*, 343, 60–62.

Radajewski, S., McDonald, I. R., and Murrell, J. C. (2003). Stable-isotope probing of nucleic acids: a window to the function of uncultured microorganisms. *Current Opinion in Biotechnology*, 14, 296-302.

Ram, R. J., VerBerkmoes, N. C., Thelen, M. P., Tyson, G. W., Baker, B. J., Blake, R. C., Shah, M., Hettich, R. L., and Banfield, J. F. (2005). Community proteomics of a natural microbial biofilm. *Science*, 308, 1915-1920.

Razatos, A., Ong, Y. L., Sharma, M. M., and Georgiou, G. (1998). Molecular determinants of bacterial adhesion monitored by atomic force microscopy. *Proceedings of the NationalAcademy of Sciences of the United States of America*, 95, 11059-11064.

Reed, A. J., Lutz, R. A., and Vetriani, C. (2006). Vertical distribution and diversity of bacteria and archaea in sulfide and methane-rich cold seep sediments located at the base of the Florida Escarpment. *Extremophiles*, 10, 199-211.

Richardson, T. L., and Jackson, G. A. (2007). Small phytoplankton and carbon export from the surface ocean. *Science*, 315, 838-840.

Rief, M., Oesterhelt, F., Heymann, B., and Gaub, H. E. (1997). Single molecule force spectroscopy on polysaccharides by atomic force microscopy. *Science*, 275, 1295 - 1297.

Ringelberg, D. B., Davis, J. D., Smith, G. A., Pfiffner, S. M., Nichols, P. D., Nickels, J. S., Henson, J. M., Wilson, J. T., Yates, M., Kampbell, D. H., Read, H. W., Stocksdale, T. T., and White, D. C. (1989). Validation of signature polarlipid fatty acid biomarkers for alkane-utilizing bacteria in soils and subsurface aquifer materials. *FEMS Microbiology Ecology*, 62, 39-50.

Rocap, G., Larimer, F. W., Lamerdin, J., Malfatti, S., Chain, P., Ahlgren, N. A., Arellano, A., Coleman, M., Hauser, L., Hess, W. R., Johnson, Z. I., Land, M., Lindell, D., Post, A. F., Regala, W., Shah, M., Shaw, S. L., Steglich, C., Sullivan, M. B., Ting, C. S., Tolonen, A., Webb, E. A., Zinser, E. R., and Chisholm, S. W. (2003). Genome divergence in two *Prochlorococcus* ecotypes reflects oceanic niche differentiation. *Nature*, 424, 1042-1047.

Rodriguez, G. G., Phipps, D., Ishiguro, K., and Ridgway, H. F. (1992). Use of a fluorescent redox probe for direct visualization of actively respiring bacteria. *Applied and Environmental Microbiology*, 58, 1801-1808.

Roszack, D. B. and Colwell, R. R. (1987). Survival strategies of bacteria in the natural environment. *Microbiology and Molecular Biology Reviews*, 51, 365-379.

Rotsch, C. and Radmacher, M. (1997). Mapping local electrostatic forces with the atomic force microscope. *Langmuir*, 13, 2825-2832.

Rusch, D. B., Halpern, A. L., Sutton, G., Heidelberg, K. B., Williamson, S., Yooseph, S., Wu, D. Y., Eisen, J. A., Hoffman, J. M., Remington, K., Beeson, K., Tran, B., Smith, H., Baden-Tillson, H., Stewart, C., Thorpe, J., Freeman, J., Andrews-Pfannkoch, C., Venter, J. E., Li, K., Kravitz, S., Heidelberg, J. F., Utterback, T., Rogers, Y. H., Falcon, L. I., Souza, V., Bonilla-Rosso, G., Eguiarte, L. E., Karl, D. M., Sathyendranath, S., Platt, T., Bermingham, E., Gallardo, V., Tamayo-Castillo, G., Ferrari, M. R., Strausberg, R. L., Nealson, K., Friedman, R., Frazier, M., and Venter, J. C. (2007). The Sorcerer II Global Ocean Sampling expedition: Northwest Atlantic through Eastern Tropical Pacific. *Plos Biology*, 5, 398-431.

Sabehi, G., Loy, A., Jung, K. H., Partha, R., Spudich, J. L., Isaacson, T., Hirschberg, J., Wagner, M., and Béjà, O. (2005). New insights into metabolic properties of marine bacteria encoding proteorhodopsins. *Plos Biology*, 3, 1409-1417.

Santschi, P. H., Balnois, E., Wilkinson, K. J., Zhang, J., and Buffle, J. (1998). Fibrillar polysaccharides in marine macromolecular organic matter as imaged by atomic force microscopy and transmission electron microscopy. *Limnology and Oceanography*, 43, 896-908.

Sassen, R., Roberts, H. H., Carney, R., Milkov, A. V., DeFreitas, D. A., Lanoil, B., and Zhang, C. L. (2004). Free hydrocarbon gas, gas hydrate, and authigenic minerals in chemosynthetic communities of the northern Gulf of Mexico continental slope: relation to microbial processes. *Chemical Geology*, 205, 195-217.

Scheuring, S., Seguin, J., Marco, S., Levy, D., Robert, B., and Rigaud, J. L. (2003). Nanodissection and high-resolution imaging of the *Rhodopseudomonas viridis* photosynthetic core complex in native membranes by AFM. *Proceedings of the NationalAcademy of Sciences of the United States of America*, 100, 1690-1693.

Scheuring, S. and Sturgis, J. N. (2005). Chromatic Adaptation of Photosynthetic Membranes. *Science*, 309, 484-487.

Schleper, C., Jurgens, G., and Jonuscheit, M. (2005). Genomic studies of uncultivated archaea. *Nature Reviews Microbiology*, 3, 479-488.

Schouten, S., Hopmans, E. C., Pancost, R. D., and Damste, J. S. S. (2000). Widespread occurrence of structurally diverse tetraether membrane lipids: Evidence for the ubiquitous presence of low-temperature relatives of hyperthermophiles. *Proceedings of the NationalAcademy of Sciences of the United States of America*, 97, 14421-14426.

Schouten, S., Hopmans, E. C., Schefuss, E., and Damste, J. S. S. (2002). Distributional variations in marine crenarchaeotal membrane lipids: a new tool for reconstructing ancient sea water temperatures? *Earth and Planetary Science Letters*, 204, 265-274.

Schubert, C. J., Coolen, M. J. L., Neretin, L. N., Schippers, A., Abbas, B., Durisch-Kaiser, E., Wehrli, B., Hopmans, E. C., Damste, J. S. S., Wakeham, S., and Kuypers, M. M. M. (2006). Aerobic and anaerobic methanotrophs in the Black Sea water column. *Environmental Microbiology*, 8, 1844-1856.

Schulze, W. X., Gleixner, G., Kaiser, K., Guggenberger, G., Mann, M., and Schulze, E. D. (2005). A proteomic fingerprint of dissolved organic carbonand of soil particles. *Oecologia*, 142, 335-343.

Schwalbach, M. S. and Fuhrman, J. A. (2005). Wide-ranging abundances of aerobic anoxygenic phototrophic bacteria in the world ocean revealed by epifluorescence microscopy and quantitative PCR. *Limnology and Oceanography*, 50, 620-628.

Seki, M. P., Polovina, J. J., Brainard, R. E., Bidigare, R. R., Leonard, C. L., and Foley, D. G. (2001). Biological enhancement at cyclonic eddies tracked with GOES thermal imagery in Hawaiian waters. *Geophysical Research Letters*, 28, 1583-1586.

Shalapyonok, A., Olson, R. J., and Shalapyonok, L. S. (1998). Ultradian growth in *Prochlorococcus* spp. *Applied and Environmental Microbiology*, 64, 1066-1069.

Shepard, J. L., Olsson, B., Tedengren, M., and Bradley, B. P. (2000). Protein expression signatures identified in Mytilus edulis exposed to PCBs, copper and salinity stress. *Marine Environmental Research*, 50, 337-340.

Sherr, B. F., A., d. G. P., and Sherr, E. B. (1999). Estimating abundance and single-cell characteristics of respiring bacteria via the redox dye, CTC. *Aquatic Microbial Ecology*, 18, 117-131.

Shibata, A., Goto, Y., Saito, H., Kikuchi, T., Toda, T., and Taguchi, S. (2006). Comparison of SYBR Green I and SYBR Gold stains for enumerating bacteria and viruses by epifluorescence microscopy. *Aquatic Microbial Ecology*, 43, 223-231.

Sieracki, M. E., Gilg, I. C., Thier, E. C., Poulton, N. J., and Goericke, R. (2006). Distribution of planktonic aerobic anoxygenic photoheterotrophic bacteria in the northwest Atlantic. *Limnology and Oceanography*, 51, 38-46.

Sullivan, M. B., Waterbury, J. B., and Chisholm, S. W. (2003). Cyanophages infecting the oceanic cyanobacterium *Prochlorococcus*. *Nature*, 424, 1047-1051.

Sullivan, M. B., Coleman, M. L., Weigele, P., Rohwer, F., and Chisholm, S. W. (2005). Three *Prochlorococcus* cyanophage genomes: Signature features and ecological interpretations. *Plos Biology*, 3, 790-806.

Summons, R. E., Jane, L. L., and Roksandic, Z. (1994). Carbon isotopic fractionation in lipids from metamorphic bacteria: Relevance for interpretation of the geochemical record of biomarkers. *Geochimica. Et. Cosmochimica. Acta.*, 58, 2853-2863.

Summons R. E., Jahnke L. L., and Simoneit R. T. (1996). Lipid biomarkers for bacterial ecosystems: Studies of cultured organisms, hydrothermal environments and ancient sediments. In G. R. Bock and J. A. Goode (Eds.), *Evolution of Hydrothermal Ecosystems on Earth (and Mars?)* (pp. 174-194). New York: John Wiley and Sons.

Suttle, C. A., Chan, A. M., and Cottrell, M. T. (1990). Infection of phytoplankton by viruses and reduction of primary productivity. *Nature*, 347, 467-469.

Suttle, C. A., and Chan, A. M. (1993). Marine cyanophages infecting oceanic and coastal strains of *Synechococcus*: abundance, morphology, cross-reactivity and growth characteristics. *Marine Ecology Progress Series*, 92, 99-109.

Suttle, C. A. a. C. A. M. (1993). Marine cyanophages infecting oceanic and coastal strains of Synechococcus: abundance, morphology, cross-reactivity and growth characteristics. *Marine Ecology Progress Series*, 92, 99-109.

Suttle, C. A. (1994). The significance of viruses to mortality in aquatic microbial communities. *Microbial Ecology,* 28, 237-243.

Suttle, C. A., and Chan, A. M. (1994). Dynamics and distribution of cyanophages and their effect on marine *Synechococcus* spp. *Applied and Environmental Microbiology*, 60, 3167-3174.

Suttle, C. A. (2005). Viruses in the sea. *Nature*, 437, 356-361.

Suyama, T., Shigematsu, T., Suzuki, T., Tokiwa, Y., Kanagawa, T., Nagashima, K. V. P., and Hanada, S. (2002). Photosynthetic apparatus in *Roseateles depolymerans*61A is transcriptionally induced by carbon limitation. *Applied and Environmental Microbiology*, 68, 1665-1673.

Teira, E., Lebaron, P., van Aken, H., and Herndl, G. J. (2006). Distribution and activity of Bacteria and Archaea in the deep water masses of the North Atlantic. *Limnology and Oceanography*, 51, 2131-2144.

Teske, A. P. (2006). Microbial communities of deep marine subsurface sediments: Molecular and cultivation surveys. *Geomicrobiology Journal*, 23, 357-368.

Thingstad, T. F., Heldal, M., Bratbak, G., and Dundas, I. (1993). Are viruses important partners in pelagic food webs. *Trends in Ecology and Evolution*, 8, 209-213.

Touhami, A., Hoffmann, B., Vasella, A., Denis, F. A., and Dufrêne, Y. F. (2003). Aggregation of yeast cells: direct measurement of discrete lectin–carbohydrate interactions*Microbiology*, 149, 2873-2878.

Tringe, S. G., von Mering, C., Kobayashi, A., Salamov, A. A., Chen, K., Chang, H. W., Podar, M., Short, J. M., Mathur, E. J., Detter, J. C., Bork, P., Hugenholtz, P., and Rubin, E. M. (2005). Comparative metagenomics of microbial communities. *Science*, 308, 554-557.

Tsutomu, M. and Koichi, K. (1992). Quantum conversion and image detection by a bacteriorhodopsin-based artificial photoreceptor. *Science*, 255, 342~344.

Tyson, G. W., Chapman, J., Hugenholtz, P., Allen, E. E., Ram, R. J., Richardson, P. M., Solovyev, V. V., Rubin, E. M., Rokhsar, D. S., and Banfield, J. F. (2004). Community structure and metabolism through reconstruction of microbial genomes from the environment. *Nature*, 428, 37-43.

Valentine, D. L. and Reeburgh, W. S. (2000). New perspectives on anaerobic methane oxidation. *Environmental Microbiology*, 2, 477-484.

Vandevivere, P. and Kirchman, D. L. (1993). Attachment stimulates synthesis by a bacterium exopolysaccharide synthesis by a bacterium. *Applied and Environmental Microbiology*, 59, 3280-3286.

Vaulot, D., Marie, D., Olson, R. J., and Chisholm, S. W. (1995). Growth of *Prochlorococcus*, a photosynthetic prokaryote, in the Equatorial Pacific Ocean. *Science*, 8, 1480-1482.

Velegol, S. B., Pardi, S., Li, X., Velegol, D., and Logan, B. E. (2003). AFM Imaging Artifacts due to Bacterial Cell Height and AFM Tip Geometry. *Langmuir*, 19, 851-857.

Venter, J. C., Remington, K., Heidelberg, J. F., Halpern, A. L., Rusch, D., Eisen, J. A., Wu, D. Y., Paulsen, I., Nelson, K. E., Nelson, W., Fouts, D. E., Levy, S., Knap, A. H., Lomas, M. W., Nealson, K., White, O., Peterson, J., Hoffman, J., Parsons, R., Baden-Tillson, H., Pfannkoch, C., Rogers, Y. H., and Smith, H. O. (2004). Environmental genome shotgun sequencing of the Sargasso Sea. *Science*, 304, 66-74.

Vetriani, C., Jannasch, H. W., MacGregor, B. J., Stahl, D. A., and Reysenbach, A. L. (1999). Population structure and phylogenetic characterization of marine benthic archaea in deep-sea sediments. *Applied and Environmental Microbiology*, 65, 4375-4384.

Wakeham, S. G. (1995). Lipid biomarkers for heterotrophic alteration of suspended particulate organic matter in oxygenated and anoxic water columns of the ocean. *Deep-Sea ResearchPart I-Oceanographic Research Papers*, 42, 1749-1771.

Whelan, J., Eglinton, L., Cathles III, L., Losh, S., Roberts, H. (2005). Surface and subsurface manifestations of gas movement through a N–S transect of the Gulf of Mexico. *Marine and Petroleum Geology* 22, 479–497.

Wang, W. W., Sineshchekov, O. A., Spudich, E. N., and Spudich, J. L. (2003). Spectroscopic and photochemical characterization of a deep ocean proteorhodopsin. *Journal of Biological Chemistry*, 278, 33985-33991.

Wang, P., Xiao, X., and Wang, F. P. (2005). Phylogenetic analysis of Archaea in the deep-sea sediments of west Pacific Warm Pool. *Extremophiles*, 9, 209-217.

Wang, K. and F., Chen. (2007). Prevalence of highly host-specific cyanophages in the estuarine environment. *Environmental Microbiology*, In press.

Wasinger, V. C., Cordwell, S. J., Cerpa Poljak, A., Yan, J. X., Gooley, A. A., wilkins, M. R., Duncan, M. W., Harris, R., Williams, K. L., and Humphery-Smith, I. (1995). Progress with gene product mapping of Moli cutes: Mycoplasma genitalium. *Electrophoresis*, 16, 1090-1094.

Waterbury, J. B. and Valois, F. W. (1993). Resistance to co-occuring phages enables marine *Synechococcus* communities to coexist wth cyanophages abundant in seawater. *Applied and Environmental Microbiology*, 59, 3393-3399.

Weijers, J. W. H., Schouten, S., Hopmans, E. C., Geenevasen, J. A. J., David, O. R. P., Coleman, J. M., Pancost, R. D., and Sinninghe Damste, J. S. (2006). Membrane lipids of mesophilic anaerobic bacteria thriving in peats have typical archaeal traits. *Environmental Microbiology*, 8, 648-657.

Weinbauer, M. G. and Suttle, C. A. (1996). Potential significance of lysogeny to bacteriophage production and bacterial mortality in coastal waters of the Gulf of Mexico. *Applied and Environmental Microbiology*, 62, 4374–4380.

Weinbauer, M. G. and Rassoulzadegan, F. (2004). Are viruses driving microbial diversification and diversity? *Environmental Microbiology*, 6, 1-11.

Wilcox, R. M. and Fuhrman, J. A. (1994). Bacterial viruses in coastal seawater: lytic rather than lysogenic production. *Marine Ecology Progress Series*, 114, 35-45.

Wilhelm, S. W., Weinbauer, M. G., Suttle, C. A., and Jeffrey, W. H. (1998). The role of sunlight in the removal and repair of viruses in the sea. *Limnology and Oceanography*, 43, 586-592.

Wilhelm, S. W. and Suttle, C. A. (1999). Viruses and Nutrient Cycles in the Sea - Viruses play critical roles in the structure and function of aquatic food webs. *Bioscience*, 49, 781-788.

Williams, P. M. and Druffel, E. R. M. (1987). Radiocarbon in dissolved organic matter in the central North Pacific Ocean. *Nature*, 330, 246-248.

Williams, S. C., Hong, Y., Danavall, D. C. A., Howard-Jones, M. H., Gibson, D., Frischer, M. E., and Verity, P. G. (1998). Distinguishing between living and nonliving bacteria: Evaluation of the vital stain propidium iodide and its combined use with molecular probes in aquatic samples. *Journal of Microbiological Methods*, 32, 225-236.

Wilmes, P. and Bond, P. L. (2004). The application of two-dimensional polyacrylamide gel electrophoresis and downstream analyses to a mixed community of prokaryotic microorganisms. *Environmental Microbiology*, 6, 911-920.

Wilmes, P. and Bond, P. L. (2006). Metaproteomics: studying functional gene expression in microbial ecosystems. *Trends in Microbiology*, 14, 92-97.

Wilson, W. H., Joint, I. R., Carr, N. G., and Mann, N. H. (1993). Isolation and molecular characterization of five marine cyanophages propagated on *Synechococcus* sp. strain WH7803. *Applied and Environmental Microbiology*, 59, 3736-3742.

Wommack, K. E. and Colwell, R. R. (2000). Virioplankton: Viruses in aquatic ecosystems. *Microbiology and Molecular Biology Reviews*, 64, 69-114.

Wuchter, C., Abbas, B., Coolen, M. J. L., Herfort, L., van Bleijswijk, J., Timmers, P., Strous, M., Teira, E., Herndl, G. J., Middelburg, J. J., Schouten, S., and Damste, J. S. S. (2006). Archaeal nitrification in the ocean. *Proceedings of the NationalAcademy of Sciences of the United States of America*, 103, 12317-12322.

Xiong, J., Fischer, W. M., Inoue, K., Nakahara, M., and Bauer, C. E. (2000). Molecular evidence for the early evolution of photosynthesis. *Science*, 289, 1724-1730.

Xu, W., Mulhern, P. J., Blackford, B. L., Jericho, M. H., and Templeton, I. (1994). A new atomic force microscopy technique for the measurement of the elastic properties of biological materials. *Scanning Microscopy*, 8, 499-506.

Index

B

C

D

E

F

G

H

I

J

K

L

M

N

O

P

Q

R

S

T

U

V

W

X

Y

Z